Solved Problems in Geophysics

Solving problems is an indispensable exercise for mastering the theory underlying the various branches of geophysics. Without this practice, students often find it hard to understand and relate theoretical concepts to their application in real-world situations.

This book is a collection of nearly 200 problems in geophysics, which are solved in detail showing each step of their solution, the equations used and the assumptions made. Simple figures are also included to help students understand how to reduce a problem to its key elements. The book begins with an introduction to the equations most commonly used in solving geophysical problems. The subsequent four chapters then present a series of exercises for each of the main, classical areas of geophysics – gravity, geomagnetism, seismology and heat flow and geochronology. For each topic there are problems with different degrees of difficulty, from simple exercises that can be used in the most elementary courses, to more complex problems suitable for graduate-level students.

This handy book is the ideal adjunct to core course textbooks on geophysical theory. It is a convenient source of additional homework and exam questions for instructors, and provides students with step-by-step examples that can be used as a practice or revision aid.

Elisa Buforn is a Professor of geophysics at the Universidad Complutense de Madrid (UCM) where she teaches courses on geophysics, seismology, physics, and numerical methods. Professor Buforn's research focuses on source fracture processes, seismicity, and seismotectonics, and she is Editor in Chief of *Física de la Tierra* and on the Editorial Board of the *Journal of Seismology*.

Carmen Pro is an Associate Professor at the University of Extremadura, Spain, where she has taught geophysics and astronomy for over 20 years. She has participated in several geophysical research projects and is involved in college management.

Agustín Udías is an Emeritus Professor at UCM and is the author of a large number of papers about seismicity, seismotectonics, and the physics of seismic sources, as well as the textbook *Principles of Seismology* (Cambridge University Press, 1999). He has held positions as Editor in Chief of *Física de la Tierra* and the *Journal of Seismology* and as Vice President of the European Seismological Commission.

Solved Problems in Geophysics

ELISA BUFORN

Universidad Complutense, Madrid

CARMEN PRO

Universidad de Extremadura, Spain

AGUSTÍN UDÍAS

Universidad Complutense, Madrid

CAMBRIDGE
UNIVERSITY PRESS

University Printing House, Cambridge CB2 8BS, United Kingdom

One Liberty Plaza, 20th Floor, New York, NY 10006, USA

477 Williamstown Road, Port Melbourne, VIC 3207, Australia

314-321, 3rd Floor, Plot 3, Splendor Forum, Jasola District Centre, New Delhi - 110025, India

79 Anson Road, #06-04/06, Singapore 079906

Cambridge University Press is part of the University of Cambridge.

It furthers the University's mission by disseminating knowledge in the pursuit of
education, learning and research at the highest international levels of excellence.

www.cambridge.org
Information on this title: www.cambridge.org/9781107602717

© Elisa Buforn, Carmen Pro and Agustín Udías 2012

First published 2012
3rd printing 2013

A catalogue record for this publication is available from the British Library

Library of Congress Cataloging in Publication data
Buforn, E.
Solved problems in geophysics / Elisa Buforn, Carmen Pro, Agustín Udías.
p.   cm.
Includes bibliographical references.
ISBN 978-1-107-60271-7 (Paperback)
1. Geophysics–Problems, exercises, etc.   I. Pro, Carmen.   II. Udías Vallina, Agustín.   III. Title.
QC807.52.B84 2012
550.78–dc23
2011046101

ISBN 978-1-107-60271-7 Paperback

Contents

Preface

This book presents a collection of 197 solved problems in geophysics. Our teaching experience has shown us that there was a need for a work of this kind. Solving problems is an indispensable exercise for understanding the theory contained in the various branches of geophysics. Without this exercise, the student often finds it hard to understand and relate the theoretical concepts with their application to practical cases. Although most teachers present exercises and problems for their students during the course, the hours allotted to the subject significantly limit how many exercises can be worked through in class. Although the students may try to solve other problems outside of class time, if there are no solutions available this significantly reduces the effectiveness of this type of study. It helps, therefore, both for the student and for the teacher who is explaining the subject if they have problems whose solutions are given and whose steps can be followed in detail. Some geophysics textbooks, for example, F.D. Stacey, *Physics of the Earth*; G.D. Garland, *Introduction to Geophysics*; C.M. Fowler, *The Solid Earth: An Introduction to Global Geophysics*; and W. Lowrie, *Fundamentals of Geophysics*, contain example problems, and, in the case of Stacey's, Fowler's, and Lowrie's textbooks, their solutions are provided on the website of Cambridge University Press. The main difference in the present text is the type of problems and the detail with which the solutions are given, and in the much greater number.

All the problems proposed in the book are solved in detail, showing each step of their solution, the equations used, and the assumptions made, so that their solution can be followed without consulting any other book. When necessary, and indeed quite often, we also include figures that allow the problems to be more clearly understood. For a given topic, there are problems with different degrees of difficulty, from simple exercises that can be used in the most elementary courses, to more complex problems with greater difficulty and more suitable for teaching at a more advanced level.

The problems cover all parts of geophysics. The book begins with an Introduction (Chapter 1) that includes the equations most used in solving the problems. The idea of this chapter is not to develop the theory, but rather to simply give a list of the equations most commonly used in solving the problems, at the same time as introducing the reader to the nomenclature. The next four chapters correspond to the division of the problems into the four thematic blocks that are classic in geophysics: gravity, geomagnetism, seismology, and heat flow and geochronology. We have not included problems in geodynamics, since this would depart too much from the approach we have taken, which is to facilitate comprehension of the theory through its application to specific cases, sometimes cases which are far from the real situation on Earth. Indeed, some of the problems may seem a bit artificial, but their function is to help the student practise with what has been seen in the theory. Neither did we want to include specific problems of geophysical prospecting as this would have considerably increased the length of the text, and moreover some of the topics that would be covered in prospecting, such as gravimetric and geomagnetic anomalies, are already included in other sections of this work.

Chapter 2 contains 68 problems in gravity divided into five sections. The first section is dedicated to the terrestrial geoid and ellipsoid, proposing calculations of the parameters that define them in order to help better understand these reference surfaces. The second corresponds to calculating the gravitational field and potential for various models of the Earth, including the existence of internal structures. Gravity anomalies are dealt with in the third section, with a variety of problems to allow students to familiarize themselves with the corrections to the observed gravity, with the concept of isostasy, and with the Airy and Pratt hypotheses. The fourth section studies the phenomenon of the Earth's tides and their influence on the gravitational field. The last section is devoted to the observations of gravity from measurements made with different types of gravimeters and the corrections necessary in each case. We also include the application of these observations to the accurate determination of different types of height.

Chapter 3 contains 42 problems in geomagnetism divided into five sections. The first is devoted to the main (internal) field generated by a tilted dipole at the centre of the Earth. It includes straightforward problems that correspond to the calculation of the geomagnetic coordinates of a point and the theoretical components of the magnetic field. This section also introduces the student to the use of the principal units used in geomagnetism. The second considers the magnetic anomalies generated by different magnetized bodies and their influence on the internal field. The third section is devoted to the external field and its variation with time. In the fourth section, we propose problems of greater complexity involving the internal field, the external field, and anomalous magnetized bodies at the same time. The last section is devoted to problems in paleomagnetism.

Chapter 4 contains 69 problems in seismology divided into seven sections. The first presents some simple exercises on the theory of elasticity. The second addresses the problem of the propagation of seismic energy in the form of elastic waves, resolving the problems on the basis of potentials, and calculating the components of their displacements. We study the reflection and refraction of seismic waves in the third section. The fourth is devoted to the problem of wave propagation using the theory of ray paths in a plane medium of constant and variable velocity of propagation. The fifth studies the problem of the propagation of rays in a spherical medium of either constant or variable propagation velocity, with the calculation of the travel-time curves for both plane and spherical media. The sixth section contains problems in the propagation of surface waves in layered media. The seventh section is devoted to problems of calculating the focal parameters and the mechanism of earthquakes.

Chapter 5 includes 11 problems in heat flow with the propagation of heat in plane and spherical media, and seven problems in geochronology involving the use of radioactive elements for dating rocks.

Finally, we provide a bibliography of general textbooks on geophysics and of specific textbooks for the topics of gravity, geomagnetism, and seismology. We have tried to include only those most recent and commonly used textbooks which are likely to be found in university libraries.

In sum, the book is a university text for students of physics, geology, geophysics, planetary sciences, and engineering at the undergraduate or Master's degree levels. It is intended to be an aid to teaching the subjects of general geophysics, as well as the specific topics of gravity, geomagnetism, seismology, and heat flow and geochronology contained in university curricula.

The teaching experience of the authors in the universities of Barcelona, Extremadura, and the Complutense of Madrid highlighted the need for a work of this kind. This text is the result of the teaching work of its authors for over 20 years. Thanks are due to the generations of students over those years who, with their comments, questions, and suggestions, have really allowed this work to see the light. We are also especially grateful to Prof. Greg McIntosh who provided us with some problems on paleomagnetism, to Prof. Ana Negredo for her comments on heat flow and geochronology problems, and to Dr R.A. Chatwin who worked on translating our text into English.

The text is an extension of the Spanish edition published by Pearson (Madrid, 2010).

E. BUFORN, C. PRO AND A. UDÍAS

Gravity

As a first approximation the Earth's gravity is given by that of a rotating sphere. The gravitational potential of a sphere of mass M is:

$$V = \frac{GM}{r}$$

where r is the position vector (Fig. A) and G the universal gravitational constant.

If the sphere is rotating with angular velocity ω the centrifugal potential at a point on the surface is given by

$$\Phi = \frac{1}{2}\omega^2 r^2 \sin^2\theta$$

where θ is the angle that r forms with the axis of rotation.

The gravity potential is their sum $U = V + \Phi$.

The value of the acceleration due to gravity (the gravity 'force') is given by the gradient of the potential:

$$\boldsymbol{g} = \nabla U$$

The radial component of the gravity force is given by

$$g_r = -\frac{GM}{r^2} + r\omega^2 \sin^2\theta$$

The potential of the Earth to a first-order approximation corresponds to that of a rotating ellipsoid, and is given by

$$U = \frac{GM}{a}\left[\frac{a}{r} - \frac{J_2}{2}\left(\frac{a}{r}\right)^3 (3\sin^2\varphi - 1) + \frac{m}{2}\left(\frac{r}{a}\right)^2 \cos^2\varphi\right]$$

where $\varphi = 90° - \theta$ is the geocentric latitude and a the equatorial radius.

The coefficient m is the ratio between the centrifugal and gravitational forces on the sphere of radius a at the equator:

$$m = \frac{a^3\omega^2}{GM}$$

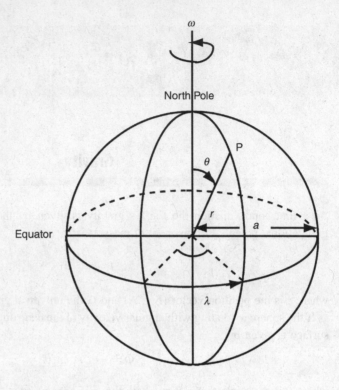

The dynamic form factor J_2 is defined as

$$J_2 = \frac{C - A}{a^2 M}$$

where C and A are the moments of inertia about the axis of rotation and an equatorial axis.

The flattening of the ellipsoid (the shape of the Earth to a first-order approximation) of equatorial and polar radius a and c is:

$$\alpha = \frac{a - c}{a}$$

In terms of J_2 and m,

$$\alpha = \frac{3}{2}J_2 + \frac{m}{2}$$

The dynamic ellipticity is

$$H = \frac{C - A}{C}$$

The gravity flattening is

$$\beta = \frac{\gamma_p - \gamma_e}{\gamma_e}$$

where γ_p and γ_e are the normal values of gravity at the pole and the equator, respectively.

The gravity at a point of geocentric latitude $\varphi = 90° - \theta$ is

$$\gamma = \gamma_e \left(1 + \beta \sin^2 \varphi \right)$$

The geocentric latitude of a point is the angle between the equator and the radius vector of the point. The geodetic latitude is defined as the angle between the equatorial plane and the normal to the ellipsoid surface at a point. Astronomical latitude is the angle between the equatorial plane and the observed vertical at a point.

The normal or theoretical gravity at a point of geocentric latitude φ referred to the GRS1980 reference ellipsoid is

$$\gamma = 9.780327 \left(1 + 0.0053024 \sin^2 \varphi - 0.0000059 \sin^2 2\varphi \right) \, \text{m s}^{-2}$$

The effect of the Sun and Moon on the Earth is to produce the phenomenon of the tides. If one considers more generally the tidal effect due to an astronomical body of mass M at a distance R from the centre of the Earth, one must add the corresponding potential, which, in the first-order approximation, is given by

$$\psi = \frac{GMr^2}{2R^3} \left(3\cos^2 \vartheta - 1 \right)$$

where r is the geocentric radius vector of the point, and ϑ is the angle the position vector r forms with the distance vector R.

Gravity anomalies, defined as $\Delta g = g - \gamma$, are the effects of the existence of anomalous masses inside the Earth. The gravity anomaly along the Z (vertical) axis at a point distance x along the horizontal axis produced by a sphere of radius R, density contrast $\Delta \rho$, and buried at a depth d, is given by

$$\Delta g(x, z) = \frac{\partial V_a}{\partial z} = \frac{G\Delta M(z + d)}{\left[x^2 + (z + d)^2 \right]^{3/2}}$$

where V_a is the potential produced by the anomalous spherical mass $\Delta M = 4/3\pi R^3 \, \Delta \rho$.

For problems in two dimensions, one uses the anomaly produced by an infinite horizontal cylinder at depth d, perpendicular to the plane under consideration. The anomalous potential is given by

$$V_a = 2\pi G\Delta \rho a^2 \ln \left(\frac{1}{\sqrt{x^2 + (z + d)^2}} \right)$$

and the anomaly by

$$\Delta g(x, z) = -\frac{\partial V_a}{\partial z} = \frac{2\pi G\Delta \rho a^2 (z + d)}{x^2 + (z + d)^2}$$

To correct for the height above sea level at which measurements are made, one uses the concepts of the free-air and Bouguer anomalies. The free-air anomaly is

$$\Delta g^{FA} = g - \gamma + 3.086h$$

where g is the observed gravity, h the height in metres, and the anomaly is obtained in gu (gravity units) $\mu m\,s^{-2}$.

The Bouguer anomaly is

$$\Delta g^B = g - \gamma + (3.086 - 0.419\rho)h$$

with ρ being the density of the plate of thickness h.

To account for isostatic compensation at height in mountainous areas, one adds an isostatic correction which can be calculated assuming either the Airy or Pratt hypotheses.

With the Airy hypothesis, the root t of a mountain is given by

$$t = \frac{\rho_c}{\rho_M - \rho_c} h$$

where ρ_c and ρ_M are the densities of the crust and mantle, and h is the height of the mountain. For an ocean zone, with water density ρ_a, the anti-root is

$$t' = \frac{\rho_c - \rho_a}{\rho_M - \rho_a} h'$$

With the Pratt hypothesis, the density contrast in a mountainous area is

$$\Delta\rho = \rho - \rho_0 = \frac{-h}{D + h}\rho_0$$

where D is the level of compensation, h the height of the mountain, and ρ_0 the density at sea level. For an oceanic zone of depth h':

$$\rho' = \frac{\rho_0 D - \rho_a h'}{D - h'}$$

$$\Delta\rho = \rho' - \rho_0$$

The isostatic correction can be calculated using a cylinder of radius a and height b, whose base is located at a distance c beneath the point, and with density contrast $\Delta\rho$:

$$C^I = 2\pi G\Delta\rho \left(b + \sqrt{a^2 + (c - b)^2} - \sqrt{a^2 + c^2} \right)$$

For mountainous zones, with the Airy hypothesis: $b = t$, $c = h + H + t$ (H = crustal thickness, h = height of the point); and with the Pratt hypothesis: $b = D$, $c = D + h$.

Geomagnetism

To a first approximation, the internal magnetic field of the Earth can be approximated by a centred dipole inclined at $11.5°$ to the axis of rotation. The potential created by a magnetic dipole at a point distant r from its centre and forming an angle θ with the axis of the dipole is

$$\Phi = \frac{-Cm\cos\theta}{r^2}$$

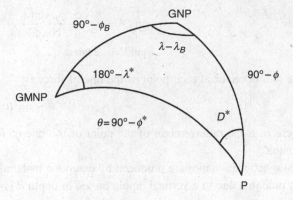

Fig. B

where $C = \mu_0/4\pi$ with $\mu_0 = 4\pi \times 10^{-7}$ H m^{-1}, and m is the dipole moment in units of A m². The product Cm is given in T m³.

The components of the magnetic dipole field $\boldsymbol{B}$ are:

$$B_r = -\frac{\partial \Phi}{\partial r} = -\frac{2Cm \cos \theta}{r^3}$$

$$B_\theta = -\frac{1}{r}\frac{\partial \Phi}{\partial \theta} = -\frac{2Cm \sin \theta}{r^3}$$

In the centred dipole approximation for the Earth's magnetic field, the geomagnetic coordinates (ϕ^*, λ^*) of a point ($\theta = 90° - \phi^*$) in terms of its geographic coordinates (ϕ, λ) and those of the Geomagnetic North Pole (GMNP) (ϕ_B, λ_B) can be calculated using the expressions of spherical trigonometry (Fig. B):

$$\sin \phi^* = \sin \phi_B \sin \phi + \cos \phi_B \cos \phi \cos(\lambda - \lambda_B)$$

$$\sin \lambda^* = \frac{\sin(\lambda - \lambda_B) \cos \phi}{\cos \phi^*}$$

The vertical and horizontal components of the field, the geomagnetic constant B_0, and the total field are given by:

$$Z^* = 2B_0 \sin \phi^*$$
$$H^* = B_0 \cos \phi^*$$
$$B_0 = \frac{Cm}{a^3}$$
$$F^* = \sqrt{H^{*2} + Z^{*2}} = B_0\sqrt{1 + 3\sin^2 \phi^*}$$

The units used for the components of the magnetic field are the tesla T and the nanotesla nT $= 10^{-9}$ T. The NS (X^*) and EW (Y^*) components are

$$X^* = H \cos D^*$$
$$Y^* = H \sin D^*$$

and the declination and inclination are given by

$$\sin D^* = \frac{-\cos \phi_B \sin(\lambda - \lambda_B)}{\cos \phi^*}$$

$$\tan I^* = 2 \tan \phi^*$$

The radius vector at each point of the line of force is:

$$r = r_0 \cos^2 \phi^* = r_0 \sin^2 \theta$$

where r_0 is the radius vector of the point of the line of force located at the geomagnetic equator.

Magnetic anomalies are produced by magnetic materials within the Earth. The anomalous potential due to a vertical dipole buried at depth d is

$$\Phi_A = \frac{Cm\cos\theta}{r^2} = \frac{Cm(z + d)}{\left[x^2 + (z + d)^2\right]^{3/2}}$$

The vertical (z) and the horizontal (x) components of the magnetic anomaly at the surface ($z = 0$) produced by a vertical magnetic dipole at depth d are:

$$\Delta Z = \frac{Cm(2d^2 - x^2)}{(x^2 + d^2)^{5/2}}$$

$$\Delta X = \frac{3Cmxd}{(x^2 + d^2)^{5/2}}$$

The Earth is affected by an external magnetic field produced mainly by the activity of the Sun. This field is variable in time, with distinct periods of variation. The most noticeable is the diurnal variation (Sq) with a maximum at 12 noon local time. The most important non-periodic variations are the so-called magnetic storms.

Seismology

Earthquakes produce elastic waves which propagate through the interior and along the surface of the Earth. Using the plane-wave approximation, the displacements of the internal P- and S-waves (u_i^P and u_i^S) can be obtained from a scalar potential and a vector potential:

$$u_i = u_i^P + u_i^S = (\nabla \varphi)_i + \left(\nabla \times \psi_j\right)_i$$

$$\varphi = A \exp ik_\alpha (\gamma_j x_j - \alpha t)$$

$$\psi_j = B_j \exp ik_\beta (\gamma_j x_j - \beta t)$$

where A and B_j are the amplitudes, x_j the coordinates of the observation point, k_α and k_β the wavenumbers, γ_j are the direction cosines defined from the azimuth a_z and angle of incidence i of the ray as:

$$\gamma_1 = \sin i \cos a_z$$
$$\gamma_2 = \sin i \, \sin a_z$$
$$\gamma_3 = \cos i$$

and α and β are the P- and S-wave velocities of propagation, respectively, defined from the Lamé coefficients (λ and shear modulus μ) and the density ρ:

$$v^P = \alpha = \sqrt{\frac{\lambda + 2\mu}{\rho}}$$

$$v^S = \beta = \sqrt{\frac{\mu}{\rho}}$$

Units used are: displacement amplitudes (u) in μm; potential amplitudes (A, B_i) in 10^{-3} m^2; wavenumber (k) in km^{-1}; and wave velocity (α, β) in km s^{-1}.

Poisson's ratio is defined in terms of the Lamé coefficients as

$$\sigma = \frac{\lambda}{2(\lambda + \mu)}$$

The angle of polarization of S-wave ε is defined as

$$\varepsilon = \tan^{-1} \frac{u^{SH}}{u^{SV}}$$

where u^{SH} is the amplitude of the SH component, and u^{SV} that of the SV component. SH and SV are the horizontal and vertical components of the S-wave on the wavefront plane.

The coefficients of reflection V and transmission W are given by the respective ratios between the amplitudes of the reflected or transmitted potentials and the incident potential:

$$V = \frac{A}{A_0}$$

$$W = \frac{A'}{A_0}$$

where A_0 is the amplitude of the incident wave potential, A that of the reflected potential, and A' of the transmitted potential.

Snell's law for plane media is expressed as

$$p = \frac{\sin i}{v}$$

and for spherical media

$$p = \frac{r \sin i}{v}$$

where p is the ray parameter, i the angle of incidence, v the propagation velocity of the medium, and r the position vector along the ray.

In the case of plane media with propagation velocity varying with depth $v(z)$, the epicentral distance and the travel time of a ray for a surface focus are given by

$$x = 2 \int_0^h \frac{p\, dz}{\sqrt{\eta^2 - p^2}}$$

$$t = 2 \int_0^h \frac{\eta^2\, dz}{\sqrt{\eta^2 - p^2}}$$

where $\eta = v^{-1}$ and h is the depth of maximum penetration of the ray. The variation of the epicentral distance x with the ray parameter p is given by

$$\frac{dx}{dp} = -\frac{2}{\varsigma_0 \sqrt{\eta_0^2 - p^2}} + 2 \int_0^\varsigma \frac{\frac{d\varsigma}{dz}\, dz}{\varsigma^2 \sqrt{\eta^2 - p^2}}$$

where

$$\varsigma = \frac{1}{v}\frac{dv}{dz}$$

In spherical media with velocity varying with depth $v(r)$, the epicentral distance, trajectory along the ray, and travel time are given by

$$\Delta = 2 \int_{r_p}^{r_0} \frac{p}{r}\frac{dr}{\sqrt{\eta^2 - r^2}}$$

$$s = 2 \int_{r_p}^{r_0} \frac{\eta\, dr}{\sqrt{\eta^2 - r^2}}$$

$$t = 2 \int_{r_p}^{r_0} \frac{dr}{v\sqrt{\eta^2 - r^2}}$$

where $\eta = rv^{-1}$, r_0 is the radius at the surface of the Earth, and r_p is the radius at the point of maximum penetration of the ray.

The variation of the distance from the epicentre Δ with the ray parameter p in a spherical medium is

$$\frac{d\Delta}{dp} = -\frac{2}{(1 - \varsigma_0)\sqrt{\eta_0^2 - p^2}} + 2 \int_0^\varsigma \frac{\frac{d\varsigma}{dr}\, dr}{(1 - \varsigma^2)\sqrt{\eta^2 - p^2}}$$

where

$$\varsigma = \frac{r}{v}\frac{dv}{dr}$$

The radial and vertical components (u_1 and u_3) of surface waves can be obtained from the potentials φ and ψ. The transverse component (u_2) is kept apart

$$u_1 = \frac{\partial \varphi}{\partial x_1} - \frac{\partial \psi}{\partial x_3} = \varphi_{,1} - \psi_{,3}$$

$$u_2 = C \exp[-iksx_3 + ik(x_1 - ct)]$$

$$u_3 = \frac{\partial \varphi}{\partial x_3} + \frac{\partial \psi}{\partial x_1} = \varphi_{,3} + \psi_{,1}$$

where c is the wave propagation velocity and

$$\varphi = A \exp[-ikrx_3 + ik(x_1 - ct)]$$
$$\psi = B \exp[-iksx_3 + ik(x_1 - ct)]$$

$$r = \sqrt{\frac{c^2}{\alpha^2} - 1}$$

$$s = \sqrt{\frac{c^2}{\beta^2} - 1}$$

For surface waves, $c < \beta < \alpha$, and hence r and s are imaginary.

For dispersive waves, the relationship between the phase velocity c and the group velocity U is

$$U = c + k\frac{dc}{dk}$$

where k is the wavenumber.

The position of the seismic focus is given by the coordinates of the epicentre (φ_0, λ_0) and the depth h. The time is that of the origin of the earthquake t_0. The size is given by the magnitude which is proportional to the logarithm of the amplitude of the recorded waves. For surface waves this is:

$$M_s = \log\frac{A}{T} + 1.66 \log \Delta + 3.3$$

where A is the amplitude of ground motion in microns, T is the period in seconds, and Δ the epicentral distance in degrees.

The magnitude of the moment is given by

$$M_w = \frac{2}{3}\log M_0 - 6.1$$

where M_0 is the seismic moment in N m (newton metres). The seismic moment is related to the displacement of the fault Δu and its area S:

$$M_0 = \mu \Delta u S$$

The mechanism of earthquakes is given by the orientation of the fracture plane (fault) defined by the angles φ (azimuth), δ (dip), and λ (slip angle or rake), or by the vectors $\boldsymbol{n}$ (the normal to the fault plane) and $\boldsymbol{l}$ (the direction of slip).

The elastic displacement of the waves produced by a point shear fault is

$$u_k(x_s, t) = \mu \Delta u(t) S (l_i n_j + l_j n_i)\frac{\partial G_{ki}}{\partial x_j}$$

where G_{ki} is the medium's Green's function which, for an isotropic, homogeneous, infinite medium, and P-waves in the far-field regime, is given by

$$G_{ki}^P = \frac{1}{4\pi\rho\alpha^2 r}\gamma_i\gamma_k\delta\left(t - \frac{r}{\alpha}\right)$$

The P-wave displacements are given by:

$$u_k^P(x_s, t) = \frac{\Delta \dot{u}(t) S}{4\pi\rho\alpha^3 r} \mu \left(l_i n_j + l_j n_i \right) \gamma_i \gamma_j \gamma_k$$

This equation can be expressed also in terms of the moment tensor M_{ij}

$$u_k^P(x_s, t) = \frac{\dot{M}_{ij}(t)}{4\pi\rho\alpha^3 r} \gamma_i \gamma_j \gamma_k$$

M_{ij} is a more general representation of a point source.

Heat flow

The Fourier law of heat transfer by diffusion states that the heat flux $\dot{q}$ is proportional to the gradient of the temperature T:

$$\dot{q} = -K\nabla T$$

where K is the thermal conductivity coefficient. The units of heat flow are $W\,m^{-2}$.

The heat diffusion equation, assuming that K is constant, is given by

$$\kappa\nabla^2 T + \frac{\varepsilon}{\rho C_v} = \frac{\partial T}{\partial t}$$

where C_v is the specific heat, ρ the density, ε the heat generated per unit volume and unit time (heat sources), and κ the thermal diffusivity:

$$\kappa = \frac{K}{\rho C_v}$$

If there are no heat sources, the diffusion equation is

$$\kappa\nabla^2 T = \frac{\partial T}{\partial t}$$

In the case of one-dimensional flow with periodic variation of temperature over time, one has:

$$T(z, t) = T_0 \exp\left[-\sqrt{\frac{\omega}{2\kappa}} z + i\left(-\sqrt{\frac{\omega}{2\kappa}} z + \omega t \right) \right]$$

where z is the vertical direction (positive towards the nadir) and ω the angular frequency.

In the case of stationary one-dimensional solutions (T constant in time) one obtains from the diffusion equation:

$$T = -\frac{\varepsilon}{2K} z^2 + \frac{\dot{q}_0}{K} z + T_0$$

where T_0 and $\dot{q}_0$ are the temperature and flow at the surface ($z = 0$).

For a spherical Earth, assuming that the thermal conductivity is constant, and that the amount of heat per unit volume depends only on time, the diffusion equation takes the form:

$$K\left(\frac{\partial^2 T}{\partial r^2} + \frac{2}{r}\frac{\partial T}{\partial r}\right) + \varepsilon(t) = \rho C_v \frac{\partial T}{\partial t}$$

where r is the radial direction.

For the stationary case, the above equation reduces to

$$\frac{1}{r^2}\frac{d}{dr}\left(r^2 \frac{dT}{dr}\right) = -\frac{\varepsilon}{K}$$

Integrating twice, one has

$$T = T_0 + \frac{\varepsilon}{6K}\left(R^2 - r^2\right)$$

where T_0 is the temperature at the surface ($r = R$).

Geochronology

Geochronology is based on determining the age of a rock by measuring the decay of its radioactive elements. In a sample of radioactive material, the number of atoms that have yet to disintegrate after time t is given by

$$n_t = n_0 e^{-\lambda t}$$

where n_0 is the initial number of atoms, and λ the decay constant. The rate of decay dn/dt is the activity R, so that

$$R = R_0 e^{-\lambda t}$$

where R_0 is the initial activity (at $t = 0$).

The half-life (or period) of the sample is the time it takes for the activity R to fall to half its initial value. It is given by:

$$T_{1/2} = \frac{0.693}{\lambda}$$

The mean life-time $\bar{t}$ of one of the atoms that existed at the start is given by:

$$\bar{t} = \frac{1}{\lambda}$$

If a sample consists of NR radioactive nuclei and NE stable nuclei, the time to arrive at the propotion NE/NR is given by

$$t = \frac{1}{\lambda}\left(1 + \frac{NE}{NR}\right)$$

If the rubidium–strontium (Rb-Sr) method is used to date a sample, a correction must be made for the contamination of the stable ^{86}Sr isotope relative to the radioisotope ^{87}Sr:

$$\left.\frac{^{87}\text{Sr}}{^{86}\text{Sr}}\right|_{\text{total}} = \left.\frac{^{87}\text{Sr}}{^{86}\text{Sr}}\right|_{\text{initial}} + \left.\frac{^{87}\text{Rb}}{^{86}\text{Sr}}\right|\left(e^{\lambda t} - 1\right)$$

This expression corresponds to a straight line (isochrone) of slope $\left(e^{-\lambda t} - 1\right)$ and intercept corresponding to the initial content $\left.\dfrac{^{87}\text{Sr}}{^{86}\text{Sr}}\right|_{\text{initial}}$.

2

Gravity

Terrestrial geoid and ellipsoid

1. Calculate the geodetic and geocentric latitudes of a point P on the ellipsoid whose radius vector is 6370.031 km, given that $m = 3.4425 \times 10^{-3}$, $GM = 39.86005 \times 10^{13}$ m^3 s^{-2}, and 6356.742 km is the polar radius. Determine J_2 and β.

The major semi-axis (equatorial radius) is a and the minor (polar radius) is c, the geocentric latitude is φ, and the geodetic latitude is φ_d (Fig. 1).

The coefficient m is given by the equation

$$m = \frac{\omega^2 a^3}{GM}$$

where $G = 6.67 \times 10^{-11}$ m^3 kg^{-1} s^{-2} is the gravitational constant, M the Earth's mass, and the angular velocity is $\omega = 2\pi/T$, where T is the rotation period ($T = 24$ h). We obtain for the semi-axis a the value

$$a = \left(\frac{mGMT^2}{4\pi^2}\right)^{1/3} = 6378.127 \, \text{km}$$

The Earth's flattening α can be obtained directly since we already know a and c so

$$\alpha = \frac{a - c}{c} = \frac{6378.127 - 6356.742}{6378.127} = 3.3529 \times 10^{-3}$$

The radius vector to the point P is given by the equation $r = a(1 - \alpha \sin^2 \varphi)$
From this equation we can calculate the geocentric latitude φ:

$$6370.031 = 6378.127\left(1 - 3.3539 \times 10^{-3} \sin^2 \varphi\right)$$

$$\varphi = \pm 37° \, 58' \, 22''$$

The relation between the geocentric φ and geodetic φ_d latitudes is given by

$$\tan \varphi_d = \frac{1}{(1 - \alpha)^2} \tan \varphi$$

Substituting the already obtained values for α and φ

$$\varphi_d = \pm 38° \, 09' \, 35''$$

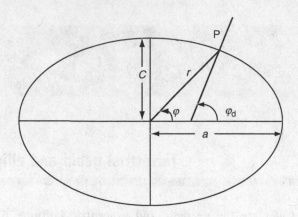

Fig. 1

The dynamic form factor J_2 can be obtained from the equation

$$\alpha = \frac{3}{2}J_2 + \frac{m}{2}$$

Then

$$J_2 = \frac{2}{3}\left(\alpha - \frac{m}{2}\right) = 1.0878 \times 10^{-3}$$

From this value we can determine the gravity flattening β using the equation

$$\beta = \frac{5}{2}m - \alpha = 5.2533 \times 10^{-3}$$

2. Taking the first-order approximation, let two points of the ellipsoid at 45° N and 30° S be situated at distances of 6367.444 km and 6372.790 km from the centre, respectively. If the normal gravity values are 9.806193 m s^{-2} for the first and 9.793242 m s^{-2} for the second, calculate: the flattening, gravity flattening, coefficient m, equatorial radius, polar radius, dynamic form factor, and the Earth's mass.

Data

$r_1 = 6367.444$ km $\varphi_1 = 45°$ N $\gamma_1 = 9.806193$ m s^{-2}
$r_2 = 6372.790$ km $\varphi_2 = 30°$ S $\gamma_2 = 9.793242$ m s^{-2}

The normal or theoretical gravity at a point can be expressed in terms of the normal gravity at the equator γ_e, the gravity flattening β, and the latitude of the point φ:

$$\gamma_1 = \gamma_e\left(1 + \beta \sin^2 \varphi_1\right) \qquad \gamma_2 = \gamma_e\left(1 + \beta \sin^2 \varphi_2\right)$$

If we divide both expressions we obtain:

$$\frac{\gamma_1}{\gamma_2} = \frac{1 + \beta \sin^2 \varphi_1}{1 + \beta \sin^2 \varphi_2}$$

From this expression we can obtain the gravity flattening, since we already know $\gamma_1, \gamma_2, \varphi_1, \varphi_2$:

$$\beta = \frac{\gamma_1 - \gamma_2}{\gamma_2 \sin^2 \varphi_1 - \gamma_1 \sin^2 \varphi_2} = 5.297 \times 10^{-3}$$

The distance r from the centre of the ellipsoid to points on its surface can be given as a function of the flattening α, the equatorial radius a, and the latitude φ:

$$r_1 = a(1 - \alpha \sin^2 \varphi_1)$$
$$r_2 = a(1 - \alpha \sin^2 \varphi_2)$$

If we divide both expressions

$$\frac{r_1}{r_2} = \frac{1 - \alpha \sin^2 \varphi_1}{1 - \alpha \sin^2 \varphi_2}$$

Thus we obtain the value of the flattening,

$$\alpha = 3.353 \times 10^{-3}$$

From this value we find the equatorial radius,

$$a = \frac{r_1}{1 - \alpha \sin^2 \varphi_1} = 6378.137 \text{ km}$$

The polar radius c can be found from this value and the flattening:

$$\alpha = \frac{a - c}{a} \quad \text{and} \quad c = a(1 - \alpha) = 6356.751 \text{ km}$$

The coefficient m is obtained from α and β:

$$\alpha + \beta = \frac{5}{2} m \quad m = \frac{2}{5}(\alpha + \beta) = 3.460 \times 10^{-3}$$

From this value we can obtain the value of the Earth's mass M from

$$m = \frac{\omega^2 a^3}{GM}$$

with $\omega = \frac{2\pi}{T} \pi$ where $T = 24$ hours.

Therefore

$$M = \frac{4\pi^2 a^3}{T^2 Gm} = 5.946 \times 10^{24} \text{kg}$$

3. Obtain the value of the terrestrial flattening in the first-order approximation, given that the normal gravity values for two points of the ellipsoid are:

Point 1: $\varphi_1 = 42° 20'$ $\gamma_1 = 980.389\ 063$ Gal
Point 2: $\varphi_2 = 47° 30'$ $\gamma_2 = 980.854\ 830$ Gal

Take the equatorial radius to be 6378.388 km.

For this problem we use the equations of Problems 1 and 2:

$$\gamma_1 = \gamma_e(1 + \beta \sin^2 \varphi_1)$$
$$\gamma_2 = \gamma_e(1 + \beta \sin^2 \varphi_2)$$

If we divide these expressions:

$$\frac{\gamma_1}{\gamma_2} = \frac{1 + \beta \sin^2 \varphi_1}{1 + \beta \sin^2 \varphi_2}$$

and we can solve for the gravity flattening, β:

$$\beta = 5.288\,2675 \times 10^{-3}$$

Using the gravity flattening β we can determine the value of gravity at the equator, γ_e:

$$\gamma_e = \frac{\gamma_1}{1 + \beta \sin^2 \varphi_1} = 978.043\,614 \, \text{Gal}$$

Using the following equations

$$\alpha + \beta = \frac{5}{2}m$$

$$\alpha = \frac{3}{2}J_2 + \frac{m}{2}$$

$$m = \frac{\omega^2 a^3}{GM}$$

we derive the expression

$$\gamma_e = \frac{GM}{a^2}(1 - \beta) + \omega^2 a$$

and substituting the values we obtain $GM = 3.986\,5415 \times 10^{14}$ m^3 s^{-2}

From this value, taking $T = 24$ hours, we obtain

$$m = \frac{\omega^2 a^3}{GM} = \frac{4\pi^2 a^3}{T^2 GM} = 3.442\,5698 \times 10^{-3}$$

And finally the Earth's flattening is

$$\alpha = \frac{5}{2}m - \beta = 3.318\,1575 \times 10^{-3}$$

4. P is a point of the terrestrial ellipsoid at latitude 60 °S and distance to the centre of 6362.121 km. The Earth's mass is 5.9761 $\times$ 10^{24} kg and the ratio between the polar and equatorial semi-axes is 0.9966. Taking the first-order approximation, calculate:

(a) The flattening and the coefficient J_2.
(b) The value of normal gravity in mGal at P.

As in the previous problems we use the equations given in Problems 1 and 2.

(a) From the equation for the flattening (Problem 2)

$$\alpha = 1 - \frac{c}{a} = 3.4 \times 10^{-3}$$

The value of the equatorial radius a is obtained from the equation

$$a = \frac{r}{1 - \alpha \sin^2 \varphi} = 6\,378\,386\,\text{m}$$

The coefficient m is given by

$$m = \frac{\omega^2 a^3}{GM} = \frac{4\pi^2 a^3}{T^2 GM} = 3.4429 \times 10^{-3}$$

The gravity flattening β is given by

$$\beta = \frac{5}{2}m - \alpha = 5.2072 \times 10^{-3}$$

The dynamic form factor J_2 is found from the relation

$$J_2 = \frac{2\alpha - m}{3} = 1.1190 \times 10^{-3}$$

(b) The normal gravity at that point is given by

$$\gamma = \gamma_e(1 + \beta \sin^2 \varphi) = 981\,856.3\,\text{mGal}$$

5. At a point P on the ellipsoid at latitude 50 °S, the value of normal gravity is 9.810 752 m s^{-2} and the distance to the centre of the Earth is 6365.587 km. Given that the mass of the Earth is 5.976 × 10^{24} kg and the ratio between the minor and major semi-axes is c/a = 0.996 6509, calculate:

(a) The flattening, equatorial radius, gravity flattening, dynamic form factor, and coefficient m.
(b) The normal gravity at the equator.
(c) The centrifugal force at P.

Data

$\varphi = 50°\,\text{S}$ $r = 6365.587\,\text{km}$ $\gamma = 9.810\,752\,\text{m s}^{-2}$.

(a) According to Problem 1, the flattening is given by

$$\alpha = 1 - \frac{c}{a} = 3.349 \times 10^{-3}$$

and the equatorial radius a is

$$a = \frac{r}{1 - \alpha \sin^2 \varphi} = 6378.122\,\text{km}$$

Taking $T = 24$ hours, the coefficient m is then given by

$$m = \frac{4\pi^2 a^3}{T^2 GM} = 3.4425 \times 10^{-3}$$

β and J_2 can be obtained from the equations

$$\beta = \frac{5}{2}m - \alpha = 5.2571 \times 10^{-3} \qquad J_2 = \frac{2}{3}\left(\alpha - \frac{m}{2}\right) = 1.0852 \times 10^{-3}$$

(b) The normal gravity γ_e at the equator is given by

$$\gamma_e = \frac{\gamma}{1 + \beta \sin^2 \varphi} = 9.780\,579\,\mathrm{m\,s^{-2}}$$

(c) The centrifugal force at point P is given by its radial and transverse components

$$f = f_r e_r + f_\theta e_\theta$$

where, since $90° - \theta = \varphi$,

$$f_r = \omega^2 r \sin^2 \theta = \omega^2 r \cos^2 \varphi = 0.013\,909\,\mathrm{m\ s^{-2}}$$

$$f_\theta = \omega^2 r \sin \theta \cos \theta = \omega^2 r \cos \varphi \sin \varphi = -0.016\,576\,\mathrm{m\ s^{-2}}$$

6. Taking the first-order approximation, calculate the Earth's flattening α, gravity flattening β, dynamic form factor J_2, and polar radius c, given that:
   $\gamma_e = 978.032$ **Gal (normal gravity at the equator)**
   $a = 6378.136$ **km (equatorial radius)**

   $GM = 39.8603 \times 10^{13}\,\mathrm{m^3\,s^{-2}}$

and that for a point on the ellipsoid at latitude 60 °N the normal gravity value is 981 921 mGal.
   Calculate also the radius vector of this point and the gravitational potential.

Assuming a first-order approximation, the expression for the normal gravity is

$$\gamma = \gamma_e (1 + \beta \sin^2 \varphi)$$

The value of β is given by

$$\beta = \frac{\gamma - \gamma_e}{\gamma_e \sin^2 \varphi} = 5.302 \times 10^{-3}$$

The coefficient m, taking $T = 24$ hours, is

$$m = \frac{4\pi^2 a^3}{T^2 GM} = 3.442 \times 10^{-3}$$

α and J_2 are determined from the equations

$$\alpha = \frac{5}{2}m - \beta = 3.303 \times 10^{-3} \qquad J_2 = \frac{2}{3}\left(\alpha - \frac{m}{2}\right) = 1.055 \times 10^{-3}$$

The polar radius c is determined from the flattening

$$\alpha = \frac{a - c}{a} \Rightarrow c = a(1 - \alpha) = 6357.069 \, \text{km}$$

The radius vector at the point of latitude 60 °N is given by

$$r = a(1 - \alpha \sin^2 \varphi) = 6362.335 \, \text{km}$$

The gravity potential at the same point, in the first-order approximation, is found using Mac Cullagh's formula,

$$U = \frac{GM}{r} \left[1 - \frac{J_2}{2} \left(\frac{a}{r} \right)^2 (3 \sin^2 \varphi - 1) + \left(\frac{r}{a} \right)^3 \frac{m}{2} \cos^2 \varphi \right] = 6.263\,57 \times 10^7 \, \text{m}^2 \, \text{s}^{-2}$$

7. Assuming that the Moon is an ellipsoid of equatorial radius 1738 km and polar radius 1737 km, with $J_2 = 3.8195 \times 10^{-4}$ and a mass of 7.3483×10^{22} kg, calculate its period of rotation.

First, we calculate the lunar flattening α,

$$\alpha = \frac{a - c}{a} = 5.7537 \times 10^{-4}$$

The coefficient m is found from the values of α and J_2:

$$\alpha = \frac{3}{2} J_2 + \frac{m}{2} \Rightarrow m = 2\alpha - 3J_2 = 7.5900 \times 10^{-6}$$

From the value of m we find the period of rotation T:

$$m = \frac{4\pi^2 a^3}{T^2 GM} \Rightarrow T = \left(\frac{4\pi^2 a^3}{mGM} \right)^{\frac{1}{2}} = 27.32 \, \text{days}$$

8. Calculate, in the first-order approximation, the latitude and radius vector of a point P of the terrestrial ellipsoid for which the value of normal gravity is 979.992 Gal, given that the Earth's mass is 5.976×10^{24} kg and that normal gravity for another point Q at 50° S latitude is 981.067 Gal, for the equator is 978.032 Gal, and that J_2 is 1.083×10^{-3}.

The gravity flattening β is found from the normal gravity at Q with latitude $\varphi_1 = -50°$:

$$\beta = \frac{\frac{\gamma_1}{\gamma_e} - 1}{\sin^2 \varphi_Q} = 5.288 \times 10^{-3}$$

Using the same expression and the normal gravity at P we calculate its latitude

$$\sin^2 \varphi_P = \frac{\frac{\gamma_A}{\gamma_e} - 1}{\beta} ; \qquad \varphi_P = \pm 37° \, 59' \, 46.57''$$

The Earth's flattening is given by

$$\alpha = \frac{15}{8} J_2 + \frac{\beta}{4} = 3.353 \times 10^{-3}$$

and the value of the coefficient m by

$$m = 2\alpha - 3J_2 = 3.457 \times 10^{-3}$$

From the value of m we obtain the equatorial radius a:

$$a = \left(\frac{mGM}{\omega^2}\right)^{\frac{1}{3}} = 6\,387\,062.758\,\text{m}$$

where we have substituted $\omega = 2\pi/T$, taking $T = 24$ hours.

Finally, we find the radius vector of point P

$$r = a(1 - \alpha\sin^2\varphi) = 6\,378\,946.678\,\text{m}$$

9. At a point P on the terrestrial ellipsoid of latitude 70 °S and radius vector 6359.253 km, the value of normal gravity is 982.609 Gal. If the mass of the Earth is 5.9769×10^{24} kg and the equatorial radius is 6378.136 km, calculate the value of normal gravity at the Pole, the dynamic form factor, and the centrifugal force at the Pole and the equator.

The flattening α is given by

$$\alpha = \frac{1}{\sin^2\varphi}\left(1 - \frac{r}{a}\right) = 3.3528 \times 10^{-3}$$

Putting $T = 24$ hours, the coefficient m is obtained from the equation

$$m = \frac{\omega^2 a^3}{GM} = 3.4425 \times 10^{-3}$$

From α and m we find the gravity flattening β:

$$\beta = \frac{5}{2}m - \alpha = 5.2535 \times 10^{-3}$$

Normal gravity at the equator is found from the value of the gravity at point P:

$$\gamma_e = \frac{\gamma}{1 + \beta\sin^2\varphi} = 9.780\,72\,\text{m s}^{-2}$$

From this value we find the normal gravity at the Pole:

$$\gamma_p = \gamma_e(1 + \beta) = 9.832\,10\,\text{m s}^{-2}$$

The dynamic form factor is found from the values of α and m:

$$J_2 = \frac{2}{3}\left(\alpha - \frac{m}{2}\right) = 1.0877 \times 10^{-3}$$

The centrifugal force is given by the expression:

$$\boldsymbol{f} = \omega^2 r\sin^2\theta\,\boldsymbol{e}_r + \omega^2 r\sin\theta\cos\theta\,\boldsymbol{e}_\theta$$

At the pole, $\theta = 90° - \varphi = 0 \to \boldsymbol{f} = 0$.
At the equator, $\theta = 90° \to \boldsymbol{f} = \omega^2 a\,\boldsymbol{e}_r = 0.033\,73\,\boldsymbol{e}_r\,\text{m s}^{-2}$

10. Let two points of the ellipsoid be of latitudes φ_1 and φ_2, with radius vectors 6372.819 km and 6362.121 km, respectively. The ratio of the normal gravities is 0.997 37, the flattening 3.3529×10^{-3}, and the gravity flattening 5.2884×10^{-3}. Calculate:

(a) The Earth's mass.
(b) The latitude of each point and the dynamic form factor.

(a) The equatorial radius a can be obtained from the ratio of the two normal gravities

$$\frac{\gamma_1}{\gamma_2} = \frac{1 + \beta\sin^2\varphi_1}{1 + \beta\sin^2\varphi_2} = \frac{1 + \beta\left(\dfrac{a - r_1}{a\alpha}\right)}{1 + \beta\left(\dfrac{a - r_2}{a\alpha}\right)} = l$$

where $l = 0.997\,37$. We solve for a and obtain

$$a = \frac{\beta(lr_2 - r_1)}{(\alpha + \beta)(l - 1)} = 6382.94\,\text{km}$$

We calculate the mass of the Earth from the flattening and gravity flattening as in Problem 2:

$$m = \frac{2}{5}(\alpha + \beta) = 3.4565 \times 10^{-3} \quad M = \frac{4\pi^2 a^3}{T^2 Gm} = 5.9653 \times 10^{24}\,\text{kg}$$

(b) The latitudes at each point are calculated from the radius vectors

$$\sin^2\varphi_1 = \frac{a - r_1}{a\alpha} \rightarrow \varphi_1 = \pm 43°\,27'\,58''$$

$$\sin^2\varphi_2 = \frac{a - r_2}{a\alpha} \rightarrow \varphi_2 = \pm 80°\,33'\,46''$$

The dynamic form factor is obtained from the values of α, β, and m:

$$\beta = 2\alpha + m - \frac{9}{2}J_2$$

$$J_2 = (2\alpha + m - \beta)\frac{2}{9} = 1.0831 \times 10^{-3}$$

11. Let a point A have a value of gravity of 9793 626.8 gu and a geopotential number of 32.614 gpu. Calculate the gravity at a point B, knowing that the increments in dynamic and Helmert height over point A are 271.116 m and 271.456 m, respectively. Take $\gamma_{45} = 9.806\,2940$ m s^{-2}. Give the units for each parameter.

The dynamic heights at points A and B are given by:

$$H_D^A = \frac{C_A}{\gamma_{45}}$$

$$H_D^B = \frac{C_B}{\gamma_{45}}$$

where C is the value of the geopotential at each point $\left(\sum_{j=1}^{N} g_j dh_j\right)$ and γ_{45} the normal gravity for a point on the ellipsoid at 45° latitude.

Subtracting both equations,

$$H_D^B - H_D^A = \frac{(C_B - C_A)}{\gamma_{45}}$$

Solving for C_B,

$$C_B = C_A + \gamma_{45}\left(H_D^B - H_D^A\right) = 298.478 \text{ gpu}$$

If heights are given in km and normal gravity in Gal, geopotentials are in gpu (geopotential units)
    1 gpu = 1 kGal m = 1 Gal km
    The Helmert orthometric height H is given by

$$H = \frac{C}{g + 0.0424H} \tag{11.1}$$

where C is in gpu, g in Gal, and H in km.
    Solving for H:

$$H = \frac{-g \pm \sqrt{g^2 + 4 \times 0.0424C}}{2 \times 0.0424}$$

Since the point A is above the geoid ($C_A > 0$), we take the positive solution,

$$H_A = 33.301 \text{ m}$$

Then, the Helmert height at point B is

$$H_B = H_A + \Delta H_A^B = 304.757 \text{ m}$$

The gravity at point B is calculated using Equation (11.1)

$$g_B = \frac{C_B}{H_B} - 0.0424H_B = 979.382\,75 \text{ Gal}$$

12. Calculate the value of gravity in gravimetric units and mGal of a point on the Earth's surface whose orthometric (Helmert) and dynamic heights are 678.612 m and 679.919 m, respectively, taking $\gamma_{45} = 9.806\,294$ m s^{-2}.

The geopotential is calculated from the dynamic height,

$$H_D = \frac{C}{\gamma_{45}} \Rightarrow C = H_D\gamma_{45} = 666.748 \text{ gpu}$$

where H_D is given in km and γ_{45} in Gal.
    Knowing the geopotential, we calculate the gravity from the orthometric (Helmert) height H, using its definition,

$$H = \frac{C}{g + 0.0424H} \Rightarrow g = \frac{C}{H} - 0.0424H = 982.489\,34 \text{ Gal} = 9\,824\,893.4 \text{ gu}$$

13. If at a point on the surface of the Earth of Helmert height 1000 m one observes a value of gravity of 9.796 235 m s^{-2}, calculate the average value of gravity

between that point and the geoid along the direction of the plumb-line, and the point's geopotential number.

The mean value of gravity between a height H and the surface of the geoid is given by

$$\bar{g} = \frac{1}{H} \int_0^H g(z)dz$$

where $g(z)$ is the value of gravity at a distance z from the geoid along the vertical path to a point of height H. This value can be obtained using the Poincaré and Prey reduction from the value of g observed at the Earth's surface at a point of height H,

$$g(z) = g + 0.0848(H - z)$$

Then

$$\bar{g} = \frac{1}{H} \int_0^H g(z)dz = \frac{1}{H} \int_0^H [g + 0.0848(H - z)]dz$$

$$= \frac{1}{H} \left[gz + 0.0848\,Hz - 0.0424z^2 \right]_0^H$$

$$\bar{g} = g + 0.0424\,H = 979.6659\,\text{Gal}$$

where g is given in Gal and H in km.

The geopotential C can be obtained from the formula for the Helmert height,

$$H = \frac{C}{g + 0.0424H} \Rightarrow C = (g + 0.0424\,H)\,H = 979.666\,\text{gpu}$$

14. For two points A and B belonging to a gravity measurement levelling line, one obtained:
$g_A = 9.801\,137\,6\ \text{m s}^{-2}$
$C_A = 933.316\ \text{gpu}$
Gross increment elevation: $\Delta h_A^B = -20.340$
Increment in dynamic height: $H_D^A - H_D^B = -20.340\ \text{m}$.
Given that the normal gravity at 45 ° latitude is 9806 294 gu, calculate the Helmert heigth of point B.

As in Problem 11, the dynamic heights at A and B are given by

$$H_D^A = \frac{C_A}{\gamma_{45}}$$

$$H_D^B = \frac{C_B}{\gamma_{45}}$$

Subtracting both equations:

$$H_D^B - H_D^A = \frac{(C_B - C_A)}{\gamma_{45}}$$

Solving for C_B:

$$C_B = C_A + \gamma_{45}\left(H_D^B - H_D^A\right) = 913.371\ \text{gpu}$$

The geopotential at B can be obtained from the gross increment in elevation between A and B,

$$C_B = C_A + \left(\frac{g_A + g_B}{2}\right) \Delta h_A^B$$

and, solving for g_B,

$$g_B = \frac{2(C_B - C_A)}{\Delta h_A^B} - g_A = 980.103\,08 \text{ Gal}$$

Finally we calculate the orthometric Helmert height at point B,

$$H = \frac{C}{g + 0.0424H}$$

Substituting the values for point B, and solving for H we obtain

$$H = \frac{-g_B \pm \sqrt{g_B^2 + 4 \times 0.0424C_B}}{2 \times 0.0424} = 931.875 \text{ m}$$

15. A, B, and C are points connected by a geometric levelling line. Given that the normal gravity at a latitude of 45° is 980.6294 Gal, complete the following table:

Station	Gravity (Gal)	Height Increment (m)	Geopotential Number (gpu)	Dynamic Height (m)	Helmert Height (m)
A	979.88696	–	664.982	?	?
B	?	−0.541	?	677.577	?
C	979.88665	?	?	?	657.134

Station A

Dynamic height:

$$H_D^A = \frac{C_A}{\gamma_{45}} = 678.118 \text{ m}$$

Helmert height:

$$H = \frac{C}{g + 0.0424H}$$

$$H = \frac{-g \pm \sqrt{g^2 + 4 \times 0.0424 \times C}}{2 \times 0.0424} = 678.611 \text{ m} \tag{15.1}$$

Station B

The geopotential number is found from the dynamic height:

$$C_B = \gamma_{45} H_D^B = 664.452 \text{ gpu}$$

From this value and the difference in height with respect to station A we find the gravity at B:

$$C_B = C_A + \left(\frac{g_A + g_B}{2}\right)\Delta h_B^A$$

from which we get $g_B = 979.877\,84$ Gal

The Helmert height of B is found as in station A:

$$H_B = 678.077\,\text{m}$$

Station C

From the known values of gravity and Helmert height we find the geopotential number (Equation 15.1)

$$C_C = gH_C + 0.0424H_C^2 = 661.574\,\text{gpu}$$

To calculate the difference in height of C with respect to B we begin with the expression

$$C_C = C_B + \left(\frac{g_B + g_C}{2}\right)\Delta h_B^C$$

from which

$$\Delta h_B^C = \frac{2(C_C - C_B)}{g_B + g_C} = -2.937\,\text{m}$$

The dynamic height is found directly from the geopotential number:

$$H_D^C = \frac{C_C}{\gamma_{45}} = 674.642\,\text{m}$$

The complete table is:

Station	Gravity (Gal)	Height increment (m)	Geopotential number (gpu)	Dynamic height (m)	Helmert height (m)
A	979.88696	–	664.982	678.118	678.611
B	979.87784	−0.541	664.452	677.577	678.077
C	979.88665	−2.937	661.574	674.642	657.134

Earth's gravity field and potential

16. Suppose an Earth is formed by a sphere of radius a and density ρ, and within it there are two spheres of radius $a/2$ with centres located on the axis of rotation. The density of that of the northern hemisphere is 5ρ and of that of the southern hemisphere is $\rho/5$. The value of the rotation is such that $m = 0.1$. Determine:

(a) The potential U in the r^{-3} approximation.
(b) The values of g_r and g_θ for a point on the equator in the r^{-2} approximation.

(c) The error made in (b) with respect to the exact solution.

(d) The deviation of the vertical from the radial at the equator.

(a) The gravitational potential is the sum of the potentials of the three spheres

$$V = V_1 + V_2 + V_3 = \frac{GM}{r} + \frac{GM_1}{q} + \frac{GM_2}{q'} \qquad (16.1)$$

where r is the distance from a point P to the centre of the sphere of radius a and mass M, where M is given by

$$M = \frac{4}{3}\pi\rho a^3$$

q and q' are the distances to the centres of the two spheres in its interior in the northern and southern hemispheres which have differential masses M_1 and M_2, respectively (Fig. 16).

The differential masses are those corresponding to the difference in density in each case with respect to the large sphere:

$$M_1 = \frac{4}{3}\pi(5\rho - \rho)\frac{a^3}{8} = \frac{M}{2}$$

differential mass of the sphere in the northern hemisphere

$$M_2 = \frac{4}{3}\pi\left(\frac{\rho}{5} - \rho\right)\frac{a^3}{8} = -\frac{M}{10}$$

differential mass of the sphere in the southern hemisphere

The distance q can be calculated using the cosine law

$$q = \sqrt{r^2 + \left(\frac{a}{2}\right)^2 - 2\frac{a}{2}r\cos\theta} = r\sqrt{1 + \left(\frac{a}{2r}\right)^2 - 2\left(\frac{a}{2r}\right)\cos\theta}$$

Considering this expression, $1/q$ corresponds to one of the generating functions of the Legendre polynomials. Then $1/q$, in the first-order approximation, is given by

$$\frac{1}{q} = \frac{1}{r}\left[1 + \frac{a}{2r}\cos\theta + \frac{1}{2}\left(\frac{a}{2r}\right)^2(3\cos^2\theta - 1)\right]$$

Since $\cos\theta' = -\cos\theta$, $1/q'$ is given by

$$\frac{1}{q'} = \frac{1}{r}\left[1 - \frac{a}{2r}\cos\theta + \frac{1}{2}\left(\frac{a}{2r}\right)^2(3\cos^2\theta - 1)\right]$$

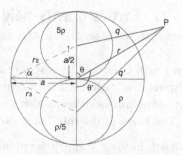

Fig. 16

If we substitute in Equation (16.1), the potentials for each sphere are given by

$$V_1 = \frac{GM}{r} \qquad V_2 = \frac{GM}{2}\left(\frac{1}{r} + \frac{a}{2r^2}\cos\theta + \frac{a^2}{8r^3}\left(3\cos^2\theta - 1\right)\right)$$

$$V_3 = -\frac{GM}{10}\left(\frac{1}{r} - \frac{a}{2r^2}\cos\theta + \frac{a^2}{8r^3}\left(3\cos^2\theta - 1\right)\right)$$

Then, the total gravity potential is the sum of the three gravitational potentials plus the potential of the centrifugal force due to the rotation:

$$U = GM\left[\left(1 + \frac{1}{2} - \frac{1}{10}\right)\frac{1}{r} + \left(\frac{1}{4} + \frac{1}{20}\right)\frac{a}{r^2}\cos\theta \right.$$
$$\left. + \left(\frac{1}{16} - \frac{1}{80}\right)\frac{a^2}{r^3}\left(3\cos^2\theta - 1\right)\right] + \frac{1}{2}r^2\omega^2\sin^2\theta\right]$$

In terms of the coefficient m, given here by $m = \dfrac{\omega^2 a^3}{GM}$,

$$U = \frac{GM}{a}\left[\frac{7}{5}\frac{a}{r} + \frac{3}{10}\frac{a^2}{r^2}\cos\theta + \frac{1}{20}\frac{a^3}{r^3}\left(3\cos^2\theta - 1\right) + \frac{1}{2}m\left(\frac{r}{a}\right)^2\sin^2\theta\right]$$

(b) Using this first-order approximation of the potential, the radial and tangential components of gravity at the equator, $r = a$ and $\theta = 90°$, putting $m = 0.1$, are

$$g_r = \frac{\partial U}{\partial r} = -1.3\frac{GM}{a^2}$$

$$g_\theta = \frac{1}{r}\frac{\partial U}{\partial \theta} = -0.3\frac{GM}{a^2}$$

(c) To calculate exactly the value of g_r at the equator we have to calculate the exact contribution of each of the three spheres plus the centrifugal force ($m = 0.1$):

$$g_r = g_r^1 + g_r^2 + g_r^3 - m\frac{GM}{a^2}$$

$$g_r^1 = -\frac{GM}{a^2}$$

$$g_r^2 = -\frac{GM}{2r_2^2}\cos\alpha$$

$$g_r^3 = -\frac{GM}{10r_3^2}\cos\alpha$$

where

$$r_2 = \sqrt{a^2 + \frac{a^2}{4}} = r_3$$

and α is the angle which forms r_2 and r_3 with the equator (Fig. 16)

$$\sin \alpha = \frac{a/2}{r_2} = \frac{1}{\sqrt{5}}$$

Then

$$g_r = -\frac{GM}{a^2}\left(1 + \frac{4}{5\sqrt{5}} - \frac{4}{25\sqrt{5}} - m\right) = -1.19\frac{GM}{a^2}$$

$$g_\theta = -g_\theta^1 + g_\theta^2 = \left(-\frac{GM}{2\frac{5}{4}a^2}\frac{1}{\sqrt{5}} + \frac{GM}{10\frac{5}{4}a^2}\frac{1}{\sqrt{5}}\right)$$

$$= \frac{GM}{a^2}\left(-\frac{2}{5\sqrt{5}} + \frac{2}{25\sqrt{5}}\right) = -0.14\frac{GM}{a^2}$$

The error made in the first-order approximation with respect to the exact solution is

$$\Delta g_r = (-1.19 + 1.3)\frac{GM}{a^2} = 0.11\frac{GM}{a^2}$$

$$\Delta g_\theta = (-0.3 + 0.14)\frac{GM}{a^2} = -0.16\frac{GM}{a^2}$$

(d) The deviation of the vertical with respect to the radial direction is given by the angle i which is determined from the gravity components g_r and g_θ. At the equator this angle is:

- Using the first order approximation

$$\tan i = \frac{g_\theta}{g_r} = \frac{0.3}{1.3} \Rightarrow i = 13.0°$$

- Using the exact values

$$\tan i = \frac{0.16}{1.19} \Rightarrow i = 7.6°$$

17. A spherical planet is formed by a sphere of radius a and density ρ, and inside it a sphere of radius $a/2$ and density 5ρ centred at the midpoint of the radius of the northern hemisphere. There is no rotation.

(a) Determine J_0, J_1, and J_2.
(b) What is the deviation of the vertical from the radial at the equator?

(a) The total gravitational potential is the sum of the potentials of the two spheres (Fig. 17) where g, is the attraction due to the potential V_1 and g_2 that due to the potential V_2:

$$V = V_1 + V_2 = \frac{GM}{r} + \frac{GM'}{q}$$

where r and q are the distances from a point P to the centres of the large and small spheres, respectively.

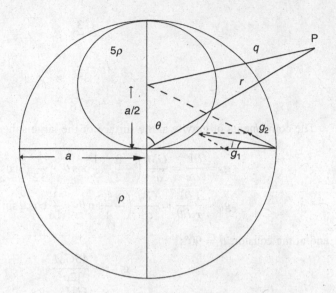

Fig. 17

As we did in Problem 16, for the small sphere of radius $a/2$ we take the differential mass M'

$$M' = \frac{4}{3}\pi\frac{a^3}{8}(5\rho - \rho) = \frac{16}{24}\pi\rho a^3 = \frac{M}{2}$$

where M is the mass of the sphere of radius a and density ρ.

For $1/q$ we take the first-order approximation of the Legendre polynomial, as we did in Problem 16:

$$\frac{1}{q} = \frac{1}{r}\left(1 + \frac{a}{2r}\cos\theta + \frac{1}{2}\left(\frac{a}{2r}\right)^2(3\cos^2\theta - 1)\right)$$

Then, the expression for the gravitational potential V is:

$$V = \frac{GM}{r} + \frac{GM}{2}\left(\frac{1}{r} + \frac{a}{2r^2}\cos\theta + \frac{1}{4}\frac{a^2}{4r^3}(3\cos^2\theta - 1)\right)$$

$$= GM\left(\frac{3}{2r} + \frac{a}{4r^2}\cos\theta + \frac{a^2}{32r^3}(3\cos^2\theta - 1)\right)$$

(b) We know that the potential can be expressed by an expansion in zonal spherical harmonics (Legendre polynomials) given in the first-order approximation by

$$V = \frac{GM}{a}\left(J_0\left(\frac{a}{r}\right) + \left(\frac{a}{r}\right)^2 J_1\cos\theta + \left(\frac{a}{r}\right)^3 J_2\frac{1}{2}(3\cos^2\theta - 1)\right)$$

Comparing the two expressions we obtain,

$$J_0 = \frac{3}{2}$$

$$J_1 = \frac{1}{4}$$

$$J_2 = \frac{1}{16}$$

The components of gravity at the surface of the large sphere ($r = a$) are:

$$g_r = \frac{\partial V}{\partial r} = \frac{GM}{a^2}\left(-\frac{3}{2} - \frac{1}{2}\cos\theta - \frac{3}{32}\left(3\cos^2\theta - 1\right)\right)$$

$$g_\theta = \frac{1}{r}\frac{\partial V}{\partial \theta} = \frac{GM}{a^2}\left(-\frac{1}{4}\sin\theta - \frac{3}{16}\cos\theta\sin\theta\right)$$

and at the equator, $\theta = 90°$:

$$g_r = -\frac{45GM}{32a^2}$$

$$g_\theta = -\frac{GM}{4a^2}$$

At the equator the deviation of the vertical with respect to the radial direction is

$$\tan i = \frac{g_\theta}{g_r} = \frac{8}{45}$$

$$i = 10.08°$$

18. Suppose an Earth is formed by a sphere of radius a and density ρ, and within it there are two spheres of radius $a/2$ and density 2ρ with centres located on the axis of rotation in each hemisphere. If M is the mass of the sphere of radius a, calculate:

(a) **The potential $U(r,\theta)$ and the form of the equipotential surface passing through the Poles.**
(b) **The component g_r of gravity in the first-order approximation for points on the surface.**
(c) **Calculate g_r directly at the Pole and the equator, and compare with the first-order approximation.**

(a) This problem is similar to Problem 16, but now the density of the two spheres is the same. The total gravity potential is the sum of the gravitational potentials of the three spheres (V, V_1 and V_2) plus the potential due to the rotation Φ:

$$U = V + V_1 + V_2 + \Phi \tag{18.1}$$

where

$$V = \frac{GM}{r}$$

$$V_1 = \frac{GM'}{q_1}$$

$$V_2 = \frac{GM'}{q_2}$$

$$\Phi = \frac{1}{2}\omega^2 r^2\sin^2\theta$$

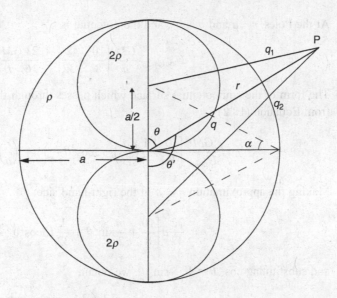

Fig. 18

and where M is the mass of the large sphere of radius a and M' the differential mass of each of the small spheres of radius $a/2$, r is the distance from a point P to the centre of the large sphere, and q_1 and q_2 the distances from P to the centres of the small spheres (Fig. 18). As in Problem 16 the differential mass is given by the difference in density between the large and the small spheres:

$$M' = \frac{4}{3}\pi(2\rho - \rho)\left(\frac{a}{2}\right)^3 = \frac{M}{8}$$

The inverse of the distance $1/q$ can be approximated by

$$\frac{1}{q_1} = \frac{1}{r}\left[1 + \frac{a}{2r}\cos\theta + \left(\frac{a}{2r}\right)^2\frac{1}{2}\left(3\cos^2\theta - 1\right)\right]$$

and since $\cos\theta' = -\cos\theta$

$$\frac{1}{q_2} = \frac{1}{r}\left[1 - \frac{a}{2r}\cos\theta + \left(\frac{a}{2r}\right)^2\frac{1}{2}\left(3\cos^2\theta - 1\right)\right]$$

The potential of the rotation can be written in terms of the coefficient $m = a^3\omega^2/GM$,

$$\Phi = \frac{GM}{r}\left(\frac{1}{2}\frac{1}{a^3}\frac{a^3\omega^2}{GM}r^3\right)\sin^2\theta = \frac{GM}{r}\frac{m}{2}\left(\frac{r}{a}\right)^3\sin^2\theta$$

Substituting in Equation (18.1)

$$U = \frac{GM}{r}\left[\frac{10}{8} + \frac{1}{8}\left(\frac{a}{2r}\right)^2(3\cos^2\theta - 1) + \left(\frac{r}{a}\right)^3\frac{m}{2}\sin^2\theta\right] \qquad (18.2)$$

At the Poles, $r = a$ and $\theta = 0°$, and the potential is

$$U_{\text{poles}} = \frac{GM}{a}\left[\frac{10}{8} + \frac{2}{32}\right] = \frac{21}{16}\frac{GM}{a}$$

The form of the equipotential surface which passes through the Poles ($r = a$) is obtained from Equation (18.2)

$$r = \frac{GM}{U_{\text{poles}}}\left[\frac{10}{8} + \frac{1}{8}\left(\frac{a}{2r}\right)^2(3\cos^2\theta - 1) + \left(\frac{r}{a}\right)^3\frac{m}{2}\sin^2\theta\right]$$

Making the approximation $r = a$ in the right-hand side:

$$r = \frac{32}{42}a\left[\frac{10}{8} + \frac{m}{2}\sin^2\theta + \frac{1}{32}(3\cos^2\theta - 1)\right]$$

and substituting $\cos^2\theta = 1 - \sin^2\theta$, we obtain

$$r = a\left[1 - \frac{32}{42}\left(\frac{3}{32} - \frac{m}{2}\right)\sin^2\theta\right]$$

This is the equation of an ellipse with flattening $\alpha = (32/42)(3/32 - m/2)$. Since there is symmetry with respect to the axis of rotation, the equipotential surface is an ellipsoid of revolution.

At the poles: $\theta = 0° \Rightarrow r_p = a$.
At the equator: $\theta = 90° \Rightarrow r_e = a\left[1 + \frac{32}{42}\left(\frac{m}{2} - \frac{3}{32}\right)\right]$

Depending on the value of m, we have the following cases,

$$\frac{m}{2} = \frac{3}{32} \Rightarrow r_e = a \Rightarrow \text{sphere}$$

$$\frac{m}{2} < \frac{3}{32} \Rightarrow r_e < a \Rightarrow \text{prolate ellipsoid}$$

$$\frac{m}{2} > \frac{3}{32} \Rightarrow r_e > a \Rightarrow \text{oblate ellipsoid}$$

$$m = 0 \Rightarrow r = a\left(1 - \frac{3}{42}\right) < a \Rightarrow \text{prolate ellipsoid}$$

(b) For the gravity at the Pole, in the first-order approximation, we take the derivative of the potential (18.2) and substitute $\theta = 0° \Rightarrow r_p = a$:

$$g_r = \frac{\partial U}{\partial r} = \frac{GM}{a^2}\left[-\frac{10}{8} - \frac{6}{32}\right] = -1.4375\frac{GM}{a^2}$$

(c) The exact solution for the gravity at the pole is the sum of the attractions of the three spheres:

$$g_r = -\frac{GM}{a^2} - \frac{GM}{2a^2} - \frac{GM}{18a^2} = -1.5555\frac{GM}{a^2}$$

At the equator we take the derivative of the potential and substitute $r = a$ and $\theta = 90°$:

$$g_r = \frac{\partial U}{\partial r} = \frac{GM}{a^2}\left[-\frac{10}{8} + \frac{3}{32} + m\right]$$

$$= -\frac{GM}{a^2}\left[\frac{37}{32} - m\right] = -\frac{GM}{a^2}[1.1562 - m]$$

For the exact solution we write

$$g_r = -\frac{GM}{a^2} - \frac{2GM}{8q^2}\cos\alpha + \omega^2 a$$

From Fig. 18 the distance q is given by

$$q^2 = \frac{a^2}{4} + a^2 = \frac{5}{4}a^2$$

$$\cos\alpha = \sqrt{\frac{4}{5}}$$

Therefore

$$g_r = -\frac{GM}{a^2}\left[1 + \frac{8}{40}\sqrt{\frac{4}{5}} - m\right] = -\frac{GM}{a^2}[1.1789 - m]$$

The approximated values are smaller than the exact solutions.

19. For the case of Problem 18, if $GM = 4 \times 10^{-3}$ m^3 s^{-2}, $a = 6 \times 10^3$ km, and $\omega = 7 \times 10^{-5}$ s^{-1}, calculate the values of J_2, α, m, H, and β.

From the definition of m we obtain

$$m = \frac{a^3\omega^2}{GM} = \frac{216 \times 10^{18} \times 49 \times 10^{-10}}{4 \times 10^{14}} = 2.6 \times 10^{-3}$$

The value of J_2 is obtained by comparing the two expressions for the potential U (Problem 18):

$$U = \frac{GM}{r}\left[1 + \left(\frac{a}{r}\right)^2 J_2 \frac{1}{2}\left(3\cos^2\theta - 1\right) + \frac{m}{2}\left(\frac{r}{a}\right)^3\sin^2\theta\right]$$

$$U = \frac{5}{4}\frac{GM}{r}\left[1 + \frac{1}{40}\left(\frac{a}{r}\right)^2\left(3\cos^2\theta - 1\right) + \frac{m}{2}\left(\frac{r}{a}\right)^3\sin^2\theta\right]$$

Then, $J_2 = 0.05$.

The flattening is obtained from the relation

$$\alpha = \frac{3}{2}J_2 + \frac{m}{2}$$

$$\alpha = \frac{3}{2} \times 50 \times 10^{-3} + \frac{2.3 \times 10^{-3}}{2} = 0.0765$$

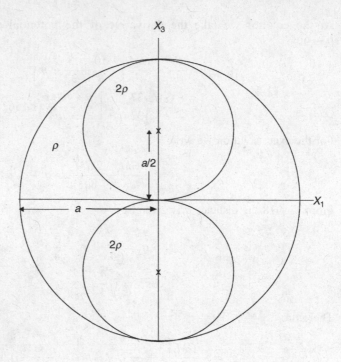

The gravity flattening is given by

$$\beta = \frac{g_p - g_e}{g_e} = \frac{1.555 - 1.175}{1.175} = 0.323$$

where we have used the values of gravity at the Pole and equator obtained in Problem 18, and in the latter we have substituted the value obtained for the coefficient m.

The dynamic ellipticity H is defined as the ratio of the moments of inertia with respect to the polar and equatorial radius (Fig. 19a):

$$H = \frac{C - A}{C}$$

where A and C are the moments of inertia of a sphere respect to the polar and equatorial radi: (axes x_1 and x_3). The moment of inertia of a sphere of radius R is

$$I_{sph} = \frac{2}{5}MR^2$$

We have to add to the moment of inertia of the sphere of radius a the moments of inertia of the two internal spheres of radius $a/2$. For the C-axis (x_3) we have

$$I_C = I_{sph\,a} + 2I_{sph\frac{a}{2}} = \frac{2}{5}Ma^2 + 2\frac{2}{5}\frac{M}{8}\frac{a^2}{4} = 0.425Ma^2$$

For the A-axis (x_1) the moment of inertia of each of the small spheres is given by (Fig. 19b)

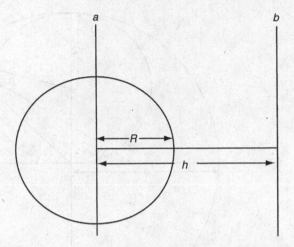

$$I = I_{CM} + Mh^2$$

since in this case the A axis does not coincide with the centre of mass, where $R = a/2$ and $h = a/2$:

$$I_{\text{sph }a/2} = \frac{2}{5}\frac{M}{8}\frac{a^2}{4} + \frac{M}{8}\frac{a^2}{4} = 0.044Ma^2$$

$$I_A = I_{\text{sph }a} + 2I_{\text{sph }a/2} = \frac{2}{5}Ma^2 + 2 \times 0.044Ma^2 = 0.488Ma^2$$

Finally

$$H = \frac{C - A}{C} = \frac{I_C - I_A}{I_C} = \frac{0.425 - 0.488}{0.425} = -0.147$$

20. Suppose an Earth is formed by a sphere of radius a and density ρ, and within it there is a sphere of radius $a/2$ and density 5ρ centred at the midpoint of the northern-hemisphere polar radius. If $m = 1/8$ and M is the mass of the sphere of radius a, determine:

(a) The form of the equipotential surface passing through the North Pole.

(b) For latitude 45°, the astronomical latitude and the deviation of the vertical from the radial.

(a) The gravitational potential is the sum of the potentials for the sphere of radius a and that of the sphere of radius $a/2$ (Fig. 20):

$$V = V_1 + V_2 = \frac{GM}{r} + \frac{GM_1}{q}$$

As in the previous problems the potential of the small sphere is given in terms of differential mass M_1:

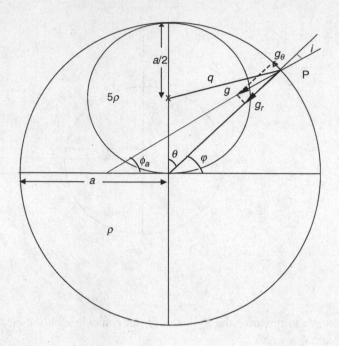

Fig. 20

$$M_1 = \frac{4}{3}\pi(5\rho - \rho)\left(\frac{a}{2}\right)^3 = \frac{M}{2}$$

and for the inverse of the distance $1/q$ we use the approximation

$$\frac{1}{q} = \frac{1}{r}\left[1 + \frac{a}{2r}\cos\theta + \left(\frac{a}{2r}\right)^2\frac{1}{2}\left(3\cos^2\theta - 1\right)\right]$$

Then, the total gravitational potential is

$$V = \frac{GM}{r}\left[\frac{3}{2} + \frac{a}{4r}\cos\theta + \frac{1}{16}\left(\frac{a}{r}\right)^2\left(3\cos^2\theta - 1\right)\right]$$

The total potential U is the sum of the gravitational potential V plus the potential of rotation Φ, where

$$\Phi = \frac{1}{2}r^2\omega^2\sin^2\theta$$

and using the coefficient $m = \omega^2 a^3/GM = 1/8$, we have

$$U = \frac{GM}{r}\left[\frac{3}{2} + \frac{a}{4r}\cos\theta + \frac{1}{16}\left(\frac{a}{r}\right)^2\left(3\cos^2\theta - 1\right) + \left(\frac{r}{a}\right)^3\frac{m}{2}\sin^2\theta\right]$$

At the North Pole, $\theta = 0°$ and $r = a$, and the value of the potential is

$$U_\text{p} = \frac{GM}{a}\left[\frac{3}{2} + \frac{1}{4} + \frac{1}{8}\right] = \frac{15GM}{8a}$$

The form of the equipotential surface is found by putting $U = U_p$:

$$\frac{15GM}{8a} = \frac{GM}{r}\left[\frac{3}{2} + \frac{a}{4r}\cos\theta + \frac{1}{16}\left(\frac{a}{r}\right)^2(3\cos^2\theta - 1) + \left(\frac{r}{a}\right)^3\frac{m}{2}\sin^2\theta\right]$$

Putting $r = a$ inside the square brackets and solving for r we find

$$r = a\frac{4}{5}\left[1 + \frac{1}{6}\cos\theta + \frac{1}{12}\cos^2\theta\right]$$

(b) The deviation of the vertical with respect to the radial direction is given by the angle i:

$$\tan i = \frac{g_\theta}{g_r}$$

To find this value we have to calculate the two components of gravity

$$g_r = \frac{\partial U}{\partial r} = GM\left[-\frac{3}{2}\frac{1}{r^2} - \frac{2a}{4}\frac{1}{r^3}\cos\theta - \frac{3a^2}{16}\frac{1}{r^4}[3\cos^2\theta - 1] + \frac{2r}{16a^3}\sin^2\theta\right]$$

$$g_\theta = \frac{1}{r}\frac{\partial U}{\partial\theta} = \frac{GM}{r}\left[-\frac{a}{4r^2}\sin\theta - \frac{a^2}{16r^3}6\cos\theta\sin\theta + \frac{r^2}{8a^3}\sin\theta\cos\theta\right]$$

For a point on the surface we put $r = a$:

$$g_r = -\frac{GM}{a^2}\left[\frac{19}{16} + \frac{1}{2}\cos\theta + \frac{11}{16}\cos^2\theta\right]$$

$$g_\theta = -\frac{GM}{a^2}\left[\frac{1}{4}\sin\theta + \frac{1}{4}\sin\theta\cos\theta\right]$$

and for latitude 45°

$$g_r = -\frac{GM}{a^2}1.88$$

$$g_\theta = -\frac{GM}{a^2}0.30$$

Then the angle i is given by $\tan i = \dfrac{0.30}{1.88} \Rightarrow i = 9.0°$.
   The astronomical latitude is

$$\phi_a = 90° - \theta - i = 36.0°$$

21. If the internal sphere of Problem 20 is located on the equatorial radius at longitude zero, find expressions for the components of gravity: g_r, g_θ, g_λ.

As in the previous problem the differential mass of the small sphere M_1 is (Fig. 21a):

$$M_1 = \frac{M}{2}$$

The total potential U is the sum of the gravitational potentials V and V_1, and the potential due to rotation Φ. According to Fig. 21b, using the relations of spherical triangles, if φ and λ are the coordinates of the point where the potential is evaluated, then

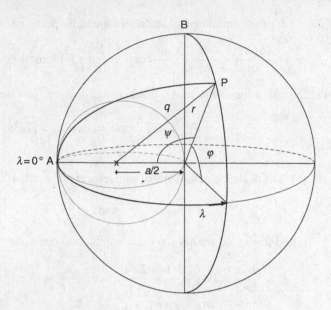

Fig. 21a

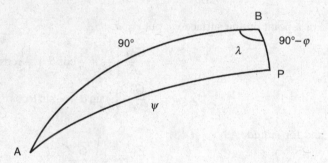

Fig. 21b

$$\cos\psi = \cos 90° \cos(90°-\varphi) + \sin 90° \sin(90°-\varphi) \cos\lambda$$
$$\cos\psi = \cos\varphi \cos\lambda$$

Using the expression for $1/q$ as in Problem 16,

$$\frac{1}{q} = \frac{1}{r}\left[1 + \frac{a}{2r}\cos\psi + \left(\frac{a}{2r}\right)^2 \frac{1}{2}\left(3\cos^2\psi - 1\right)\right]$$

The gravitational potential of the small sphere is given by

$$V_1 = \frac{GM}{2r}\left[1 + \frac{a}{2r}\cos\varphi\cos\lambda + \frac{a^2}{8r^2}\left(3\cos^2\varphi\cos^2\lambda - 1\right)\right]$$

The total potential U is given by:

$$U = \frac{GM}{r}\left[\frac{3}{2} + \frac{a}{4r}\cos\varphi\cos\lambda + \frac{a^2}{16r^2}\left(3\cos^2\varphi\cos^2\lambda - 1\right) + \frac{1}{2}r^2\omega^2\cos^2\varphi\right]$$

Using the coefficient m and $\sin \theta = \cos \varphi$, we obtain

$$U = GM \left[\frac{3}{2r} + \frac{a^2}{4r^2} \sin \theta \cos \lambda + \frac{a^2}{16r^3} (3\sin^2\theta\cos^2\lambda - 1) + \frac{r^2}{a^3} \frac{m}{2} \sin^2\theta \right]$$

The three components of gravity are found by differentiating U with respect to r, θ, and λ:

$$g_r = \frac{\partial U}{\partial r} = GM \left[-\frac{3}{2r^2} - \frac{a^2}{2r^3} \sin \theta \cos \lambda - \frac{3a^2}{16r^4} (3\sin^2\theta\cos^2\lambda - 1) + \frac{r}{a^3} m\sin^2\theta \right]$$

$$g_\theta = \frac{1}{r} \frac{\partial U}{\partial \theta} = GM \left[\frac{a^2}{4r^3} \cos \theta \cos \lambda + \frac{a^2}{16r^4} 6 \cos \lambda \sin \theta \cos \theta + \frac{r}{a^3} m \sin \theta \cos \theta \right]$$

$$g_\lambda = \frac{1}{r \sin \theta} \frac{\partial U}{\partial \lambda} = GM \left[-\frac{a^2}{4r^3} \sin \lambda - \frac{3a^2}{8r^4} \sin \theta \cos \lambda \sin \lambda \right]$$

22. A planet is formed by a sphere of radius a and density ρ, with a spherical core of density 5ρ and radius $a/2$ centred on the axis of rotation in the northern hemisphere and tangential to the equator. The planet rotates with $m = 1/4$. For the point at coordinates (45° N, 45° E), calculate:

(a) The astronomical latitude.
(b) The deviation of the vertical from the radial.
(c) The angular velocity of rotation that would be required for this deviation to be zero.

(a) The gravitational potential is the sum of the potentials of the two spheres (Fig. 22):

$$V = V_1 + V_2 = \frac{GM}{r} + \frac{GM'}{q} \tag{22.1}$$

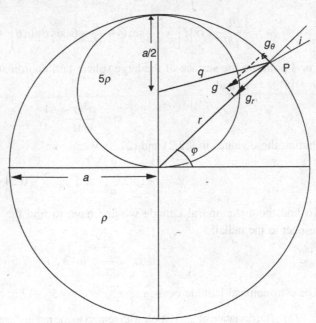

Fig. 22

As in Problem 16, the inverse of the distance from a point P to the centre of the small sphere, $1/q$, can be approximated by

$$\frac{1}{q} = \frac{1}{r}\left[1 + \frac{a}{2r}\cos\theta + \frac{1}{2}\left(\frac{a}{2r}\right)^2(3\cos^2\theta - 1)\right]$$

As in previous problems we use the differential mass of the small sphere,

$$M' = \frac{4}{3}\pi\frac{a^3}{8}(5\rho - \rho) = \frac{2}{3}\pi\rho a^3$$

and since $M = \frac{4}{3}\pi\rho a^3$, then $M' = M/2$.

Substituting in Equation (22.1) we obtain

$$V = \frac{GM}{r} + \frac{GM}{2}\left[\frac{1}{r} + \frac{a}{2r^2}\cos\theta + \frac{1}{2}\frac{a^2}{4r^3}(3\cos^2\theta - 1)\right]$$

$$= GM\left[\frac{3}{2r} + \frac{a}{4r^2}\cos\theta + \frac{a^2}{16r^3}(3\cos^2\theta - 1)\right]$$

The total potential U is the sum of V plus the potential due to rotation Φ:

$$U = V + \Phi = \frac{GM}{r}\left[\frac{3}{2} + \frac{a}{4r}\cos\theta + \frac{a^2}{16r^2}(3\cos^2\theta - 1)\right] + \frac{1}{2}r^2\omega^2\sin^2\theta$$

The components of gravity g_r and g_θ are

$$g_r = \frac{\partial U}{\partial r} = GM\left[-\frac{3}{2r^2} - \frac{a}{2r^3}\cos\theta - \frac{3a^2}{16r^4}(3\cos^2\theta - 1)\right] + r\omega^2\sin^2\theta \qquad (22.2)$$

$$g_\theta = \frac{1}{r}\frac{\partial U}{\partial\theta} = GM\left[-\frac{a}{4r^3}\sin\theta - \frac{a^2}{16r^4}6\cos\theta\sin\theta\right] + r\omega^2\sin\theta\cos\theta \qquad (22.3)$$

For a point on the surface of the large sphere and coordinates 45° N, 45° E, we have that $\theta = 45°$, $r = a$, and

$$m = \frac{a^3\omega^2}{GM} = \frac{1}{4}$$

Putting these values in (22.2) and (22.3), we obtain

$$g_r = -1.82\frac{GM}{a^2} \text{ and } g_\theta = -0.24\frac{GM}{a^2}$$

To find the astronomical latitude we first have to find the deviation of the vertical with respect to the radial:

$$\tan i = \frac{g_\theta}{g_r} = 0.13\,, \ i = 7.5°$$

The astronomical latitude is, then, $\phi = \varphi - i = 45° - 7.5° = 37.5°$.

(b) The deviation of the vertical with respect to the radial direction, as already found, is $i = 7.5°$.
(c) If we want the deviation of the vertical to be null, $i = 0$, this implies $g_\theta = 0$.

writing Equation (22.3) in terms of the coefficient m, where

$$m = \frac{a^3 \omega^2}{GM} \tag{22.4}$$

we have

$$g_\theta = 0 = \frac{GM}{a^2} \left(\frac{2\sqrt{2}+3}{16} - \frac{m}{2} \right)$$

and solving for m gives $m = 0.73$.

Substituting in Equation (22.4) we obtain

$$\omega = \sqrt{\frac{GM}{a^3}} 0.85$$

23. A planet consists of a very thin spherical shell of mass M and radius a, within which is a solid sphere of radius $a/2$ and mass M' centred at the midpoint of the equatorial radius of the zero meridian. The planet rotates with angular velocity ω about an axis normal to the equatorial plane. Calculate:

(a) The potential at points on the surface as a function of latitude and longitude.
(b) The components of the gravity vector.
(c) If $M' = 10\,M$, what is the ratio between the tangential and radial components of gravity at the North Pole?

(a) The potential U is the sum of the gravitational potentials due to the spherical shell V_1, and to the interior sphere V_2, plus the potential due to the rotation of the planet Φ (Fig. 23):

$$U = V_1 + V_2 + \Phi$$
$$\Phi = \frac{1}{2}\omega^2 r^2 \cos^2\varphi$$
$$V_1 = \frac{GM}{r} \tag{23.1}$$
$$V_2 = \frac{GM'}{q}$$

where r is the distance from a point P on the surface of the planet to its centre, q is the distance from point P to the centre of the interior sphere, and φ the latitude of point P.

Using the cosine law,

$$q = \sqrt{r^2 + \frac{a^2}{4} - ar\cos\psi}$$

where ψ is the angle between r and the equatorial radius, and its inverse can be approximated by (Problem 16)

$$\frac{1}{q} = \frac{1}{r}\left[1 + \frac{a}{2r}\cos\psi + \frac{a^2}{8r^2}\left(3\cos^2\psi - 1\right) \right] \tag{23.2}$$

Using the relation for spherical triangles

$$\cos a = \cos b \cos c + \sin b \sin c \cos A$$

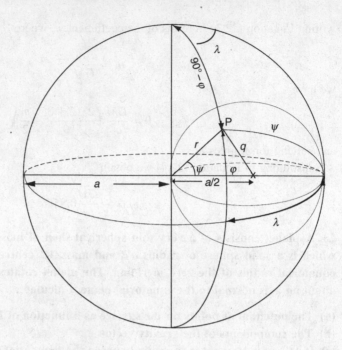

putting $b = 90° - \varphi$, $c = 90°$, $A = \lambda$, and $a = \psi$, where λ is the longitude of P, then

$$\cos \psi = \cos \varphi \cos \lambda$$

Substituting in (23.1), the potential due to the small sphere is

$$V_2 = -\frac{GM'}{r}\left[1 + \frac{a}{2r}\cos\varphi\cos\lambda + \frac{a^2}{8r^2}\left(3\cos^2\varphi\cos^2\lambda - 1\right)\right]$$

The total potential U is

$$U = \frac{GM}{r} + \frac{GM'}{r}\left[1 + \frac{a}{2r}\cos\varphi\cos\lambda + \frac{a^2}{8r^2}\left(3\cos^2\varphi\cos^2\lambda - 1\right)\right]$$
$$+ \frac{1}{2}\omega^2 r^2 \cos^2\varphi \tag{23.3}$$

(b) The components of the gravity vector are obtained from Equation (23.3):

$$g_r = \frac{\partial U}{\partial r}$$
$$= -\frac{GM}{r^2} + GM'\left[-\frac{1}{r^2} - \frac{a}{r^3}\cos\varphi\cos\lambda - \frac{3a^2}{8r^4}\left(3\cos^2\varphi\cos^2\lambda - 1\right)\right]$$
$$+ \omega^2 r\cos^2\varphi$$
$$g_\theta = \frac{1}{r}\frac{\partial U}{\partial \theta} = -\frac{1}{r}\frac{\partial U}{\partial \varphi}$$
$$= \frac{GM'}{r}\left[\frac{a}{2r^2}\sin\varphi\cos\lambda + \frac{a^2}{8r^3}6\cos\varphi\cos^2\lambda\sin\varphi\right]$$
$$+ \omega^2 r\cos\varphi\sin\varphi$$

$$g_\lambda = \frac{1}{r\cos\varphi}\frac{\partial U}{\partial \lambda} = \frac{GM'}{r\cos\varphi}\left[-\frac{a}{2r^2}\cos\varphi\sin\lambda - \frac{a^2}{8r^3}6\cos^2\varphi\cos\lambda\sin\lambda\right]$$

(c) At the North Pole, $\varphi = 90°$ and $r = a$. Putting $M' = 10M$ and substituting in the previous equations we obtain

$$g_r = -\frac{GM}{a^2} - \frac{GM'}{a^2} + \frac{3GM'a^2}{8a^4} = -7.25\frac{GM}{a^2}$$

$$g_\theta = \frac{GM'}{2a^2} = -\frac{5GM}{a^2}$$

The ratio between the radial and the tangential components of gravity at the North Pole is

$$\frac{g_r}{g_\theta} = 1.45$$

24. An Earth consists of a sphere of radius a and density ρ, within which there are two spheres of radius $a/2$ centred on the axis of rotation and tangent to each other. The density of that of the northern hemisphere is 4ρ and that of the southern hemisphere is $\rho/4$.

(a) **Express the gravitational potential in terms of M (the mass of the large sphere) up to terms of $1/r^3$.**
(b) **What astronomical latitude corresponds to points on the equator (without rotation)?**
(c) **What error is made by using the $1/r^3$ approximation in calculating the value of g_r at the equator?**

(a) The total gravitational potential V is the sum of the potentials of the sphere of radius a (V_0) and of the two spheres of radius $a/2$ situated in the northern (V_1) and southern (V_2) hemispheres (Fig. 24):

$$V = V_0 + V_1 + V_2$$

As in previous problems the large sphere is considered to have uniform density ρ and the effect of the two interior spheres is calculated using their differential masses

$$M = \frac{4}{3}\pi\rho a^3$$

$$M_1 = \frac{4}{3}\pi(4\rho - \rho)\frac{a^3}{8} = \frac{3M}{8}$$

$$M_2 = \frac{4}{3}\pi\left(\frac{\rho}{4} - 1\right)\frac{a^3}{8} = -\frac{3M}{32}$$

The potentials are

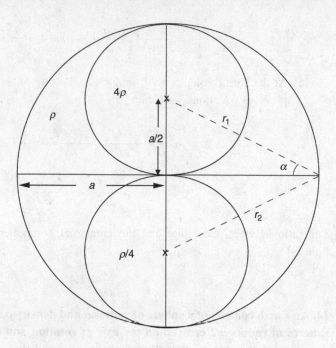

$$V_0 = \frac{GM}{r}$$

$$V_1 = \frac{GM_1}{r_1} = \frac{3GM}{8}\left[\frac{1}{r} + \frac{a}{2r^2}\cos\theta + \frac{a^2}{8r^3}\left(3\cos^2\theta - 1\right)\right]$$

$$V_2 = \frac{GM_2}{r_2} = -\frac{3GM}{32}\left[\frac{1}{r} - \frac{a}{2r^2}\cos\theta + \frac{a^2}{8r^3}\left(3\cos^2\theta - 1\right)\right]$$

where r_1 and r_2 have been calculated as in Problem 16. Then, the total gravitational potential in the $1/r^3$ approximation is

$$V = GM\left[\frac{41}{32r} + \frac{15}{64}\frac{a}{r^2}\cos\theta + \frac{9}{256}\frac{a^2}{r^3}\left(3\cos^2\theta - 1\right)\right]$$

(b) The components of the gravity vector, taking into account that there is no rotation, are

$$g_r = \frac{\partial V}{\partial r} = GM\left[-\frac{41}{32r^2} - \frac{15a}{32r^3}\cos\theta - \frac{27a^2}{256r^4}\left(3\cos^2\theta - 1\right)\right]$$

$$g_\theta = \frac{1}{r}\frac{\partial V}{\partial\theta} = GM\left[-\frac{15a}{64r^3}\sin\theta - \frac{27a^2}{128r^3}\cos\theta\sin\theta\right]$$

(24.1)

At the equator, $r = a$ and $\theta = 90°$ and we obtain

$$g_r = -1.175 \frac{GM}{a^2}$$

$$g_\theta = -0.243 \frac{GM}{a^2}$$

The astronomical latitude (φ_a) is the angle between the vertical and the equatorial plane. In our case at the equator this is given by the deviation of the vertical from the radial direction:

$$\tan \varphi_a = \frac{g_\theta}{g_r} = 0.207$$

Then

$$\varphi_a = 11.68° \, \text{N}$$

(c) If we want to calculate the exact value of $\overset{\text{!}}{g}_r$ at the equator, we calculate the exact attractions of each sphere and add them:

$$g_r^0 = -\frac{GM}{a^2} \qquad g_r^1 = -\frac{3GM}{8r_1^2} \cos \alpha \qquad g_r^2 = -\frac{3GM}{32r_2^2} \cos \alpha \qquad (24.2)$$

where r_1 and r_2 are the distances from the centre of each of the two interior spheres (Fig. 24):

$$r_1 = r_2 = \sqrt{a^2 + \frac{a^2}{4}} = \frac{a\sqrt{5}}{2}$$

and α is the angle which r_1 and r_2 form with the equatorial plane:

$$\sin \alpha = \frac{a/2}{r_1} = \frac{1}{\sqrt{5}}$$

The radial component of gravity is given by

$$g_r^T = g_r^0 + g_r^1 + g_r^2 = -1.335 \frac{GM}{a^2}$$

The error we make using the approximation is

$$g_{\text{approx}} - g_{\text{exact}} = 0.160 \frac{GM}{a^2}, \qquad \text{that is, } 16\%.$$

25. An Earth consists of a sphere of radius a and density ρ within which there are two spheres of radius $a/2$ centred on the axis of rotation and tangent to each other. The density of that of the northern hemisphere is 2ρ and that of the southern hemisphere is $\rho/2$.

(a) Express the potential V in terms of M (the mass of the large sphere), G, and r up to terms in $1/r^3$.

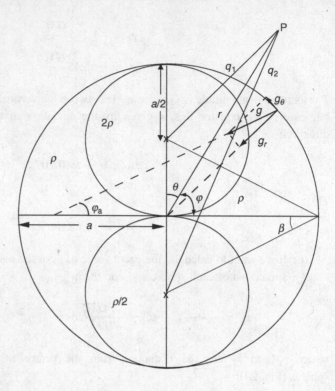

Fig. 25

(b) **According to the value of this potential V, which astronomical latitudes corres-pond to the geocentric latitudes 45° N and 45° S?**

(c) **What must the rotation period be for the astronomical and geocentric latitudes to coincide?**

(d) **What error is made by the $1/r^3$ approximation in calculating the value of g_r at the equator? And at the North Pole?**

(a) As in previous problems the effect of the interior spheres is given in terms of their differential masses (Fig. 25):

$$M_1 = \frac{4}{3}\pi(2\rho - \rho)\left(\frac{a}{2}\right)^3 = \frac{M}{8}$$

$$M_2 = \frac{4}{3}\pi\left(\frac{\rho}{2} - \rho\right)\left(\frac{a}{2}\right)^3 = -\frac{M}{16}$$

The distances q_1 and q_2 from the centre of each sphere to an arbitrary point P are found using the cosine law:

$$q_1 = r^2 + \frac{a^2}{4} - 2\frac{ar}{2}\cos\theta$$

$$q_2 = r^2 + \frac{a^2}{4} + 2\frac{ar}{2}\cos\theta$$

Using the approximation for $1/q$ (Problem 16), the total gravitational potential V is the sum of the potentials of the three spheres:

$$V = \frac{GM}{r} + \frac{GM_1}{q_1} + \frac{GM_2}{q_2}$$

$$= \frac{GM}{r} \left[\frac{17}{16} + \frac{3}{32} \frac{a}{r} \cos\theta + \frac{1}{128} \left(\frac{a}{r}\right)^2 (3\cos^2\theta - 1) \right]$$

(b) The components of the gravity vector are given by

$$g_r = \frac{\partial V}{\partial r} = -GM \left[\frac{1}{r^2} \frac{17}{16} + \frac{6}{32} \frac{a}{r^3} \cos\theta + \frac{3}{128} \frac{a^2}{r^4} (3\cos^2\theta - 1) \right]$$

$$g_\theta = \frac{1}{r} \frac{\partial V}{\partial \theta} = -\frac{GM}{r} \left[\frac{3}{32} \frac{a}{r^2} \sin\theta + \frac{1}{128} \frac{a^2}{r^3} 6 \cos\theta \sin\theta \right]$$

(25.1)

If the point P is at the surface, $r = a$, then

$$g_r = -\frac{GM}{a^2} \left[\frac{17}{16} + \frac{3}{16} \cos\theta + \frac{3}{128} (3\cos^2\theta - 1) \right]$$

$$g_\theta = -\frac{GM}{a^2} \left[\frac{3}{32} \sin\theta + \frac{6}{128} \cos\theta \sin\theta \right]$$

At geocentric latitude 45° N, $\theta = 45°$,

$$g_r = -1.21 \frac{GM}{a^2}$$

$$g_\theta = -0.09 \frac{GM}{a^2}$$

The deviation of the vertical with respect to the radial direction i is given by

$$\tan i = \frac{g_\theta}{g_r} = 0.074 \Rightarrow i_{45} = 4.2°$$

According to Fig. 25, the astronomical latitude φ_a can be determined from the deviation of the vertical i,

$$\varphi_a + i + (180° - \varphi) = 180° \Rightarrow \varphi_a = \varphi - i = 45° - 4.2° = 40.8° \, \text{N}$$

In the same way, for geocentric latitude 45° S ($\theta = 135°$)

$$g_r = -0.94 \frac{GM}{a^2}$$

$$g_\theta = -0.04 \frac{GM}{a^2}$$

Then $\tan i_{135} = \frac{g_\theta}{g_r} = 0.043 \Rightarrow i_{135} = 2.5°$

Then the astronomical latitude is

$$\varphi_a = -45° - 2.5° = -47.5° = 47.5° \, \text{S}$$

(c) If we want the astronomical and geocentric latitudes to coincide, then the deviation of the vertical must be null, $i = 0°$. This implies that g_θ^{total} must be zero. To do this by means of the rotation, we have to make the tangential component of the centrifugal force g_θ^R be equal and of opposite sign to that of the gravitational potential g_θ^V:

$$g_\theta^{\text{total}} = g_\theta^V + g_\theta^R = 0 \Rightarrow g_\theta^V = -g_\theta^R$$

The tangential component due to rotation is

$$g_\theta^R = \frac{1}{r}\frac{\partial \Phi}{\partial \theta}$$

where $\Phi = \frac{1}{2}\omega^2 r^2 \sin^2\theta$. Then

$$g_\theta^R = -\omega^2 r \cos\theta \sin\theta$$

For a point on the surface at latitude 45° N, $r = a$ and $\theta = 45°$, so

$$g_\theta^V = -g_\theta^R \Rightarrow -0.09\, GM/a^2 = -\omega^2\, a/2$$

From here we can calculate the period of rotation

$$T = \frac{2\pi}{\omega} = \frac{2\pi}{\sqrt{0.18}}\sqrt{\frac{a^3}{GM}}$$

For a point at latitude 45° S, $r = a$ and $\theta = 135°$, so

$$g_\theta^R = g_\theta^V \Rightarrow -\frac{1}{2}\omega^2 a = -0.04\frac{GM}{a^2} \Rightarrow T = \frac{2\pi}{\sqrt{0.08}}\sqrt{\frac{a^3}{GM}}$$

(d) The value of the radial component of gravity at the equator, $r = a$, $\theta = 90°$, by substitution in (25.1), is

$$g_r = -1.04\frac{GM}{a^2}$$

If we calculate the exact value by adding the contributions of the three spheres (Fig. 25)

$$g_r^{\text{exact}} = g_r^M + g_r^1 + g_r^2$$

$$g_r^M = -\frac{GM}{a^2}$$

$$g_r^1 = -g_1 \cos\beta$$

$$g_r^2 = g_2 \cos\beta$$

where

$$\cos \beta = \frac{a}{\sqrt{\frac{5}{4}a}} = \frac{2}{\sqrt{5}}$$

$$g_1 = -\frac{GM}{8}\frac{4}{5a^2}$$

$$g_2 = \frac{GM}{16}\frac{4}{5a^2}$$

$$g_r^{exact} = -\frac{GM}{a^2}0.96$$

The error in the approximation is:

$$g_r^{error} = -0.96\frac{GM}{a^2} - \left(-1.04\frac{GM}{a^2}\right) = -0.08\frac{GM}{a^2}$$

In a similar way, for a point at the North Pole, $r = a$, $\theta = 0°$:

$$g_r = -1.30\,GM/a^2$$

$$g_r^{exact} = g_r^M + g_r^1 + g_r^2$$

$$g_r^M = -\frac{GM}{a^2}$$

$$g_r^1 = -\frac{GM}{8\left(\frac{a}{2}\right)^2} = -\frac{GM}{2a^2}$$

$$g_r^2 = \frac{GM}{16\left(a+\frac{a}{2}\right)^2} = \frac{GM}{36a^2}$$

$$g_r^{exact} = -\frac{GM}{a^2}1.47$$

The error in the approximation is

$$g_r^{error} = -1.47\frac{GM}{a^2} - \left(-1.30\frac{GM}{a^2}\right) = -0.17\frac{GM}{a^2}$$

26. A spherical Earth of radius a has a core of radius $a/2$ whose centre is displaced $a/2$ along the axis of rotation towards the North Pole. The core density is twice that of the mantle.

(a) **What should the period of rotation of the Earth be for the direction of the plumb-line to coincide with the radius at a latitude of 45° S?**

(b) **What are the values of J_0, J_1, J_2, and m?**

(a) As in previous problems we calculate the gravitational potential by the sum of the potentials of the two spheres, using for the core the differential mass (Fig. 26):

$$V = V_1 + V_2 = \frac{GM}{r} + \frac{GM'}{q}$$

$$M' = \frac{4}{3}\pi(2\rho - \rho)\frac{a^2}{8} = \frac{M}{8}$$

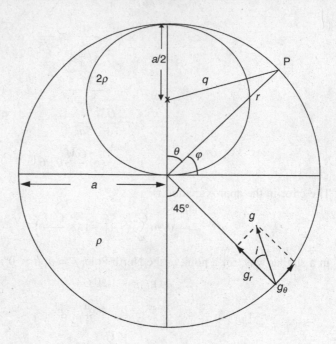

Fig. 26

As we saw in Problem 16, we use for $1/q$ the first-order approximation

$$V = \frac{GM}{r} + \frac{GM}{8}\left[\frac{1}{r} + \frac{a}{2r^2}\cos\theta + \frac{1}{2}\frac{a^2}{4r^3}\left(3\cos^2\theta - 1\right)\right]$$

(26.1)

The total potential U is the sum of the gravitational potential V and the potential due to rotation

$$\Phi = \frac{1}{2}r^2\omega^2\sin^2\theta$$

$$U = GM\left[\frac{9}{8}\frac{1}{r} + \frac{a}{16r^2}\cos\theta + \frac{a^2}{64r^3}\left(3\cos^2\theta - 1\right)\right] + \frac{1}{2}r^2\omega^2\sin^2\theta$$

In order that the direction of the plumb-line coincides with the radial direction, the tangential component of gravity, g_θ, must be null:

$$g_\theta = \frac{1}{r}\frac{\partial U}{\partial \theta} = \frac{GM}{r}\left(-\frac{a}{16}\frac{1}{r^2}\sin\theta - \frac{a^2}{64}\frac{1}{r^3}6\cos\theta\sin\theta\right)$$
$$+ r\omega^2\sin\theta\cos\theta$$

For a point on the surface at latitude $45°\,$S, the tangential component of gravity is, with $r = a$, $\theta = 135°$,

$$g_\theta = 0.003\frac{GM}{a^2} - \frac{a\omega^2}{2}$$

Putting this component equal to zero, we find the value of the period of rotation T:

$$g_\theta = 0.003 \frac{GM}{a^2} - \frac{a\omega^2}{2} = 0 \Rightarrow \omega^2 = 0.006 \frac{GM}{a^3} \Rightarrow T = \frac{2\pi}{0.077} \sqrt{\frac{a^3}{GM}}$$

(b) The gravitational potential V of Equation (26.1) can be written as

$$V = \frac{GM}{r} \frac{9}{8} \left[1 + \frac{a}{18r} \cos\theta + \frac{a^2}{72r^2} \left(3\cos^2\theta - 1 \right) \right]$$

We obtain the values of J_1 and J_2 by comparison with the equation

$$V = \frac{GM}{r} \left[J_0 + J_1 \frac{a}{r} P_1 + J_2 \left(\frac{a}{r} \right)^2 P_2 \right]$$

Since the total mass is $(9/8)M$, we obtain

$$J_0 = 1$$

$$J_1 = \frac{1}{18}$$

$$J_2 = \frac{1}{72}$$

27. Within a spherical planet of radius a and density ρ there are two spherical cores of radius $a/2$ and density ρ' with centres located on the axis of rotation at $a/2$ from the planet's centre, one in the northern hemisphere and the other in the southern hemisphere.

(a) **Neglecting rotation of the planet, calculate what the ratio ρ'/ρ should be for the gravity flattening to be 1/8.**

(b) **If the planet rotates so that $m = 1/16$, and the ratio of the densities is that found in part (a), calculate the astronomical latitude which corresponds to the geocentric latitude 45° N.**

(a) Since there is no rotation the total potential U is the sum of the gravitational potentials of the three spheres (Fig. 27). As in previous problems we use the mass M of the planet with uniform density ρ and for the two cores the differential masses M'. For $1/q$ we use the approximation as in Problem 16:

$$M' = \frac{4}{3}\pi \frac{a^3}{8} (\rho' - \rho) = \frac{4}{3}\pi \frac{a^3}{8} (\rho' - \rho) \frac{\rho}{\rho} = \frac{M}{8} \left(\frac{\rho'}{\rho} - 1 \right) \tag{27.1}$$

The potential U is

$$U = \frac{GM}{r} + \frac{GM'}{r} \left(2 + \left(\frac{a}{2r} \right)^2 \left(3\cos^2\theta - 1 \right) \right)$$

The radial components of gravity at the equator and the Pole are found by taking the derivative of the potential U:

$$g_r = \frac{\partial U}{\partial r} = -\frac{GM}{r^2} + GM' \left(-\frac{2}{r^2} - \frac{3a^2}{4r^4} \left(3\cos^2\theta - 1 \right) \right)$$

On the surface $r = a$, and at the equator $\theta = 90°$ and at the Pole $\theta = 0°$, so

$$g_r^e = -\frac{GM}{a^2} - \frac{GM'}{a^2}\frac{5}{4}$$

$$g_r^p = -\frac{GM}{a^2} - \frac{GM'}{a^2}\frac{14}{4}$$

The gravity flattening is given by

$$\beta = \frac{g_p - g_e}{g_e} = \frac{1}{8}$$

By substituting the values of gravity we find the relation between M and M':

$$\frac{1}{8} = \frac{-M - \frac{7}{2}M' + M + \frac{5}{4}M'}{-M - \frac{5}{4}M'} \Rightarrow M = \frac{67}{4}M'$$

Putting M' in terms of M from Equation (27.1) we find the ratio of the densities:

$$M' = \frac{4}{67}M = \frac{M}{8}\left(\frac{\rho'}{\rho} - 1\right) \Rightarrow \frac{\rho'}{\rho} = 1.48$$

(b) For a rotating planet we add to the potential U the rotational potential, Φ:

$$U = \frac{GM}{r} + \frac{GM'}{r}\left(2 + \left(\frac{a}{2r}\right)^2 (3\cos^2\theta - 1)\right) + \frac{GM}{r}\left(\frac{r}{a}\right)^3 \frac{m}{2}\sin^2\theta$$

The radial and tangential components of gravity are now

$$g_r = \frac{\partial U}{\partial r} = -\frac{GM}{r^2} + GM'\left(-\frac{2}{r^2} - \frac{3a^2}{4r^4}(3\cos^2\theta - 1)\right)$$

$$+ GM\frac{rm}{a^3}\sin^2\theta$$

$$= -1.11\frac{GM}{r^2}$$

$$g_\theta = \frac{1}{r}\frac{\partial U}{\partial \theta} = \frac{GM'}{r^2}\left[-\left(\frac{a}{2r}\right)^2 6\cos\theta\sin\theta\right]$$

$$+ \frac{GM}{r^2}\left(\frac{r}{a}\right)^3 m\sin\theta\cos\theta$$

$$= -0.013\frac{GM}{r^2}$$

From Fig. 27 we see that the relation between the geocentric and astronomical latitudes is

$$\varphi_a = \varphi - i$$

where i is the deviation of the vertical with respect to the radial direction, which is given by

$$\tan i = \frac{g_\theta}{g_r} = 0.012 \Rightarrow i = 0.7°$$

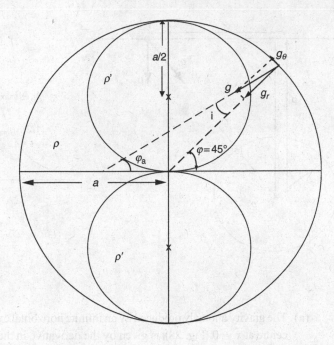

Fig. 27

Then the astronomical latitude for geocentric latitude 45° is

$$\varphi_a = 45° - 0.7° = 44.3°$$

Gravity anomalies. Isostasy

28. For two-dimensional problems, the gravitational potential of an infinite horizontal cylinder of radius a is

$$V = 2\pi G\rho a^2 \ln\left(\frac{1}{r}\right)$$

where r is the distance measured perpendicular to the axis. Assume that a horizontal cylinder is buried at depth d as measured from the surface to the cylinder's axis.

(a) Calculate the anomaly along a line of zero elevation on the surface perpendicular to the axis of the cylinder.
(b) At what point on this line is the anomaly greatest?
(c) What is the relationship between the distance at which the anomaly is half the maximum and the depth at which the cylinder is buried?
(d) For a sphere of equivalent mass to produce the same anomaly, would it be at a greater or lesser depth?

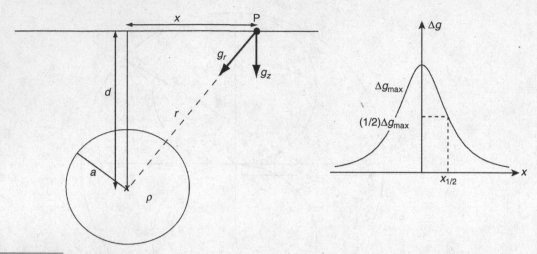

Fig. 28

(a) The gravity anomaly produced by an infinite horizontal cylinder buried at depth d, with centre at $x = 0$ (Fig. 28), is given by the derivative in the vertical direction (z-axis) of the gravitational potential V:

$$V = 2\pi G \Delta \rho a^2 \ln \left(\frac{1}{\sqrt{x^2 + (z+d)^2}} \right)$$

$$\Delta g = g_z = -\frac{\partial V}{\partial z} = \frac{2\pi \Delta \rho G a^2 d}{x^2 + (z+d)^2} \qquad (28.1)$$

For points on the surface ($z = 0$):

$$\Delta g = \frac{2\pi \Delta \rho G a^2 d}{x^2 + d^2}$$

(b) To find the point at which the anomaly has its maximum value, we take the derivative with respect to x and put it equal to zero:

$$\frac{\partial \Delta g}{\partial x} = 0 \Rightarrow -2\pi \Delta \rho G a^2 d 2x = 0 \Rightarrow x = 0$$

Substituting $x = 0$ in (28.1):

$$\Delta g_{max} = \frac{2\pi G \Delta \rho a^2}{d}$$

(c) The distance at which the anomaly has a value equal to half its maximum value gives us the depth d at which the cylinder is buried:

$$\frac{\Delta g_{max}}{2} = \Delta g \Rightarrow \frac{2\pi G \Delta \rho a^2}{2d} = \frac{2\pi G \Delta \rho a^2 d}{x^2 + d^2} \Rightarrow x_{1/2} = \pm d$$

(d) The gravitational potential produced by a sphere of differential mass ΔM buried at depth d under $x = 0$ is given by

$$V = \frac{G\Delta M}{r} = \frac{G\Delta M}{\left(x^2 + (z+d)^2\right)^{1/2}}$$

The gravity anomaly is

$$\Delta g_z = \frac{\partial V}{\partial z} = \frac{G\Delta M(z+d)}{\left(x^2 + (z+d)^2\right)^{3/2}}$$

and for a point on the surface $z = 0$,

$$\Delta g_z = \frac{G\Delta M d}{(x^2 + d^2)^{3/2}}$$

The maximum value for $x = 0$ is

$$\Delta g_{max} = \frac{G\Delta M}{d^2}$$

The distance at which the anomaly has half its maximum value is

$$\frac{G\Delta M d}{\left(x_{1/2}^2 + d^2\right)^{3/2}} = \frac{G\Delta M}{2d^2}$$

$$x_{1/2} = \sqrt{2^{2/3} - 1}\, d = 0.766d$$

Therefore, the sphere is at a greater depth than the cylinder.

29. At a point at latitude 42° 29′ 19″ and height 378.7 m the value of gravity is observed to be 980 252.25 mGal. Calculate in gravimetric units (gu):

(a) The free-air anomaly.

(b) The Bouguer anomaly if the density of the crust is 2.65 g cm⁻³.

(a) We first calculate the normal or theoretical value of gravity given by the expression

$$\gamma = 9.7803268 \left(1 + 0.00530244 \sin^2 \varphi - 0.0000058 \sin^2 2\varphi\right) \mathrm{m\, s^{-2}}$$

We substitute for φ its value 42° 29′ 19″ and obtain

$$\gamma = 9.803\,9299 \,\mathrm{m\, s^{-2}}$$

The free-air anomaly, using the free-air correction, is

$$\Delta g^{FA} = g + 3.086h - \gamma = -238.7 \,\mathrm{gu}$$

(b) The Bouguer anomaly is calculated from the free-air anomaly using the Bouguer correction with a crust density of 2.65 g cm^{-3}:

$$\Delta g^{\text{B}} = g + 3.086h - 2\pi G\rho h - \gamma = \Delta g^{\text{FA}} - 2\pi G\rho h = -659.3 \text{ gu}$$

30. An anomalous mass is formed by two equal tangent spheres of radius R, with centres at the same depth d ($d \gg R$) and density contrast $\Delta\rho$.

(a) Calculate the Bouguer anomaly at the surface ($z = 0$) produced by the mass anomaly along a profile passing through the centres of the two spheres.

(b) Represent it graphically for $x = 0$ (above the tangent point), 500, 1000, and 2000 m taking $R = 1$ km, $d = 3$ km, and $\Delta\rho = 1$ g cm^{-3}.

(a) For one sphere the anomaly for points on the surface ($z = 0$) is (Problem 28)

$$\Delta g = \frac{G\Delta M d}{(x^2 + d^2)^{3/2}}$$

For two spheres the anomaly is the sum of the attractions of the two spheres (Fig. 30a):

$$\Delta g = \frac{G\Delta M d}{r_1^3} + \frac{G\Delta M d}{r_2^3}$$

where

$$r_1 = \sqrt{(x - R)^2 + d^2}$$
$$r_2 = \sqrt{(x + R)^2 + d^2}$$

Then

$$\Delta g = \frac{G\Delta M d}{\left[(x - R)^2 + d^2\right]^{3/2}} + \frac{G\Delta M d}{\left[(x + R)^2 + d^2\right]^{3/2}}$$

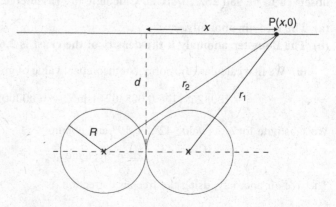

Fig. 30a

(b) To represent graphically the curve of the anomaly (Fig. 30b), we first find the point
at which it is a maximum:

$$\frac{\partial \Delta g}{\partial x} = \frac{\partial}{\partial x} \left(\frac{G \Delta M d}{\left[(x - R)^2 + d^2 \right]^{3/2}} + \frac{G \Delta M d}{\left[(x + R)^2 + d^2 \right]^{3/2}} \right) = 0$$

$$(x + R)^2 \left[(x - R)^2 + d^2 \right]^5 = (x - R)^2 \left[(x + R)^2 + d^2 \right]^5 \Rightarrow x = 0 \, \text{maximum}$$

Using the data given in the problem, we find the values of the anomaly for the five points,
with $\Delta \rho = 1 \, \text{g cm}^{-3}$, $R = 1$ km, d = 3 km

$$\Delta M = \frac{4}{3} \pi R^3 \Delta \rho = \frac{4}{3} \pi \times 10^9 \times 10^3 = 4.19 \times 10^{12} \, \text{kg}$$

x (m)	Δg (gu)
0	53.0
500	52.0
1000	48.9
1500	43.9
2000	37.5

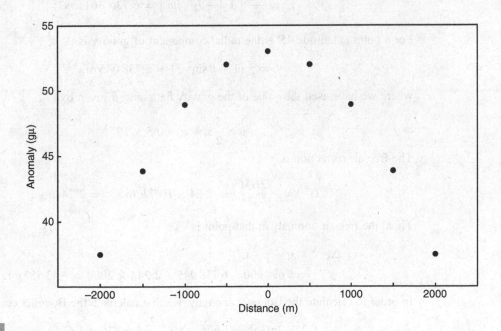

Fig. 30b

31. At a point at geocentric latitude 45° N and height 2000 m the observed value of gravity is $g = 6690\,000$ gu. Taking the approximation that the Earth is an ellipsoid of equatorial radius $a = 6000$ km, density $= 4$ g cm^{-3}, $J_2 = 10^{-3}$, and $m = 10^{-3}$, calculate for that point:

(a) The free-air and the Bouguer anomalies.
(b) The distance from the free surface to that of the sphere of radius a (precision 1 gu).

(a) The volume of an ellipsoid is:

$$V = \frac{4}{3}\pi a^3 (1 + 2\alpha)$$

The flattening is

$$\alpha = \frac{3J_2}{2} + \frac{m}{2} = 2 \times 10^{-3}$$

and the mass is

$$M_e = V\rho = \frac{4}{3}\pi a^3 (1 + 2\alpha)\rho = 3.624 \times 10^{24}\,\text{kg}$$

Using $G = 6.67 \times 10^{-11}$ m^3 kg^{-1} s^{-2}

$$\frac{GM}{a^2} = \frac{6.67 \times 10^{-11} \times 3.624 \times 10^{24}}{36 \times 10^{12}} = 6.732\,994\,\text{m s}^{-2}$$

The value of gravity at the equator in the first-order approximation is given by

$$\gamma_e = \frac{GM}{a^2}\left(1 + \frac{3}{2}J_2 - m\right) = 6.736\,361\,\text{m s}^{-2}$$

For a point at latitude 45° N the radial component of gravity is

$$\gamma_r = \gamma_e\left(1 + \beta \sin^2 \varphi\right) = 6.738\,045\,\text{m s}^{-2}$$

where we have used the value of the gravity flattening β given by

$$\beta = \frac{5}{2}m - \alpha = 0.5 \times 10^{-3}$$

The free-air correction is

$$C^{\text{FA}} = -\frac{2GM}{a^3}h = 2.24 \times 10^{-6}h \text{ m s}^{-2} = 2.244h \text{ gu}$$

Then, the free-air anomaly at that point is

$$\Delta g^{\text{FA}} = g - \gamma + C^{\text{FA}}$$
$$= 6\,690\,000 - 6\,738\,045 + 2.244 \times 2000 = -43\,557\,\text{gu}$$

In order to calculate the Bouguer anomaly, we first calculate the Bouguer correction

$$C^{\text{B}} = 2\pi G\rho h = 1.676 \times 10^{-6} \times h \text{ m s}^{-2} = 1.676h \text{ gu}$$

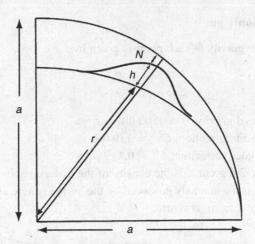

Fig. 31

Then, the Bouguer anomaly is:

$$\Delta g^B = g - \gamma + C^{AL} - 2\pi G\rho h = \Delta g^{FA} - C^B$$
$$= -45\,557 - 1.676 \times 2000 = -46\,909\,\text{gu}$$

(b) If we call N the distance at the given point between the free surface and the surface of the sphere of radius a (Fig. 31), this is given by:

$$N = a - r - h$$

where r is the radius of the ellipsoid at latitude 45° N which to a first approximation is

$$r = a\left(1 - \alpha\sin^2\varphi\right) = 6000\left(1 - 2 \times 10^{-3} \times \frac{1}{2}\right) = 5994\,\text{km}$$

Then,

$$N = 6000 - 5994 - 2 = 4\,\text{km}$$

32. Beneath a point A at height 400 m there exists an anomalous spherical mass of radius 200 m, density 3.5 g cm^{-3}, whose centre is 200 m below the reference level. A point B is located at a height of 200 m and a horizontal distance of 400 m from A, and a third point C is at a height of 0 m and at a horizontal distance of 800 m from A. The density of the medium above the reference level is 2.6 g cm^{-3}, and below the reference level it is 2.5 g cm^{-3}. The theoretical value of gravity is 980 000 mGal. Calculate:

(a) The values of gravity at A, B, and C.
(b) The Bouguer anomalies at these points.

Precision 1 gu.

(a) The gravity at each point is given by

$$g = \gamma - C^{FA} + C^{B} + C^{am}$$

where

Normal gravity: $\gamma = 9800\ 000$ gu

Free-air correction: $C^{FA} = 3.086\ h$

Bouguer correction: $C^{B} = 0.419\ \rho_1\ h$

$\rho_1 = 2.6$ g cm^{-3} is the density of the material above the reference level

C^{am} is the anomaly produced by the buried sphere at a point at height h and a horizontal distance x from its centre:

$$C^{am} = \frac{G\Delta M(h+d)}{\left(x^2 + (h+d)^2\right)^{3/2}} = \frac{G\frac{4}{3}\pi R^3 (\rho_{sph} - \rho_2)(h+d)}{\left(x^2 + (h+d)^2\right)^{3/2}} \tag{32.1}$$

where d is the depth to the centre from the reference level; and ρ_{sph} and ρ_2 are the densities of the sphere and of the medium where it is located, respectively. In our case: $d = 200$ m, $\rho_{sph} = 3.5$ g cm^{-3}, and $\rho_2 = 2.5$ g cm^{-3}.

For point A, $x = 0$, we obtain

$$C^{FA} = 1234\,\text{gu} \qquad C^{B} = 436\,\text{gu}$$

$$C^{am} = \frac{G\Delta M}{(h+d)^2} = \frac{G\frac{4}{3}\pi R^3 (\rho_{sph} - \rho_2)}{(h+d)^2} = 6\,\text{gu}$$

The value of gravity is $g_A = 9\ 799\ 208$ gu.

At point B:

$C^{FA} = 617$ gu

$C^{B} = 218$ gu

The anomaly produced by the sphere is calculated by Equation (32.1), substituting

$x = x_B = 400$ m and $h = h_B = 200$ m

$C^{am} = 5$ gu

We obtain $g_B = 9\ 799\ 606$ gu.

At point C:

The free-air and Bouguer corrections are null, because the point is at the reference level. The anomaly due to the sphere, by substitution in Equation (32.1), $x = x_C = 800$ m, and $h = h_C = 0$, is

$C^{am} = 1$ gu

The value of gravity is: $g_C = 9800\ 001$ gu.

(b) The Bouguer anomaly is given by

$$\Delta g^{B} = g + C^{FA} - C^{B} - \gamma$$

By substitution of the values for each point we obtain that the anomalies correspond to those produced by the sphere:

$$\Delta g_A^B = 6 \, \text{gu}$$

$$\Delta g_B^B = 5 \, \text{gu}$$

$$\Delta g_C^B = 1 \, \text{gu}$$

33. For a series of points in a line and at zero height which are affected by the gravitational attraction exerted by a buried sphere of density contrast 1.5 g cm^{-3}, the anomaly versus horizontal distance curve has a maximum of 4.526 mGal and a point of inflexion at 250 m from the maximum. Calculate:

(a) The depth, anomalous mass, and radius of the sphere.

(b) The horizontal distance to the centre of the sphere of the point at which the anomaly is half the maximum.

(a) We know that the inflection point of the curve of the anomaly produced by a sphere buried at depth d corresponds to the horizontal distance $d/2$. Then

$$x_{\text{inf}} = \frac{d}{2} \Rightarrow d = 2x_{\text{inf}} = 2 \times 250 = 500 \, \text{m}$$

The maximum value of the anomaly at $x = 0$ is

$$\Delta g_{\text{max}} = \frac{G\Delta M}{d^2} = 4.526 \, \text{mGal} = 45.26 \, \text{gu}$$

and solving for ΔM

$$\Delta M = \frac{45.26 \times 10^{-6} \, \text{m s}^{-2} \times 500^2 \, \text{m}^2}{6.67 \times 10^{-11} \, \text{m}^3 \, \text{s}^{-2} \, \text{kg}^{-1}} = 1.6964 \times 10^{11} \, \text{kg}$$

From this value we calculate the radius of the sphere:

$$\Delta M = \frac{4}{3}\pi R^3 \Delta \rho \Rightarrow R = \left(\frac{3\Delta M}{4\pi\Delta\rho}\right)^{1/3} = \left(\frac{3 \times 1.6964 \times 10^{11} \, \text{kg}}{4 \times 3.14 \times 1.5 \times 10^3 \, \text{kg m}^{-3}}\right)^{1/3} = 300 \, \text{m}$$

(b) In order that $\Delta g = \frac{1}{2}\Delta g_{\text{max}}$ with $z = 0$ we write

$$\frac{G\Delta M d}{\left(x_{1/2}^2 + d^2\right)^{3/2}} = \frac{1}{2}\frac{G\Delta M}{d^2} \Rightarrow x_{1/2} = 383 \, \text{m}$$

34. At a point at height 2000 m, the measured value of gravity is 9.794 815 m s^{-2}. The reference value at sea level is 9.8 m s^{-2}. The crust is 10 km thick and of density 2 g cm^{-3}, and the mantle density is 3 g cm^{-3}. Calculate:

(a) The free-air, Bouguer, and isostatic anomalies. Use the Pratt hypothesis with a cylinder of radius 10 km and a 40 km depth of compensation.

(b) If beneath this point there is a spherical anomalous mass of $G\Delta M = 160$ m^3 s^{-2} at a 2000 m depth, what should the compensatory cylinder's density be for the compensation to be total?

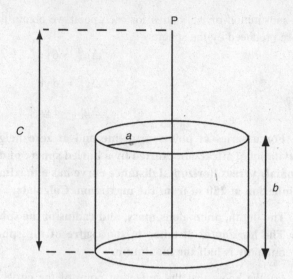

Fig. 34a

(a) The free-air anomaly is

$$\Delta g^{FA} = g - \gamma + C^{FA} = 9\,794\,815 - 9\,800\,000 + 3.086 \times 2000 = 987\,\text{gu}$$

The Bouguer anomaly is

$$\Delta g^{B} = g - \gamma + C^{FA} - C^{B} = \Delta g^{FA} - 0.4191\rho h$$
$$= 987 - 0.4191 \times 2 \times 2000 = -689\,\text{gu}$$

To calculate the isostatic anomaly (Fig. 34a) we begin with the calculation of the isostatic correction assuming Pratt's hypothesis and using only a vertical cylinder of radius 10 km under the point and the compensation level at 40 km. In this way, the correction consists of the gravitational attraction of a cylinder of radius a and height b at a point at distance c from the base of the cylinder, which is given by

$$C^{I} = 2\pi G \Delta\rho \left[b + \sqrt{a^2 + (c - b)^2} - \sqrt{a^2 + c^2} \right] \tag{34.1}$$

where $\Delta\rho$ is the contrast of densities, which according to Pratt's hypothesis is given by

$$\Delta\rho = \frac{h\rho_0}{D + h} \tag{34.2}$$

where ρ_0 is the density for a block at sea level, which in our case is formed by a crust of density 2 g cm^{-3} and thickness 10 km over a mantle of density 3 g cm^{-3} and thickness 30 km. For the whole 40 km we use a mean value of density

$$\rho_0 = \frac{2 \times 10 + 3 \times 30}{40} = 2.75\,\text{g cm}^{-3}$$

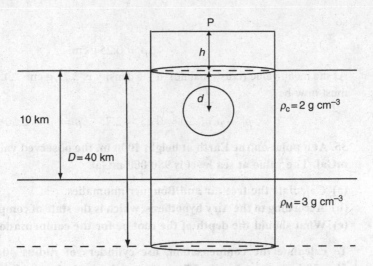

By substitution in (34.2) we obtain

$$\Delta\rho = \frac{2 \times 2.75}{42} = 0.13 \, \text{g cm}^{-3}$$

The isostatic correction (34.1) is

$$C^{\text{I}} = 2 \times 3.1416 \times 6.67 \times 10^{-11} \times 0.13 \times 10^3 \lfloor 40 + \sqrt{100 + 4} - \sqrt{100 + 1764} \rfloor \times 10^3$$
$$= 382 \, \text{gu}$$

Finally the isostatic anomaly is given by

$$\Delta g^{\text{I}} = g - \gamma + C^{\text{FA}} + C^{\text{B}} + C^{\text{I}} = \Delta g^{\text{B}} + C^{\text{I}} = -689 + 382 = -307 \, \text{gu}$$

(b) If under the point considered there is an anomalous spherical mass (Fig. 34b) at depth $d = 2$ km, the anomaly it produces is

$$\Delta g^{\text{am}} = \frac{G\Delta M}{d^2} = \frac{160}{(2000)^2} = 40 \, \text{gu}$$

The total anomaly now is the Bouguer anomaly plus the anomaly due to the sphere:

$$\Delta g = -689 - 40 = -729 \, \text{gu}$$

If the isostatic compensation is total (isostatic anomaly equal to zero), this anomaly must be compensated by the cylinder. Thus, the necessary contrast of densities $\Delta\rho$ to do this can be calculated using expression (34.1):

$$729 \times 10^6 = 2 \times 3.1416 \times 6.67 \times 10^{-11}$$
$$\times \left[40 + \sqrt{100 + 4} - \sqrt{100 + 1764} \right] \times 10^3 \times \Delta\rho$$

so

$$\Delta\rho = 0.25\,\mathrm{g\,cm}^{-3}$$

As the mean value (crust–mantle) of the density is 2.75 g cm^{-3}, the density of the cylinder must now be

$$\rho_0 - \rho = \Delta\rho = 0.25 = 2.75 - \rho \Rightarrow \rho = 2.50\,\mathrm{g\,cm}^{-3}$$

35. At a point on the Earth at height 1000 m, the observed value of gravity is 979 700 mGal. The value at sea level is 980 000 mGal.

(a) **Calculate the free-air and Bouguer anomalies.**
(b) **According to the Airy hypothesis, which is the state of compensation of that height?**
(c) **What should the depth of the root be for the compensation to be total?**

To calculate the compensation, use cylinders of radius 40 km, crustal thickness $H = 30$ km, crust density 2.7 g cm^{-3}, and mantle density 3.3 g cm^{-3}.

(a) The free-air anomaly is

$$\Delta g^{\mathrm{FA}} = g - \gamma + C^{\mathrm{FA}} = g - \gamma + 3.086h = 86\,\mathrm{gu}$$

The Bouguer anomaly is

$$\Delta g^{\mathrm{B}} = g - \gamma + C^{\mathrm{FA}} - C^{\mathrm{B}} = \Delta g^{\mathrm{FA}} - 2\pi\,G\,\rho\,h = -1046\,\mathrm{gu}$$

(b) To calculate the isostatic anomaly according to the Airy hypothesis we first need to obtain the value of the root given by the equation

$$t = \frac{\rho_{\mathrm{c}}}{\rho_{\mathrm{M}} - \rho_{\mathrm{c}}} h = 4500\,\mathrm{m}$$

The isostatic correction is given by

$$C^{\mathrm{I}} = 2\pi G \Delta\rho \left[b + \sqrt{a^2 + (c-b)^2} - \sqrt{a^2 + c^2} \right]$$

Substituting the values
$\Delta\rho = \rho_{\mathrm{M}} - \rho_{\mathrm{C}} = 600\ \mathrm{kg\ m}^{-3}$
$b = t = 4500\ \mathrm{m}$
$c = h + H + t = 35\ 500\ \mathrm{m}$
$a = 40\ \mathrm{km}$
we obtain:
$C^{\mathrm{I}} = 409\ \mathrm{gu}$
the isostatic anomaly is:

$$\Delta g^{\mathrm{I}} = g - \gamma + C^{\mathrm{FA}} + C^{\mathrm{B}} + C^{\mathrm{I}} = \Delta g^{\mathrm{B}} + C^{\mathrm{I}} = -637\,\mathrm{gu}$$

The negative value of the anomaly indicates that the zone is overcompensated.

(c) If we want the compensation to be total, the value of the isostatic correction must be

$$\Delta g^{\mathrm{I}} = \Delta g^{\mathrm{B}} + C^{\mathrm{I}} = 0 \Rightarrow C^{\mathrm{I}} = -\Delta g^{\mathrm{B}} = 1046 \text{ gu}$$

Since the isostatic correction under the Airy hypothesis is

$$C^{\mathrm{I}} = 2\pi G \Delta\rho \left[t + \sqrt{a^2 + (h+H)^2} - \sqrt{a^2 + (h+H+t)^2} \right]$$

substituting and solving for t, we obtain

$$t = 13\,068 \text{ m}$$

For a total isostatic compensation the value of the root (13 068 m) must be much larger than that corresponding to the 1000 m height, which is only 4500 m.

36. Gravity measurements are made at two points A and B of altitude 1000 m and − 1000 m above the reference level, respectively, 2 km apart along a W-E profile at latitude 38.80° N. Below a point C located in the direction AB and 1 km from A is buried a sphere of radius 1 km and centre 3 km below the reference level, of density $\rho = 1.76$ g cm^{-3}. Calculate:

(a) The value of gravity at A and B.
(b) Using the Airy assumption and neglecting the sphere, calculate the root at A and B.

Crustal density $\rho_{\mathrm{C}} = 2.76$ g cm^{-3}, mantle density $\rho_{\mathrm{M}} = 3.72$ g cm^{-3}, $a = 10$ km, and $H = 30$ km.

(a) The gravity observed at points A and B is given by

$$g_{\mathrm{A}} = \gamma - C^{\mathrm{FA}} + C^{\mathrm{B}} + C^{\mathrm{am}}$$
$$g_{\mathrm{B}} = \gamma + C^{\mathrm{FA}} - C^{\mathrm{B}} + C^{\mathrm{am}}$$

where γ is the theoretical gravity, C^{FA} the free-air correction, C^{B} the Bouguer correction, and C^{am} the attraction due to the anomalous mass.

The theoretical gravity at the observation point at latitude 38.80° N is

$$\gamma = 9.780\,32 \left(1 + 0.005\,3025 \sin^2\varphi - 0.000\,0058 \sin^2 2\varphi \right) = 9.800714 \text{ m s}^{-2}$$

The free-air and Bouguer corrections are:

$$C^{\mathrm{FA}} = 3.806h = 3.806 \times 1000 = 3806 \text{ gu}$$
$$C^{\mathrm{B}} = 0.419\rho_{\mathrm{C}} h = 0.419 \times 2.76 \times 1000 = 1156 \text{ gu}$$

The attraction due to the spherical anomalous mass (Fig. 36) is given by

$$C^{\mathrm{am}} = \frac{G \Delta M (z + d)}{\left(x^2 + (z+d)^2 \right)^{3/2}}$$

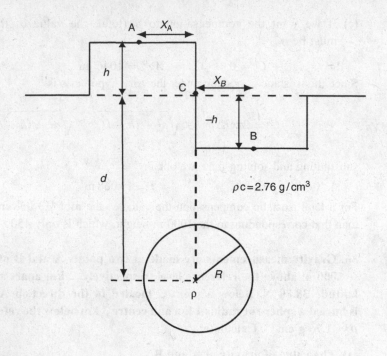

Fig. 36

For points A and B, by substitution of the values

$$\Delta M = \frac{4}{3}\pi R^3 \Delta\rho$$

$$R = 1000\,\text{m}, \qquad \Delta\rho = -1000\,\text{kg m}^{-3}$$

$$z_A = h = 1000\,\text{m}$$

$$z_B = -h = -1000\,\text{m}$$

$$x_A = x_B = 1000\,\text{m}$$

$$d = 3000\,\text{m}$$

we find

$$C_A^{am} = -16\,\text{gu}$$

$$C_B^{am} = -50\,\text{gu}$$

Then the values of gravity at both points are

$$g_A = 9800\,627.9 - 3806 + 1156.4 - 16 = 9.798\,048\,\text{m s}^{-2}$$

$$g_B = 9800\,627.9 + 3806 - 1156.4 - 50 = 9.803\,314\,\text{m s}^{-2}$$

(b) To calculate the value of the root under A and B according to the Airy hypothesis we use the equation

$$t = \frac{\rho_C}{\rho_M - \rho_C}h$$

where ρ_C and ρ_M are the crust and mantle densities. Then we find

$$t_A = \frac{2.76}{3.72 - 2.76} \times 1000 = 2875\,\text{m}$$

$$t_B = \frac{2.76}{3.72 - 2.76} \times (-1000) = -2875\,\text{m}$$

37. At a point at latitude 43° N, the observed value of gravity is 9800 317 gu, and the free-air anomaly is 1000 gu.

(a) Calculate the Bouguer anomaly. Take $\rho_C = 2.67$ g cm^{-3}.

(b) If the isostatic compensation is due to a cylinder of radius 10 km which is beneath the point of measurement, what percentage of the Bouguer anomaly is compensated by the classical models of Airy and Pratt?

(c) According to the Pratt hypothesis, what density should the cylinder have for the compensation to be total?

(a) First we calculate the normal gravity at latitude 43° N:

$$\gamma = 978.0320\left(1 + 0.005\,3025\sin^2\varphi - 0.000\,0058\sin^2 2\varphi\right)\,\text{Gal}$$

$$\gamma = 9804\,385\,\text{gu}$$

The height of the point is determined from the free-air anomaly,

$$\Delta g^{FA} = g - \gamma + 3.086h = 9\,800\,317 - 9\,804\,385 + 3.086h = 1000\,\text{gu}$$

and solving for h,

$$h = 1642\,\text{m}$$

From this value we calculate the Bouguer anomaly

$$\Delta g^{B} = g - \gamma + 1.967h = -838\,\text{gu}$$

(b) To apply the isostatic compensation using the Airy hypothesis we first calculate the root corresponding to the height $h = 1642$ m:

$$t = 4.45h = 7307\,\text{m}$$

The isostatic correction is determined using Equation (34.1) of Problem 34:

$$C^I = 2\pi G\Delta\rho\left(b + \sqrt{a^2 + (c - b)^2} - \sqrt{a^2 + c^2}\right) \tag{37.1}$$

where

$$a = 10\,\text{km}, b = t = 7307\,\text{m}; c = t + 30\,000 + h = 38\,949\,\text{m}$$

$$\Delta\rho = \rho_M - \rho_C = 3.27 - 2.67 = 0.6\,\text{g cm}^{-3}$$

which results in $C^I = 70$ gu. This represents 8% of the observed Bouguer anomaly.

If we use the Pratt hypothesis, the contrast of densities corresponding to $h = 1642$ m is given by

$$\Delta\rho = \frac{h\rho_0}{D+h} = 0.043 \text{ g cm}^{-3}$$

where we have used $\rho_0 = \rho_C = 2.67$ g cm^{-3} as the density of the crust. We now substitute in Equation (37.1), $b = D = 100$ km, $c = D + h = 101.642$ km and the obtained value of $\Delta\rho = 0.043$ g cm^{-3}, and obtain

$$C^I = 148 \text{ gu}$$

We have to determine again the Bouguer anomaly using the density according to the Pratt hypothesis

$$\rho = \frac{D\rho_0}{D+h} = 2.63 \text{ g cm}^{-3}$$
$$\Delta g^B = \Delta g^{AL} - 2\pi G\rho h = -810 \text{ gu}$$

The isostatic correction corresponds now to 18% of the Bouguer anomaly.

(c) If the compensation is total the isostatic correction must be equal to the Bouguer anomaly with changed sign:

$$C^I = -\Delta g^B$$

Using the Pratt hypothesis in order to calculate the density ρ of the cylinder under the point, we have to take into account that this density must also be the density used in the determination of the Bouguer anomaly. Then we write

$$C^I = -\Delta g^B = -\Delta g^{FA} + 2\pi \, G \, \rho \, h$$

$$2\pi G(\rho_0 - \rho)\left(b + \sqrt{a^2 + (c-b)^2} - \sqrt{a^2 + c^2}\right) = -\Delta g^{FA} + 2\pi G\rho h$$

and putting

$$N = \left(b + \sqrt{a^2 + (c-b)^2} - \sqrt{a^2 + c^2}\right)$$

and solving for ρ, we obtain

$$\rho = \frac{\Delta g^{FA} + 2\pi G\rho_0 N}{2\pi G(h+N)} = 2.46 \text{ g cm}^{-3}$$

where we have used the values $\rho_0 = 2.67$ g cm^{-3} and $\Delta g^{FA} = 1000$ gu.

38. At a point on the Earth's surface, a measurement of gravity gave a value of 9795 462 gu. The point is 2000 m above sea level. At sea level the crust is 20 km thick and of density $\rho_C = 2$ g cm^{-3}. The density of the mantle is $\rho_M = 4$ g cm^{-3}.

(a) Calculate the free-air and Bouguer anomalies.

(b) **Calculate the isostatic anomaly according to the Airy and Pratt assumptions. Use cylinders of 10 km radius and compensation depth of 60 km.**

(c) **Beneath the point, there is an anomalous spherical mass of $G\Delta M = 1200$ m^3 s^{-2}. How deep is it?**

Take $\gamma = 9.8$ m s^{-2}.

(a) The free-air anomaly is given by

$$\Delta g^{FA} = g - \gamma + 3.086h = 9\,795\,462 - 9\,800\,000 + 3.086h = 1634\,\text{gu}$$

For the Bouguer anomaly we first calculate the Bouguer correction

$$C^B = 0.419\rho h = 0.419 \times 2 \times 2000 = 1676\,\text{gu}$$

Then we obtain

$$\Delta g^B = g - \gamma + C^{FA} - C^B = \Delta g^{FA} - C^B = -42\,\text{gu}$$

(b) To calculate the isostatic anomaly according to the Airy hypothesis we determine first the value of the root corresponding to the height 2000 m:

$$t = \frac{\rho_c}{\rho_M - \rho_c}h = \frac{2}{4-2}2000 = 2000\,\text{m}$$

The isostatic correction, using a single cylinder under the point, is given by

$$C^I = 2\pi G\Delta\rho\left(b + \sqrt{a^2 + (c-b)^2} - \sqrt{a^2 + c^2}\right) \tag{38.1}$$

where (Fig. 38a)

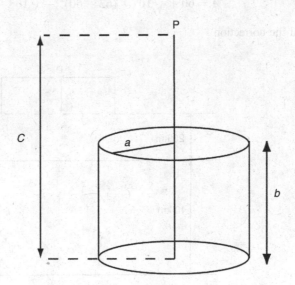

$$a = 10 \, \text{km}, \qquad b = t = 2 \, \text{km}, \qquad c = H + t + h = 20 + 2 + 2 = 24 \, \text{km}$$

Calling A the term inside the brackets in Equation (38.1)

$$A = 2 + \sqrt{100 + 484} - \sqrt{100 + 576} = 0.166 \, \text{km}$$

The isostatic correction is, then, given by

$$C^{\text{I}} = 2 \times 3.1416 \times 6.67 \times 10^{-8} \, \text{cm}^3/\text{gs}^{-2} \times 2 \, \text{g}/\text{cm}^{-3} \times 166 \times 10^2 \, \text{cm} = 139 \, \text{gu}$$

Finally, the isostatic anomaly using the Airy hypothesis is

$$\Delta g^{\text{I}} = g - \gamma + C^{\text{FA}} - C^{\text{B}} + C^{\text{I}} = 97 \, \text{gu}$$

According to the Pratt hypothesis, the regional density is given by

$$\rho = \frac{D}{D + h} \rho_0 = \frac{60}{60 + 2} 3.33 = 3.22 \, \text{g} \, \text{cm}^{-3}$$

where D is the compensation depth (in this problem 60 km) and for ρ_0 (Fig. 38b) we have used the mean value of the density of the crust ($2 \, \text{g} \, \text{cm}^{-3}$) and of the mantle ($4 \, \text{g} \, \text{cm}^{-3}$) along the compensation depth

$$\rho_0 = \frac{1}{3} 2 + \frac{2}{3} 4 = 3.33 \, \text{g} \, \text{cm}^{-3}$$

The contrast of densities is

$$\Delta \rho = 3.33 - 3.22 = 0.11 \, \text{g} \, \text{cm}^{-3}$$

For the isostatic correction, using the Pratt hypothesis, the term A is now

$$A = 60 + \sqrt{10^2 + (62 - 60)^2} - \sqrt{10^2 + 62^2} = 7.4 \, \text{km}$$

and the correction

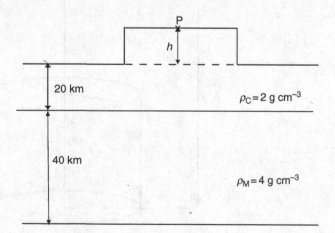

$$C^I = 2 \times 3.1416 \times 6.67 \times 10^{-11} \, \text{m}^3/\text{kg}^{-1}\text{s}^{-2} \times 0.11 \times 10^3 \, \text{kg/m}^{-3} \times 7.4 \times 10^3 \, \text{m}$$
$$= 341 \, \text{gu}$$

Since according to the Pratt hypothesis, the density of the compensating cylinder extends to the surface of the height 2000 m, we have to calculate again the Bouguer anomaly using this density (3.33 g cm^{-3}). We find for the Bouguer and isostatic anomalies the values

$$C^B = 2 \, \pi \, G \, \rho \, h = 2699 \, \text{gu}$$
$$\varDelta g^B = \varDelta g^{FA} - C^B = -1065 \, \text{gu}$$
$$\varDelta g^I = \varDelta g^B + C^I = -724 \, \text{gu}$$

(c) If we assume that the isostatic anomaly is produced by a spherical anomalous mass buried under the point at a depth d under sea level its gravitational effect is given by

$$\Delta g_{\text{max}} = \frac{G \Delta M}{(h+d)^2} = \Delta g^I = 724 \, \text{gu}$$

Solving for d we obtain

$$h + d = 3517 \, \text{m} \Rightarrow d = 1517 \, \text{m}$$

39. At a point P at height 2000 m above sea level, a measurement is made of gravity. The crust at sea level, where gravity is 9.8 m s^{-2}, is 20 km thick and of density 3 g cm^{-3}, and the density of the mantle is 4 g cm^{-3}. Below the point P, at 2000 m depth under sea level, is an anomalous spherical mass of $G\Delta M = 1200$ m^3 s^{-2}.

(a) Neglecting the isostatic compensation, what would be the value of gravity at the point P?

(b) With isostatic compensation, what now is the value of gravity at that point?

Use the Airy and Pratt assumptions for the isostatic compensation (Pratt depth of compensation, 100 km) with single cylinders of 20 km radius under the point.

(a) Without isostatic compensation, the gravity observed at point P is equal to the sum of the normal gravity plus the free-air and Bouguer corrections and the effect of the anomalous mass. Remember that the free-air correction has negative sign:

$$g_P = \gamma - C^{FA} + C^B + C^{am} \tag{39.1}$$

where,

$$\gamma = 9800\,000 \, \text{gu}$$
$$C^{FA} = 3.086 \times h = 3.086 \times 2000 = 6172 \, \text{gu}$$
$$C^B = 0.419 \rho h = 0.419 \times 3 \times 2000 = 2514 \, \text{gu}$$

The anomaly due to the anomalous mass is given by

$$C^{am} = \Delta g_{\text{max}} = \frac{G \Delta M}{(h+d)^2} \tag{39.2}$$

By substitution of the values,

$$C^{am} = \frac{G\Delta M}{(h+d)^2} = 75 \text{ gu}$$

The gravity observed a P is, then, given by

$$g_P = 9\,800\,000 - 6172 + 2514 + 75 = 9\,796\,417\,\text{gu}$$

(b) If there is isostatic compensation, according to the Airy hypothesis, we determine first the depth of the root, using the density of the crust ρ_C and of the mantle ρ_M:

$$t = \frac{\rho_C}{\rho_M - \rho_C} h = \frac{3}{4-3} \times 2000 = 6000 \text{ m}$$

The isostatic correction is

$$C^I = 2\pi G\Delta\rho \left(b + \sqrt{a^2 + (c-b)^2} - \sqrt{a^2 + c^2} \right) \tag{39.3}$$

where $a = 20$ km, $b = 6$ km, $c = 2 + 20 + 6 = 28$ km, $\Delta\rho = 1$ g cm^{-3}, so

$$C^I = 2 \times 3.14 \times 6.67 \times 10^{-11} \times 1 \times 10^3 \times \left(6 + \sqrt{20^2 + (28-6)^2} - \sqrt{20^2 + 28^2} \right) \times 10^3$$

$$= 553 \text{ gu}$$

Then the observed gravity at point P is

$$g_P = 9\,796\,417 - 553 = 9\,795\,864\,\text{gu}$$

According to the Pratt hypothesis we first determine the contrast of densities

$$\Delta\rho = \frac{h\rho_0}{D+h}$$

where D is the level of compensation (100 km) and ρ_0 is the mean density for the crust and mantle down to depth 100 km:

$$\rho_0 = \frac{20 \times 3 + 80 \times 4}{100} = 3.8 \text{ g cm}^{-3}$$

Substituting we find

$$\Delta\rho = \frac{2 \times 3.8}{100 + 2} = 0.074 \text{ g cm}^{-3}$$

The isostatic correction is

$$C^I = 2 \times 3.14 \times 6.67 \times 10^{-11} \times 0.074 \times 10^3$$

$$\times \left(100 + \sqrt{20^2 + (102-100)^2} - \sqrt{20^2 + 102^2} \right) \times 10^3$$

$$= 501 \text{ gu}$$

where we have used in Equation (39.3) the values

$$a = 20\,\text{km}, \qquad b = 100\,\text{km}, \qquad c = 100 + 2 = 102\,\text{km}$$

The value of gravity at point P is now

$$g_P = 9796\,417 - 501 = 9795\,916\,\text{gu}$$

which is larger by 52 gu than using the Airy hypothesis

40. A point P is at altitude 1000 m above sea level. Beneath this point is a sphere of 1 km radius and $G\Delta M = 650\ \text{m}^3\ \text{s}^{-2}$, with its centre 4 km vertically below the point P. Given that the density of the sphere is twice that of the crust and 3/2 that of the mantle, calculate:

(a) The density of the sphere, crust, and mantle.
(b) The value of gravity that would be observed at P for the isostatic compensation to be total including the sphere.
(c) The radius of the sphere for the root to be null. Comment on the result.

Use the Airy hypothesis for the isostatic compensation with $H = 30\ \text{km}$, 20 km radius of the cylinder, and theoretical gravity $\gamma = 980\ \text{Gal}$.

(a) If we know $G\Delta M$ we can calculate the contrast of densities between the anomalous mass and the crust

$$G\Delta M = \frac{4}{3}\pi\Delta\rho R^3 G \Rightarrow \Delta\rho = \frac{3G\Delta M}{4\pi R^3 G} = \frac{3 \times 650}{4 \times 3.1416 \times 10^9 \times 6.67 \times 10^{-11}}$$
$$= 2.326\,\text{g cm}^{-3}$$

Since the density of the sphere ρ_{sph} is double that of the crust ρ_{c}, the densities of the crust and mantle are

$$\rho_{\text{sph}} = 2\rho_{\text{C}} \Rightarrow \Delta\rho = \rho_{\text{sph}} - \rho_{\text{C}} = 2\rho_{\text{C}} - \rho_{\text{C}} = \rho_{\text{C}} = 2.326\,\text{g cm}^{-3}$$
$$\rho_{\text{sph}} = 4.652\,\text{g cm}^{-3}$$
$$\rho_{\text{M}} = \frac{2}{3}\rho_{\text{sph}} = 3.101\,\text{g cm}^{-3}$$

(b) If isostatic compensation is total we have

$$\Delta g^I = 0 = -\gamma + g^P + C^{\text{FA}} + C^{\text{B}} + C^I + C^{\text{am}} \tag{40.1}$$

where γ is the normal gravity, g^P the observed gravity at point P, C^{FA} the free-air correction, C^{B} the Bouguer correction, C^I the isostatic correction and C^{am} the gravitational effect of the anomalous mass.

The free-air and Bouguer corrections are given by

$$C^{\text{FA}} = 3.806h = 3.806 \times 1000 = 3086\,\text{gu}$$
$$C^{\text{B}} = 0.419\rho\,h = 0.419 \times 2.326 \times 1000 = 975\,\text{gu}$$

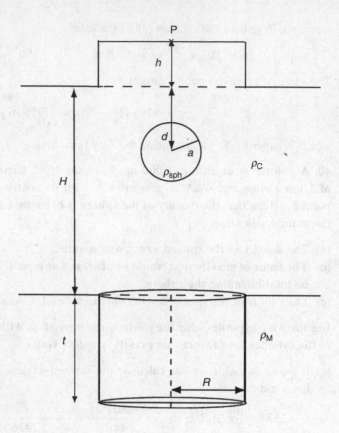

Fig. 40

The effect of the spherical mass is

$$C^{am} = \frac{G\Delta M}{(h+d)^2} = \frac{650}{(4)^2 \times 10^6} = 41 \text{ gu}$$

We calculate the isostatic correction using the Airy hypothesis and taking into account the presence of the spherical anomalous mass. Thus, according to Fig. 40, the equilibrium between the gravity at P and at sea level far from P is given by

$$\pi a^2 \rho_C H + \pi a^2 \rho_M t = \pi a^2 \rho_C (h + H + t) + \frac{4}{3}\pi R^3 \left(\rho_{sph} - \rho_C\right)$$

and solving for t:

$$t = \frac{a^2 h \rho_C + \frac{4}{3} R^3 \left(\rho_{sph} - \rho_C\right)}{a^2 (\rho_M - \rho_C)} = 3011 \text{ m}$$

As in previous problems we calculate the isostatic correction using a cylinder under point P

$$C^I = 2\pi G \Delta\rho \left[b + \sqrt{a^2 + (c-b)^2} - \sqrt{a^2 + c^2} \right]$$

where $a = 20$ km, $b = t = 3011$ m, $c = h + H + t = 34\,011$ m and $\Delta\rho = \rho_M - \rho_C = 0.775$ g cm^{-3}, resulting in

$$C^I = 145 \text{ gu}$$

If compensation is total, the isostatic anomaly must be null. This implies that the Bouguer anomaly is equal, with opposite sign, to the isostatic correction

$$\Delta g^I = 0 = \Delta g^B - C^I \Rightarrow \Delta g^B = -145 \text{gu}$$

But the Bouguer correction can be obtained from Equation (40.1):

$$\Delta g^I = 0 = -\gamma + g^P + C^{FA} + C^B + C^I + C^{am}$$

Solving for C^B we obtain

$$C^B = 975 \text{ gu}$$

From the definition of the Bouguer anomaly we can find the value g^P of gravity at P:

$$\Delta g^B = g^P - \gamma + C^{FA} - C^B + C^{am}$$
$$-145 = g^P - 9800\,000 + 3086 - 975 + 41 \Rightarrow g^P = 9797\,703 \text{ gu}$$

(c) Since the density of the sphere is greater than the density of the crust, there is an excess of gravity at P with respect to other points at sea level far from P, which must be compensated by a root of crustal material inside the mantle with negative gravitational influence. In this situation the root can never be null.

41. At 10 km beneath sea level vertically under a point P of height 2000 m there exists an anomalous spherical mass $G\Delta M = 10^4$ m^3 s^{-2}. At sea level, gravity is 9800 000 gu and the crustal thickness 20 km. The density of the crust is 2 g cm^{-3}, and of the mantle 4 g cm^{-3}. Using the Airy assumption for the isostatic compensation with a cylinder of 10 km radius, calculate for that point:

(a) The observed gravity.
(b) The free-air, Bouguer, and isostatic anomalies.

(a) For point P the Bouguer correction is

$$C^B = 0.419\rho h = 0.419 \times 2 \times h = 0.838h = 1676 \text{ gu}$$

The gravity at point P, if there is no isostatic compensation and other effects, can be obtained from the normal gravity and the free-air and Bouguer corrections

$$g^P = \gamma - C^{FA} + C^B = 9800\,000 - 3.086 \times 2000 + 0.838 \times 2000$$
$$= 9795504 \text{ gu} \tag{41.1}$$

Since there is an anomalous mass under point P we have to add its gravitational contribution to the gravity at P. For a spherical mass at depth $h + d$ under P the gravitational attraction is

$$C^{am} = \frac{G\Delta M}{(h+d)^2} = \frac{10^4 \, m^3 \, s^{-2}}{10^8 \, m^2} = 10^{-4} \, m\,s^{-2} = 100 \, gu$$

We calculate the root corresponding to the isostatic compensation, assuming the Airy hypothesis, and taking into account the presence of the anomalous mass in the same way as in Problem 40:

$$\pi a^2 \rho_C H + \pi a^2 \rho_M t = \pi a^2 \rho_C (h + H + t) + \frac{4}{3}\pi R^3 \left(\rho_{sph} - \rho_C\right)$$

$$= \pi a^2 \rho_C (h + H + t) + M_a$$

$$t = \frac{\pi a^2 \rho_C h + M_a}{\pi a^2 (\rho_M - \rho_C)} = 2239 \, m$$

The isostatic correction using a cylinder is given by

$$C^I = 2\pi G \Delta \rho \left(b + \sqrt{a^2 + (c-b)^2} - \sqrt{a^2 + c^2} \right) \tag{41.2}$$

where $a = 10$ km, $b = t = 2.239$ km, $c = 20 + 2 + 2 = 24.239$ km, $\Delta\rho = 4 - 2 = 2 \, g\,cm^{-3}$, resulting in

$$C^I = 154 \, gu$$

The gravity at point P is the value obtained in (41.1) plus the contribution of the anomalous mass and minus the isostatic correction:

$$g^P = 9795\,504 + 100 - 154 = 9795\,450 \, gu$$

(b) The free-air anomaly is equal to this observed value minus the normal gravity plus the free-air correction:

$$\Delta g^{FA} = g^P - \gamma + C^{AL}$$

Substituting the values we obtain

$$\Delta g^{FA} = 9795\,450 - 9800\,000 + 3.086 \times 2000 = 1622 \, gu$$

The Bouguer anomaly is given by

$$\Delta g^B = g^P - \gamma + C^{AL} - C^B = -54 \, gu$$

Finally the isostatic anomaly is the Bouguer anomaly plus the isostatic correction:

$$\Delta g^I = -54 + 154 = 100 \, gu$$

This value corresponds to the gravitational contribution of the anomalous mass.

42. At a point P of height 2000 m above sea level the measured value of gravity is 979.5717 Gal. Beneath P is a sphere centred at a depth of 12 km below sea level, 1 g cm^{-3} density, and radius 5 km. Assuming the Airy hypothesis ($H = 30$ km, $\rho_C = 2.5$ g cm^{-3}), $\rho_M = 3.0$ g cm^{-3}), calculate the isostatic anomaly at the point in gu and mGal.

For the compensation, assume cylinders of the same radius as the sphere. Normal gravity $\gamma = 9.8$ m s^{-2}.

We first calculate the root t, assuming the Airy hypothesis, corresponding to the height 2000 m of point P. If the situation is of total isostatic equilibrium, we have to introduce the effect produced by the sphere in the determination of the root (Fig. 40):

$$\frac{4}{3}\pi\left(\rho_{sph} - \rho_C\right)a^3 + \pi a^2 \rho_C h + \pi a^2 \rho_C H + \pi a^2 \rho_C t = \pi a^2 H \rho_C + \pi a^2 \rho_M t$$

so

$$t = \frac{\rho_C h + \frac{4}{3}a\left(\rho_{sph} - \rho_C\right)}{\rho_M - \rho_C} = -10\,000\,\text{m} \qquad (42.1)$$

The negative value of t (anti-root) is due to the deficit of mass produced by the presence of the sphere ($\rho_{sph} < \rho_C$) under point P.

The isostatic correction, as in previous problems, is calculated taking a cylinder under the point:

$$C^I = 2\pi G(\rho_M - \rho_C)\left(b + \sqrt{a^2 + (c-b)^2} - \sqrt{a^2 + c^2}\right)$$

where $b = t = 10\,000$ m, $a = 5000$, and $c = H + h = 32\,000$ m.

We obtain $C^I = 36$ gu.

The isostatic anomaly at P is equal to the observed gravity minus the normal gravity and the free-air, Bouguer, and isostatic corrections, and the attraction of the spherical mass:

$$\Delta g^I = g^P - \gamma + C^{FA} - C^B - C^I - C^{am} \qquad (42.2)$$

The effect of the anomalous mass is given by

$$C^{am} = \frac{G\frac{4}{3}\pi a^3\left(\rho_{sph} - \rho_C\right)}{(h+d)^2} = -267\,\text{gu}$$

By substitution in (42.2)

$$\Delta g^I = 9\,795\,717 - 9\,800\,000 + 3.086 \times 2000 - 0.419 \times 2.5 \times 2000 + 267 - 36$$
$$= 25\,\text{gu}$$

43. A point A on the Earth's surface is at an altitude of 2100 m above sea level. Calculate:

(a) **The value of gravity at A if the isostatic anomaly is 2.5 mGal. Assume the Airy hypothesis ($\rho_C = 2.6$ g cm^{-3}, $\rho_M = 3.3$ g cm^{-3}, $H = 30$ km).**

(b) **If the previous value had been measured with a Worden gravimeter of constant 0.301 82 mGal/division giving a reading of 630.6, calculate the value of gravity at another point B at which the device reads 510.1 (both readings corrected for drift).**

(c) **At what depth is the centre of a sphere of density 4 g cm^{-3} and radius 5 km which is buried in the crust, given that the anomaly created at a point A, 12 km from the**

centre of the sphere, not in the same vertical, is 321 gu. Also calculate the horizontal distance from the centre to point A.

Take, for compensation, cylinders of 10 km radius. $\gamma = 9.8$ m s^{-2}

(a) The isostatic anomaly is given by

$$\Delta g^{\mathrm{I}} = g_{\mathrm{A}} - \gamma + 3.086h - 0.419\rho_{\mathrm{C}}h + C^{\mathrm{I}} \tag{43.1}$$

To calculate the isostatic correction C^{I}, assuming the Airy hypothesis, we must first calculate the root t that corresponds to the height h

$$t = \frac{\rho_{\mathrm{C}}h}{\rho_{\mathrm{M}} - \rho_{\mathrm{C}}} = 7800\,\mathrm{m}$$

As in other problems the isostatic correction is calculated using

$$C^{\mathrm{I}} = 2\pi G(\rho_{\mathrm{M}} - \rho_{\mathrm{C}})\left(b + \sqrt{a^2 + (c - b)^2} - \sqrt{a^2 + c^2}\right)$$

and substituting the values

$$b = t = 7800\,\mathrm{m}, \qquad c = h + H + t = 39900\,\mathrm{m},$$
$$a = 10\,\mathrm{km}, \qquad \Delta\rho = \rho_{\mathrm{M}} - \rho_{\mathrm{C}} = 700\,\mathrm{kg\,m}^{-3}$$

we obtain

$$C^{\mathrm{I}} = 84\,\mathrm{gu}$$

Solving for g_{A} in Equation (43.1) we obtain

$$g_{\mathrm{A}} = \Delta g^{\mathrm{I}} + \gamma - 3.086h + 0.419\rho_{\mathrm{C}}h - C^{\mathrm{I}} = 9795\,748\,\mathrm{gu}$$

(b) For a Worden gravimeter the increment in gravity between two points (Δg) is proportional to the increment in the values given by the instrument (ΔL) corrected by the instrumental variations

$$\Delta g = K\Delta L$$
$$g_{\mathrm{B}} - g_{\mathrm{A}} = K(L_{\mathrm{B}} - L_{\mathrm{A}}) \Rightarrow g_{\mathrm{B}} = g_{\mathrm{A}} + K(L_{\mathrm{B}} - L_{\mathrm{A}})$$
$$= g_{\mathrm{A}} - 364\,\mathrm{gu} = 9795\,384\,\mathrm{gu}$$

where K is the constant of the gravimeter.

(c) The anomaly produced by a sphere buried at depth d under sea level at a point at height h and at a horizontal distance x from the centre of the sphere is given by

$$C^{\mathrm{am}} = \frac{G\frac{4}{3}\pi a^3\left(\rho_{\mathrm{sph}} - \rho_{\mathrm{C}}\right)(h + d)}{\left(x^2 + (h + d)^2\right)^{3/2}} \tag{43.2}$$

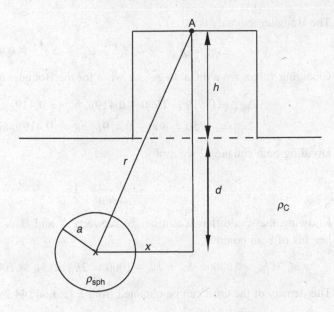

Fig. 43

For point A (Fig. 43) if r is the distance from the centre of the sphere to the point A, $x^2 + (h + d)^2 = r^2$ and solving for d in Equation (43.2) gives

$$d = \frac{C^{am} r^3}{G \frac{4}{3} \pi a^3 \left(\rho_{sph} - \rho_C \right)} - h$$

Substituting the values $r = 12$ km, $a = 5$ km, $\rho_{sph} - \rho_C = 1400$ kg m^{-3}, $h = 2100$ m, $C^{am} = 321 \times 10^{-6}$ m s^{-2}, we obtain:

$$d = 9245\,\text{m}$$

The horizontal distance is:

$$x = \sqrt{r^2 - (h + d)^2} = 3910\,\text{m}$$

44. In a gravity survey, two points A and B on the Earth's surface gave the values 159 and 80 mGal for the free-air anomaly, and −51 and −25 mGal for the Bouguer anomaly, respectively. Given that B is at an altitude 1000 m lower than A, and that the density of the mantle is 25% greater than that of the crust, calculate:

(a) The value of gravity at A and B, and the densities of the crust and mantle.

(b) The isostatic anomaly according to the hypotheses of Airy ($H = 30$ km) and Pratt ($D = 100$ km, ρ_0 the value determined in the previous part) at point A. Take, for compensation, cylinders of 10 km radius. $\gamma = 980$ Gal.

(a) The free-air anomaly at point A is given by

$$\Delta g_A^{FA} = g_A - \gamma + C_A^{FA} = g_A - \gamma + 3.086 h_A \tag{44.1}$$

The Bouguer anomaly is

$$\Delta g_A^B = g_A - \gamma + C_A^{FA} - C_A^B = \Delta g_A^{FA} - 0.419 \rho_C h_A$$

Changing values from mGal to gu, we write for the Bouguer anomalies at points A and B

$$\Delta g_A^B = -510 = 1590 - 0.419 \rho_C h_A \Rightarrow 0.419 \rho_C h_A = 2100 \, \text{gu}$$
$$\Delta g_B^B = -250 = 800 - 0.419 \rho_C h_B \Rightarrow 0.419 \rho_C h_B = 1050 \, \text{gu} \qquad (44.2)$$

Dividing both equations, we find

$$\frac{h_A}{h_B} = 2$$

Knowing that the difference in height between A and B is 1000 m, we obtain for the heights of both points,

$$h_B = h_A - 1000 \Rightarrow h_A = h_B + 1000 = 2h_B \Rightarrow h_B = 1000 \, \text{m} \Rightarrow h_A = 2000 \, \text{m}$$

The density of the crust can be obtained from Equation (44.2):

$$0.419 \rho_C h_B = 105 \times 10^{-5} \, \text{m s}^{-2} = 0.419 \rho_C \times 1000 \Rightarrow \rho_C = 2.505 \, \text{g cm}^{-3}$$

The density of the mantle is 25% more than that of the crust, so

$$\rho_M = \rho_C (1 + 0.25) = 1.2 \times 2.505 = 3.131 \, \text{g cm}^{-3}$$

The gravity at A and B is obtained using Equation (44.1):

$$g_A = \Delta g_A^{FA} - 3.086 h_A + \gamma = 1590 - 3.086 \times 2000 + 9800\,000 = 9\,795\,418 \, \text{gu}$$
$$g_B = 800 - 3.086 \times 1000 + 9800\,000 = 9797\,714 \, \text{gu}$$

(b) For the isostatic anomaly at point A, according to the Airy hypothesis, we first calculate the value of the root t corresponding to its height:

$$t = \frac{\rho_C h_A}{1.25 \rho_C - \rho_C} = \frac{h_A}{0.25} = 8000 \, \text{m}$$

For the isostatic correction we use, as in other problems, a cylinder

$$C^I = 2\pi G (\rho_M - \rho_C) \left(b + \sqrt{a^2 + (c-b)^2} - \sqrt{a^2 + c^2} \right)$$

Substituting the values

$$b = t = 8000 \, \text{m}, \qquad c = h + H + t = 40\,000 \, \text{m},$$
$$a = 10 \, \text{km}, \qquad \Delta \rho = \rho_M - \rho_C = 626 \, \text{kg m}^{-3}$$

we obtain

$$C^I = 77 \, \text{gu}$$

The isostatic anomaly is

$$\Delta g_A^I = \Delta g_A^B + C^I = -510 + 77 = -433 \, \text{gu}$$

If we use the Pratt hypothesis, we first calculate the density corresponding to the material under point A:

$$\rho = \frac{D\rho_0}{D+h} = \frac{100 \times 2.505}{100 + 2} = 2.456 \, \text{g cm}^{-3}$$

and the contrast of density

$$\Delta\rho = \rho_0 - \rho = 2.505 - 2.456 = 0.049 \, \text{g cm}^{-3}$$

The isostatic correction is determined using a cylinder,

$$C^I = 2\pi G(\rho_M - \rho_C)\left(b + \sqrt{a^2 + (c-b)^2} - \sqrt{a^2 + c^2} \right)$$

where $\Delta\rho = 0.049$ g cm^{-3}, $b = D = 100$ km, $a = 10$ km, $c = D + h = 102$ km.
   Then, we obtain

$$C^I = 158 \, \text{gu}$$

The isostatic anomaly will be the Bouguer anomaly plus the isostatic correction

$$\Delta g_A^I = \Delta g_A^B + C^I = -510 + 158 = -352 \, \text{gu}$$

In both cases the anomaly is negative, but using the Airy model the value is greater than using the Pratt model.

45. At a point P on the Earth's surface, the observed value of gravity is 9.795 636 m s^{-2}, and the Bouguer anomaly is -26 mGal. Assuming the Airy hypothesis ($\rho_C = 2.7$ g cm^{-3}, $\rho_M = 3.3$ g cm^{-3}, $H = 30$ km), calculate:

(a) The height of the point.
(b) The isostatic anomaly.
(c) The value of gravity that would be observed at the point if beneath it were a sphere at a depth of 10 km below sea level, with a density of 2.5 g cm^{-3} and a radius of 5 km, such that the compensation was total.

Compensation with cylinders of 5 km radius; $\gamma = 9.8$ m s^{-2}.

(a) We calculate the height of point P from the Bouguer anomaly:

$$\Delta g^B = g^P - \gamma + 3.086h - 0.419\rho_C h \Rightarrow h = \frac{\Delta g^B - g^P + \gamma}{3.086 - 0.419\rho_C}$$

so

$$h = \frac{-260 - 9\,795\,636 + 9\,800\,000}{3.086 - 0.419 \times 2.7} = 2099.8 \, \text{m} \cong 2100 \, \text{m}$$

(b) To calculate the isostatic anomaly, using the Airy hypothesis, we first calculate the value of the root t corresponding to the height of the point:

$$t = \frac{\rho_C}{\rho_M - \rho_C} h = \frac{2.7}{3.3 - 2.7} \times 2100 = 9450 \, \text{m}$$

Using a cylinder, the isostatic correction is given by

$$C^I = 2\pi G \Delta \rho \left(b + \sqrt{a^2 + (c - b)^2} - \sqrt{a^2 + c^2} \right) \tag{45.1}$$

where we substitute the values

$$a = 5 \, \text{km}, \qquad b = t = 9450 \, \text{m},$$
$$c = h + H + t = 2100 + 30\,000 + 9450 = 41\,550 \, \text{m},$$
$$\Delta \rho = 3.3 - 2.7 = 0.6 \, \text{g cm}^{-3}$$

and obtain

$$C^I = 22 \, \text{gu}$$

The isostatic anomaly is then

$$\Delta g^I = \Delta g^B + C^I = -260 + 22 = -238 \, \text{gu}$$

(c) If the compensation is total then the isostatic anomaly must be zero. But now we have to include the gravitational effect C^{am} produced by the presence of the anomalous mass of the sphere.

$$\Delta g^I = 0 = g^P - \gamma + 3.086h - 0.419\rho_C h - C^{\text{am}} + C^I$$

Solving for g^P:

$$g^P = \gamma - 3.086h + 0.419\rho_C h + C^{\text{am}} - C^I \tag{45.2}$$

where the effect of the sphere is given by

$$C^{\text{am}} = \frac{G \frac{4}{3} \pi a^3 \left(\rho_{\text{sph}} - \rho_C \right)}{(h + d)^2} = -48 \, \text{gu}$$

We calculate the isostatic correction according to the Airy hypothesis. First we calculate the value of root t, but now we add the effect of the sphere on point P (Fig. 45):

$$\rho_C \pi a^2 h + \rho_C \pi a^2 H + \frac{4}{3} \pi a^3 \left(\rho_{\text{sph}} - \rho_C \right) + \rho_C \pi a^2 t = \rho_C \pi a^2 H + \rho_M \pi a^2 t$$

Solving for t, we obtain

$$t = \frac{\rho_C h + \frac{4}{3} a \left(\rho_{\text{sph}} - \rho_C \right)}{\rho_M - \rho_C} = 7228 \, \text{m}$$

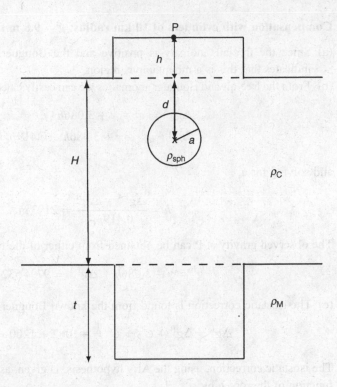

Fig. 45

We substitute this value of t in Equation (45.1) together with the other values

$$a = 5\,\text{km}, \qquad b = 7 = 7228\,\text{m},$$
$$c = h + H + t = 39\,328\,\text{m}, \qquad \Delta\rho = 3.3 - 2.7 = 0.5\,\text{kg\,m}^{-3}$$

and obtain

$$C^{\text{I}} = 18\,\text{gu}$$

By substitution in (45.2) we find the value of the gravity at P under the given conditions:

$$g^{\text{P}} = 9800000 - 3.086 \times 2100 + 0.419 \times 2.7 \times 2100 - 48 - 18 = 9795830\,\text{gu}$$

46. Consider a point on the surface of the Earth in an overcompensated region at which the values of the free-air and the Bouguer anomalies are 1300 gu and −1200 gu, respectively.

(a) **Is this a mountainous or an oceanic zone? Give reasons.**

(b) **Calculate the altitude and the value of gravity at the point given that the density of the crust is 2.72 g cm^{-3}.**

(c) **If the isostatic anomaly is -1062 gu calculate, according to the Airy hypothesis ($\rho_{\text{C}} = 2.72$ g cm^{-3}, $\rho_{\text{M}} = 3.30$ g cm^{-3}, $H = 30$ km), the value of the root responsible for this anomaly. Compare it with the value that it would have if the region were in isostatic equilibrium.**

Compensation with cylinders of 10 km radius; g $= 9.8$ m s^{-2}.

(a) Since the free-air anomaly is positive and the Bouguer anomaly is negative, this indicates that this is a mountainous region.

(b) From the free-air and Bouguer anomalies we can easily calculate the height of the point:

$$\Delta g^{FA} = g^P - \gamma + 3.086h$$
$$\Delta g^B = g^P - \gamma + 3.086h - 0.419\rho_C h \tag{46.1}$$

and solving for h,

$$h = \frac{\Delta g^{FA} - \Delta g^B}{0.419\rho_C} = 2193 \,\text{m}$$

The observed gravity at P can be obtained from either of the two equations (46.1):

$$g^P = \gamma - 3.086h + \Delta g^{FA} = 9794\,532 \,\text{gu}$$

(c) The isostatic correction is found from the known Bouguer and isostatic anomalies:

$$\Delta g^I = \Delta g^B + C^I \Rightarrow C^I = -1062 + 1200 = 138 \,\text{gu}$$

The isostatic correction, using the Airy hypothesis, is given, as in previous problems, as a function of the root t, by

$$C^I = 138 = 2\pi G \Delta\rho \left(b + \sqrt{a^2 + (c-b)^2} - \sqrt{a^2 + c^2} \right)$$

where we substitute

$$\Delta\rho = 3.3 - 2.72 = 0.58 \,\text{g cm}^{-3}, \qquad a = 10 \,\text{km},$$
$$b = t, \qquad c = h + H + t = 2.193 + 30 + t \,\text{km}$$

and obtain for t,

$$t = 19\,984 \,\text{m}$$

If the region is in equilibrium the root due to the height h would be

$$t = \frac{\rho_C h}{\rho_M - \rho_C} = \frac{2.72 \times 2193}{3.3 - 2.72} = 10\,284 \,\text{m}$$

Since we have already obtained a larger value ($t = 19\,984$ m), this indicates that the region is overcompensated.

47. In an oceanic region, gravity is measured at a point on the surface of the sea, obtaining a value of 979.7950 Gal. Calculate, using the Airy hypothesis ($H = 30$ km, $\rho_C = 2.9$ g cm^{-3}, $\rho_M = 3.2$ g cm^{-3}, $\rho_W = 1.04$ g cm^{-3}):

(a) The isostatic anomaly if the thickness of the crust is 8.4 km.

(b) **The thickness that the water layer would have to have if 15 km vertically below the point there was centred a sphere of 10 km radius such that the anti-root is null and the compensation total. Also calculate the density of the sphere.**

For the compensation, take cylinders of 10 km radius. $\gamma = 9.8$ m s^{-2}.

(a) We are in an oceanic region, therefore in the calculation of the root for the isostatic compensation according to the Airy hypothesis we have to consider the layer of water of density ρ_W. The value of the root is now given by

$$t' = \frac{\rho_C - \rho_W}{\rho_M - \rho_C} h' = \frac{2.9 - 1.04}{3.2 - 2.9} h' = 6.2h' \tag{47.1}$$

According to Fig. 47a, we have the following relation: $H - h' - t' = e \Rightarrow h' = H - t' - e$, where e is the thickness of the crust at the oceanic region, H the thickness of the normal (sea level) continental crust, h' the thickness of the water layer, and t' the negative root.

Substituting the values of t' from Equation (47.1) we obtain for h'

$$h' = 30 - 6.2h' - 8.4 \Rightarrow h' = 3\,\text{km} = 3000\,\text{m}$$

$$t' = 6.2 \times 3000 = 18\,600\,\text{m}$$

From the value of the root we calculate the isostatic correction using a cylinder of 10 km radius,

$$C^{\mathrm{I}} = 2\pi G \Delta \rho \left(b + \sqrt{a^2 + (c-b)^2} - \sqrt{a^2 + c^2} \right)$$

where $\Delta \rho = 3.2 - 2.9 = 0.3$ g cm^{-3} = 300 kg m^{-3}, $b = t'$, $c = H = 30\,000$ m, and so

$$C^{\mathrm{I}} = 269\,\text{gu}$$

To calculate the isostatic anomaly we first have to apply the Bouguer correction which in this case consists of two terms: the first to eliminate the attraction of the water layer

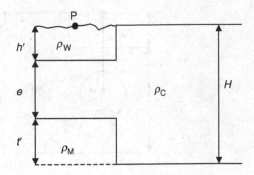

$(-2\pi G\rho_W h')$ and the second to replace this layer by one of density equal to the crustal density $(+2\pi G\rho_C h')$. Since the point is at sea level the free-air correction is null:

$$\Delta g^I = g - \gamma + C^B - C^I$$

$$C^B = 0.419(\rho_C - \rho_W)h' = 2338 \text{ gu}$$

$$\Delta g^I = 19 \text{ gu}$$

(b) If the anti-root is null and there is total compensation, then we have

$$t' = 0$$

$$\Delta g^I = 0 = g - \gamma - 2\pi G\rho_W h' + 2\pi G\rho_C h' - C^{am} \qquad (47.2)$$

Since the isostatic anomaly must be null, then the anomalous spherical mass and the water layer must compensate each other. The attraction of the anomalous mass is $C^{am} = G\Delta M/d^2$ where d is the depth of its centre below sea level. Then we can write

$$g - \gamma + 2\pi G(\rho_C - \rho_W)h' - \frac{G\Delta M}{d^2} = 0$$

where the mass of the sphere is

$$\Delta M = \frac{4}{3}\pi a^3 \Delta\rho_{sph}$$

If the point P is totally isostatically compensated and the anti-root is null, then (Fig. 47b)

$$\pi a^2 h' \rho_W + \pi a^2 (H - h')\rho_C + \frac{4}{3}\pi a^3(\rho_{sph} - \rho_C) = \pi a^2 H\rho_C$$

Solving for h' gives

$$h' = \frac{4a\left(\rho_C - \rho_{sph}\right)}{3(\rho_W - \rho_C)} \qquad (47.3)$$

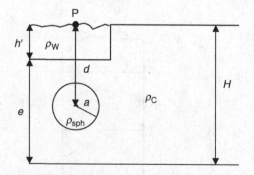

Substituting this value in (47.2) we obtain

$$\rho_{sph} = \frac{\gamma - g}{\frac{4}{3}\pi G\left(2a - \frac{a^3}{d^2}\right)} + \rho_C = 3372 \, \text{kg m}^{-3}$$

and putting this value in (47.3), $h' = 3369$ m.

48. At a point on the Earth's surface, 500 m below sea level, a gravity value is measured of 980.0991 Gal. If the region is in isostatic equilibrium calculate, using the Airy hypothesis ($H = 30$ km, $\rho_C = 2.7$ g cm^{-3}, $\rho_M = 3.2$ g cm^{-3}):

(a) The thickness of the crust.

(b) The isostatic anomaly in gu, with reasons for the sign of each correction. Take compensating cylinders of 5 km radius; g = 9.8 m s^{-2}.

(a) We calculate first the root, according to the Airy hypothesis which corresponds to the depth of the point, applying the condition of isostatic equilibrium (Fig. 48)

$$\rho_C H = \rho_C(H - h' - t') + \rho_M t'$$

so

$$t' = \frac{\rho_C}{\rho_M - \rho_C} h' = 2700 \, \text{m}$$

The thickness of the crust e at that point is

$$e = H - h' - t' = 30\,000 - 2700 - 500 = 26\,800 \, \text{m}$$

(b) The isostatic anomaly is given by

$$\Delta g^I = g^P - \gamma - 3.086h' + 0.419\rho_C h' - C^I$$

The free-air correction ($3.086h'$) is negative because the point is below sea level. For the same reason the Bouguer correction ($0.419 \, \rho_C \, h'$) is positive. The isostatic correction, using the Airy hypothesis, is calculated as in previous problems using a cylinder under the point of 5 km radius and density contrast $\Delta \rho = 3.2 - 2.7 = 0.5$ g cm^{-3}. The value of the anti-root t' has already been calculated, so

$$C^I = 2\pi G\Delta\rho\left(b + \sqrt{a^2 + (c - b)^2} - \sqrt{a^2 + c^2}\right)$$

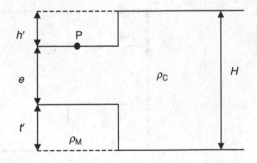

Fig. 48

Substituting $c = H - h'$ and $b = t'$, we obtain

$$C^I = 9 \, \text{gu}$$

Then the isostatic anomaly is $\Delta g^I = 5 \, \text{gu}$.

49. In an oceanic region where the density of the crust is 2.90 g cm^{-3} and that of the mantle 3.27 g cm^{-3}, the value of gravity measured at a point P on the sea floor at depth 4000 m is 9.806 341 m s^{-2}.
Calculate, according to the Airy hypothesis:

(a) The thickness of the crust.
(b) The isostatic anomaly in gravimetric units.

Data: $\rho_w = 1.04$ g cm^{-3}, $H = 30$ km, $g = 9.8$ m s^{-2}. Take, for compensation, cylinders of 10 km radius.

(a) First we calculate the value of the root according to the Airy hypothesis

$$t' = \frac{\rho_C - \rho_W}{\rho_M - \rho_C} h' = 20\,108 \, \text{m}$$

The thickness of the crust under the point is found by (Fig. 49)

$$e = H - h' - t' = 30000 - 40000 - 2018 = 5892 \, \text{m}$$

(b) Because the point is located at the bottom of the sea, to reduce the observed value of gravity to the surface of the geoid (sea level) we eliminate first the attraction of the water layer. Then we apply the free-air and the Bouguer corrections, to take into account the attraction of a layer of crustal material which replaces the water. Finally we apply the isostatic correction:

$$\Delta g^I = g^P - \gamma + 0.419\rho_W h' - 3.086 h' + 0.419\rho_C h' - C^I$$

The isostatic correction is calculated using a cylinder of 10 km radius,

$$C^I = 2\pi G \Delta\rho \left(b + \sqrt{a^2 + (c - b)^2} - \sqrt{a^2 + c^2} \right)$$

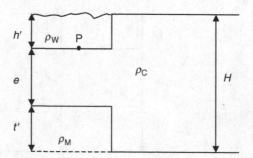

Fig. 49

where $b = t' = 20\,108$ m, $c = H - h' = 26\,000$ m, $\Delta\rho = 3.27 - 2.9 = 0.37$ g cm^{-3} = 370 kg m^{-3}

By substitution we obtain

$$C^{\mathrm{I}} = 598 \text{ gu}$$

The isostatic anomaly is:

$$\Delta g^{\mathrm{I}} = 9\,806\,341 - 9\,800\,000 - 3.086 \times 4000 + 0.419 \times (1.04 + 2.9) \times 4000 - 598$$
$$= 4 \text{ gu}$$

50. At a point with coordinates 42.78° N, 0.5° E and height 1572 m, the observed value of gravity is 980.0317 Gal.

(a) Calculate the free-air and Bouguer anomalies.
(b) If cylinders of 10 km radius beneath that point are used for the isostatic compensation, calculate the gravimetric attraction of the mass defect corresponding to the altitude of the point according to the Airy and Pratt hypotheses. Take, for the crust, $H = 30$ km, $\rho_{\mathrm{C}} = 2.67$ g cm^{-3}, for the mantle, $\rho_{\mathrm{M}} = 3.27$ g cm^{-3}, and $D = 100$ km for the Pratt level of compensation.
(c) How deep should the root of the Airy model be for the compensation to be total?

(a) We calculate first the normal gravity at the point where gravity has been observed using the expression

$$\gamma = 9.780\,32\left(1 + 0.005\,3025 \sin^2 \varphi\right) = 9.804\,243 \text{ m s}^{-2}$$

where φ is the latitude.

The free-air anomaly is given by

$$\Delta g^{\mathrm{FA}} = g^{\mathrm{P}} - \gamma + C^{\mathrm{FA}} = 9800\,317 - 9804\,243 + 3.086 \times 1572 = 925 \text{ gu}$$

and the Bouguer anomaly by

$$\Delta g^{\mathrm{B}} = g^{\mathrm{P}} - \gamma + C^{\mathrm{FA}} + C^{\mathrm{B}} = 9800\,317 - 9804\,243 + 1.967 \times 1572 = -834 \text{ gu}$$

(b) If we approximate the isostatic compensation by means of a cylinder of radius a under the point, we use the expression

$$C^{\mathrm{I}} = -2\pi G \Delta\rho \left(b + \sqrt{a^2 + (c - b)^2} - \sqrt{a^2 + c^2} \right) \qquad (50.1)$$

where $\Delta\rho$ is the density contrast, b the height of the cylinder, and c the distance from the base of the cylinder to the observation point.

Airy: We calculate first the root given by the equation

$$t = 4.45 \times h = 4.45 \times 1572 = 6995 \text{ km}$$

For the isostatic correction we substitute in (50.1) the values

$$a = 10 \, \text{km}, \qquad b = t = 6995 \, \text{km}, \qquad c = t + H + h = 38\,567 \, \text{km},$$
$$\Delta \rho = \rho_M - \rho_C = 3.27 - 2.67 = 0.6 \, \text{g cm}^{-3}$$

and obtain

$$C^{I} = 2 \times 3.1416 \times 6.67 \times 10^{-11} \, \text{m}^{-3} \, \text{s}^{-2} \, \text{kg}^{-1} \times 0.5 \times 10^{3} \, \text{kg m}^{-3} \times A$$

where:

$$A = \left(6.995 + \sqrt{10^2 + (38.567 - 6.995)^2} - \sqrt{10^2 + 38.567^2} \right) \times 10^3 = 270 \, \text{m}$$

so

$$C^{I} = 68 \, \text{gu}$$

Pratt: The contrast of densities is now given by

$$\Delta \rho = \frac{h}{D + h} \rho_0 = \frac{1575}{100\,000 + 1572} \times 2.67 = 0.04 \, \text{g cm}^{-3}$$

and substituting in Equation (50.1) with the values
$$a = 10 \, \text{km}, \qquad b = 100 \, \text{km}, \qquad c = D + h = 101\,572 \, \text{m}, \text{ we have}$$

$$C^{I} = 135 \, \text{gu}$$

(c) If the isostatic compensation is total (isostatic anomaly null) the isostatic correction, according to the Airy hypothesis, coincides with the value of the Bouguer anomaly (-834 gu):

$$\Delta g^{B} = 2\pi G \Delta \rho \left(t + \sqrt{a^2 - (H + h)^2} - \sqrt{a^2 + (t + H + h)^2} \right)$$

so

$$t + \sqrt{a^2 - (H + h)^2} - \sqrt{a^2 + (t + H + h)^2} = \frac{\Delta g^{B}}{2\pi G \Delta \rho}$$

In this expression we solve for the value of the root t:

$$\sqrt{a^2 + (t + H + h)^2} = t + \sqrt{a^2 - (H + h)^2} - \frac{\Delta g^{B}}{2\pi G \Delta \rho} = t + N$$
$$a^2 + (t + H + h)^2 = t^2 + N^2 + 2tN$$
$$a^2 + t^2 + (H + h)^2 + 2t(H + h) = t^2 + N^2 + 2tN$$

so

$$t = \frac{N^2 - a^2 - (H + h)^2}{2(H + h - N)}$$

By substitution of the values of N, a, H, and h we obtain

$$t = -58\,875\,\text{m}$$

Because this root has a negative value greater than the thickness of the crust, total compensation is not possible.

51. Calculate the free-air anomaly observed on a mountain of height 2000 m which is fully compensated by a root of depth $t = 10$ km. The compensation is by a cylinder of radius 20 km, the density of the crust is 2.67g cm^{-3}, and that of the mantle is 3.27g cm^{-3}.

The free-air anomaly is given by

$$\Delta g^{\text{FA}} = g - \gamma + 3.086h$$

Since the point is isostatically compensated, we calculate the isostatic correction using a cylinder as in previous problems:

$$C^{\text{I}} = 2\pi G \Delta \rho \left(b + \sqrt{a^2 + (c-b)^2} - \sqrt{a^2 + c^2} \right)$$

where we substitute
$b = t = 10$ km, $c = h + H + t = 42$ km, $a = 20$ km, $\Delta \rho = 600$ kg m^{-3}
and obtain

$$C^{\text{I}} = 306\,\text{gu}$$

Since the point is totally compensated the isostatic anomaly must be zero:

$$\Delta g^{\text{I}} = g - \gamma + C^{\text{FA}} - C^{\text{B}} + C^{\text{I}} = 0$$

The free-air anomaly can now be written as

$$\Delta g^{\text{FA}} = g - \gamma + C^{\text{FA}} = C^{\text{B}} - C^{\text{I}} \tag{51.1}$$

We can calculate the Bouguer correction:

$$C^{\text{B}} = 1.119h = 1.119 \times 2000 = 2238\,\text{gu}$$

and substituting in (51.1) we obtain, for the free-air anomaly,

$$\Delta g^{\text{FA}} = 2238 - 301.5 = 1932\,\text{gu}$$

52. Calculate for a point P at height 2100 m, latitude 40°N, and observed gravity 979.7166 Gal the refined Bouguer anomaly in gravimetric units. Consider a surplus mass compartment of 3000 m, mean height, 3520 m inner radius, 5240 m outer radius, with $n = 16$. The density of the crust is 2.5 g cm^{-3}.

The refined Bouguer anomaly is obtained using, besides the free-air and Bouguer corrections, the correction for the topography or topographic or terrain correction (T):

$$\Delta g^{\text{B}} = g - \gamma + C^{\text{FA}} - C^{\text{B}} + T$$

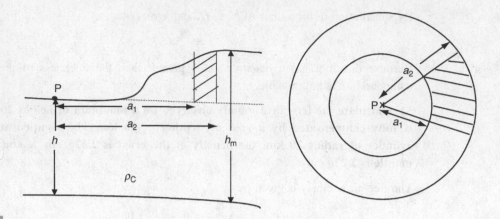

The normal gravity at latitude 40° N is given by

$$\gamma = 9.780\,32\left(1 + 0.0053025\,\sin^2 \varphi\right) = 9.801\,747\,\mathrm{m\,s^{-2}}$$

the free-air correction by

$$C^{\mathrm{FA}} = 3.086h = 6481\,\mathrm{gu}$$

and the Bouguer correction by

$$C^{\mathrm{B}} = 0.419\rho_{\mathrm{C}}h = 2200\,\mathrm{gu}$$

The topographic correction is introduced in order to correct for the topographic masses not included in the Bouguer correction, that is, in this case those above the height h (Fig. 52). Remember that the Bouguer correction corrects for an infinite layer or plate of thickness h and doesn't consider the additional masses above h or the lack of masses below h. The topographic correction is always positive because the masses above height h produce on point P an attraction of negative sign which must be added and the lack of mass under h must also be taken into account with a positive sign, since it has been subtracted in the Bouguer correction.

To calculate the attraction of the mass above height h we use the attraction of concentric cylinders (in our case two) with axis passing through point P and with height equal to the difference between the height h of the point P and the height of the mass of the topography h_{m} above h. The cylinder is divided into n sectors with radius a_1 and a_2 to approximate the topography (Fig. 52). Then the topographic correction is given by

$$T = \frac{2\pi G\rho_{\mathrm{C}}}{n}\left(\sqrt{a_2^2 + (c-b)^2} - \sqrt{a_2^2 + c^2} - \sqrt{a_1^2 + (c-b)^2} + \sqrt{a_1^2 + c^2}\right)$$

In our case we substitute the values

$$a_2 = 5240\,\mathrm{m}, \qquad a_1 = 3520\,\mathrm{m}, \qquad b = h_{\mathrm{m}} - h = 900\,\mathrm{m}, \qquad c = 0, \qquad n = 16$$

and obtain

$$T = 2\,\text{gu}$$

The refined Bouguer anomaly is

$$\Delta g^{B} = g - \gamma + C^{FA} - C^{B} + T = -298\,\text{gu}$$

53. Calculate the topographic correction for a terrestrial compartment of inner radius $a_1 = 5240$ m, outer radius $a_2 = 8440$ m, $n = 20$, mean height 120 m, with 2000 m being the height of the point P. Take $\rho = 2.65$ g cm^{-3}.

In this problem we consider the topographic correction for the case of the lack of mass in the topography at heights below that of the point P. Since in the Bouguer correction we have subtracted an infinite layer of thickness h, we have to correct for the places where the mass was not present (Fig. 53).

The topographic or terrain correction T in this case is calculated in the same way as in the previous problem. Thus we take n sectors of cylinders with axis at point P and height equal to the difference between h and h_{m} (Fig. 52b). The correction is then given by

$$T = \frac{2\pi G \rho_{C}}{n}\left(\sqrt{a_2^2 + (c-b)^2} - \sqrt{a_2^2 + c^2} - \sqrt{a_1^2 + (c-b)^2} + \sqrt{a_1^2 + c^2}\right)$$

where we substitute $a_2 = 8440$ m, $a_1 = 5240$ m, $b = c = h - h_{\mathrm{m}}$, $n = 20$ to obtain

$$T = 0.67\,\text{mGal}$$

54. Calculate the topographic correction for an oceanic sector or compartment of inner radius $a_1 = 5240$ m, outer radius $a_2 = 8440$ m, $n = 20$, mean depth 525 m, with 600 m being the height of the point P. Take $\rho_C = 2.67$ g cm^{-3}, $\rho_W = 1.03$ g cm^{-3}.

In this problem we have to correct for the lack of mass in the oceanic area near the point P, between the sea level and height h (column 1 in Fig. 54). Also we have to take into account the attraction produced by the water layer between sea level and the bottom of the sea (column 2 in Fig. 54).

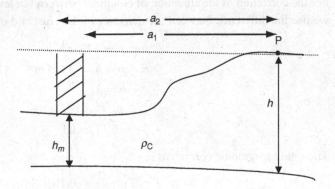

Fig. 53

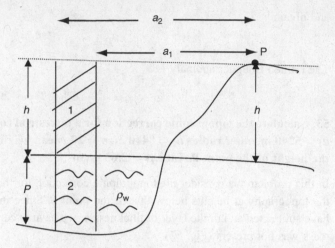

Fig. 54

To calculate the necessary topographic correction we proceed as in Problems 52 and 53, using cylindrical sectors:

$$T = \frac{2\pi G\rho}{n} \left(\sqrt{a_2^2 + (c - b)^2} - \sqrt{a_2^2 + c^2} - \sqrt{a_1^2 + (c - b)^2} + \sqrt{a_1^2 + c^2} \right)$$

where

$$a_2 = 8440 \, \text{m}, \qquad a_1 = 5240 \, \text{m}, \qquad n = 20$$

For the correction corresponding to the attraction of column 1, between height h and sea level, we substitute the values:

$$b = h = c, \qquad \rho = \rho_C$$

and obtain

$$T_1 = 0.07 \, \text{mGal}$$

For the correction of the attraction of column 2 between sea level and the bottom of the sea we use the difference between the densities of the crust and of water:

$$b = p = 525 \, \text{m}$$
$$c = p + h_s = 1125 \, \text{m}$$
$$c - b = h_s$$
$$\rho = \rho_C - \rho_W$$
$$T_2 = 0.11 \, \text{mGal}$$

The total topographic correction is

$$T = T_1 + T_2 = 0.18 \, \text{mGal}$$

Tides

55. Two spherical planets A and B of radii 2a and a and masses 3m and m are separated by a distance (centre to centre) of 6a. The only forces acting are gravitational, and the system formed by the two planets rotates in the equatorial plane.

(a) Calculate the value of the components of the acceleration of the tides at the Pole of each planet directly and using the tidal potential. On which planet are they greater?

(b) If each planet spins on its axis with the same angular velocity as the system, what, for each planet, is the ratio between the centrifugal force and the maximum of the tidal force at the equator? On which planet is this ratio greater?

   (a) From Fig. 55a we can deduce that at the Pole of planet A, the radial component of the acceleration of the tides produced by planet B is

$$g_r^T = -\frac{Gm}{q^2}\cos\alpha$$

where q is the distance from the Pole of planet A to the centre of planet B and a the angle formed by q and the radius at the Pole of planet A.

By substitution of the required values we obtain

$$q = \sqrt{(6a)^2 + (2a)^2} = \sqrt{40}a$$

$$\cos\alpha = \frac{2a}{\sqrt{40}a} \Rightarrow \alpha = 71.6°$$

$$g_r^T = -\frac{Gm}{40a^2}\cos(71.6) = -0.008\frac{Gm}{a^2}$$

The tangential component g_θ^T is given by (Fig. 55a)

$$g_\theta^T = -\frac{Gm}{36a^2} + \frac{Gm}{q^2}\sin\alpha = \frac{-Gm}{36a^2} + \frac{Gm}{40a^2}\sin(71.6)$$

$$= \frac{Gm}{a^2}(-0.028 + 0.024) = -0.004\frac{Gm}{a^2}$$

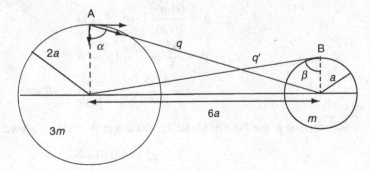

If we use the tidal potential,

$$\psi = \frac{GMr^2}{2R^3}\left(3\cos^2\vartheta - 1\right)$$

where R is the distance between the centres of planets A and B (Fig. 55a), and ϑ is the angle the position vector r forms with the distance vector R (in this case it is equal to the colatitude, $\vartheta = \theta$)

$$\psi = \frac{GMr^2}{2R^3}\left(3\cos^2\theta - 1\right)$$

The radial and tangential components of the acceleration are given by

$$g_r^T = \frac{\partial \psi}{\partial r} = \frac{\partial}{\partial r}\frac{Gmr^2}{2(6a)^3}\left(3\cos^2\theta - 1\right) = \frac{Gmr}{216a^3}\left(3\cos^2\theta - 1\right)$$

$$g_\theta^T = \frac{1}{r}\frac{\partial \psi}{\partial \theta} = -\frac{1}{r}\frac{\partial}{\partial \theta}\frac{Gmr^2}{2(6a)^3}\left(3\cos^2\theta - 1\right) = \frac{Gmr}{216a^3}3(\cos\theta\sin\theta)$$

For planet A, at the Pole, $r = 2a$ and $\theta = 90°$, so

$$g_r^T = -\frac{Gm}{108a^2} = -0.009\frac{Gm}{a^2}$$

$$g_\theta^T = 0$$

For planet B, we proceed in a similar manner:

$$q'^2 = 36a^2 + a^2 = 37a^2$$

$$\cos\alpha = \frac{a}{\sqrt{37}a} \Rightarrow \alpha = 80.5°$$

Therefore,

$$g_r^T = -\frac{G3m}{37a^2}\cos(80.5) = 0.013\frac{Gm}{a^2}$$

$$g_\theta^T = -\frac{G3m}{36a^2} + \frac{G3m}{37a^2}\sin(80.5) = -0.003\frac{Gm}{a^2}$$

Using the tidal potential, we obtain the acceleration components

$$\psi = \frac{G3mr^2}{2(6a)^3}\left(3\cos^2\theta - 1\right)$$

$$g_r^T = \frac{\partial \psi}{\partial r} = \frac{Gmr}{72a^3}\left(3\cos^2\theta - 1\right)$$

$$g_\theta^T = -\frac{1}{r}\frac{\partial \psi}{\partial \theta} = \frac{9}{216}\frac{Gmr}{(a)^3}(\cos\theta\sin\theta)$$

Substituting at the Pole of planet B, $r = a$ and $\theta = 90°$, we have

$$g_r^T = -0.014\frac{Gm}{a^2}$$

$$g_\theta^T = 0$$

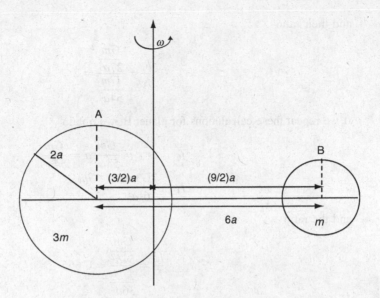

(b) First we calculate the centre of gravity of the system formed by the two planets measured from the centre of planet A (Fig. 55b):

$$x = \frac{3m \times 0 + m \times 6a}{3m + m} = \frac{3}{2}a$$

The rotation radius for planet A is $3/2a$ and for planet B

$$6a - \frac{3}{2}a = \frac{9}{2}a$$

In the rotating system the centrifugal force equals the force of gravitational attraction, which at the equator ($\theta = 0°$) is

$$f_g = \frac{Gm3m}{(6a)^2} = f_C = \omega^2 r = m\omega^2 \frac{9}{2}a$$

From this expression we obtain the value of the angular velocity ω of the rotation system:

$$\frac{3Gm^2}{36a^2} = m\omega^2 \frac{9}{2}a \Rightarrow \omega^2 = \frac{Gm}{54a^3}$$

Since the angular velocity of the spin of each planet is equal to that of the system, the spin centrifugal force at the equator of planet A is

$$f_C = \omega^2 r = \frac{Gm}{54a^3} 2a = \frac{Gm}{27a^2}$$

The tidal force is

$$f_T = g_r^T = \frac{2Gmr}{R^3} = \frac{2Gm2a}{6^3 a^3}$$

and their ratio

$$\frac{f_C}{f_T} = \frac{\dfrac{Gm}{27a^2}}{\dfrac{Gm}{54a^2}} = 2$$

If we repeat these calculations for planet B, we obtain

$$f_C = \omega^2 r = \frac{Gm}{54a^3}\, a = \frac{Gm}{54a^2}$$

$$f_T = \frac{2G3ma}{6^3 a^3} = \frac{Gm}{36a^2}$$

and the ratio

$$\frac{f_C}{f_T} = \frac{\dfrac{Gm}{54a^2}}{\dfrac{Gm}{36a^2}} = 0.666$$

The ratio is larger for planet A, as expected owing to its larger radius.

56. Two planets of mass M and radius a are separated by a centre-to-centre distance of $8a$. The planets spin on their own axes with an angular velocity such that the value of the centrifugal force at the equator is equal to the maximum of the tidal force (the equatorial plane is the plane in which the system formed by the two planets rotates around an axis normal to that plane).

(a) **Calculate the value of the components of the vector g as multiples of GM/a^2 for a point of $\lambda' = 60°$ and $\varphi = 45°$ (with $\lambda = 0°$, being the meridian in front of the other planet) including all the forces that act.**

(b) **What is the relationship between the angular velocity of each planet and that of the system?**

(a) The tidal potential is given by

$$\psi = \frac{GMr^2}{2R^3}\left(3\cos^2\vartheta - 1\right)$$

where R is the centre-to-centre distance between the planets, and ϑ the angle formed by the vector r to a point and R (Fig. 56a). From this potential we calculate the radial component of the tidal force:

$$f_r^T = \frac{\partial\psi}{\partial r} = \frac{\partial}{\partial r}\left(\frac{GMr^2}{2R^3}\left(3\cos^2\vartheta - 1\right)\right) = \frac{GMr}{R^3}\left(3\cos^2\vartheta - 1\right)$$

At the equator of one planet $\vartheta = 0°$, $r = a$, and $R = 8a$, so

$$f_r^T = \frac{GMa}{(8a)^3}\, 2 = \frac{GM}{256a^2}$$

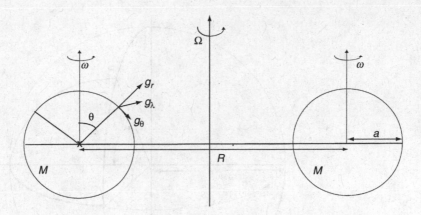

The spin centrifugal force is

$$f_C = \omega^2 a$$

Equating these two expressions we find the value of the spin angular velocity,

$$f_C = f_r^T \Rightarrow \omega^2 a = \frac{GM}{256a^2} \Rightarrow \omega = \frac{1}{16}\sqrt{\frac{GM}{a^3}}$$

At a point on the surface of one of the planets the total potential is the sum of the gravitational potential V, the spin potential Φ, and the tidal potential ψ:

$$U = V + \Phi + \psi = \frac{GM}{r} + \frac{1}{2}\omega^2 r^2 \cos^2\varphi + \frac{GMr^2}{2R^3}\left(3\cos^2\vartheta - 1\right)$$

For a point P at latitude φ and longitude λ (Fig. 56b)

$$\cos\vartheta = \cos\varphi\cos\lambda$$

and the potential U is

$$U = \frac{GM}{r} + \frac{1}{2}\omega^2 r^2 \cos^2\varphi + \frac{GMr^2}{2R^3}\left(3\cos^2\varphi\cos^2\lambda - 1\right)$$

The components of gravity including the three effects are,

$$g_r = \frac{\partial U}{\partial r} = -\frac{GM}{r^2} + \omega^2 r \cos^2\varphi + \frac{GMr}{R^3}\left(3\cos^2\varphi\cos^2\lambda - 1\right)$$

$$g_\varphi = -\frac{1}{r}\frac{\partial U}{\partial\varphi} = \omega^2 r \cos\varphi\sin\varphi + \frac{GMr}{R^3}3\cos^2\lambda\cos\varphi\sin\varphi$$

$$g_\lambda = \frac{1}{r\cos\varphi}\frac{\partial U}{\partial\lambda} = -\frac{1}{r\cos\varphi}\frac{GMr^2}{2R^3}6\cos^2\varphi\cos\lambda\sin\lambda$$

$$= -\frac{GMr}{R^3}3\cos\varphi\cos\lambda\sin\lambda$$

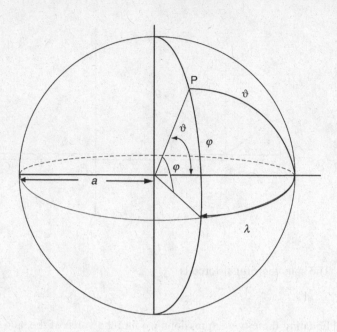

Fig. 56b

At the required point,

$$r = a, \qquad \varphi = 45°,$$

$$\lambda = 60° \Rightarrow \cos \vartheta = \cos 45° \cos 60° = 0.35 \Rightarrow \vartheta = 69.3°$$

so

$$g_r = -\frac{GM}{a^2} + \left(\frac{GM}{256a^3}\right) a \frac{1}{2} + \frac{GMa}{(8a)^3}\left(3\frac{1}{2}\frac{1}{4} - 1\right) = -0.9993\frac{GM}{a^2}$$

$$g_\varphi = \frac{GM}{256a^3} a \frac{1}{\sqrt{2}}\frac{1}{\sqrt{2}} + \frac{GMa}{(8a^3)} 3\frac{1}{4}\frac{1}{\sqrt{2}}\frac{1}{\sqrt{2}} = 0.0027\frac{GM}{a^2}$$

$$g_\lambda = -\frac{GMa}{(8a)^3} 3 \frac{\sqrt{2}}{2}\frac{1}{2}\frac{\sqrt{3}}{2} = -0.0018\frac{GM}{a^2}$$

(b) To obtain the angular velocity of the rotation of the system (Ω) we take into account that the centrifugal force due to the rotation of the system at the equatorial plane is equal to the gravitational attraction between the two planets:

$$M\Omega^2 p = G\frac{MM}{(8a)^2}$$

where p is the distance from the centre of one planet to the centre of gravity of the system. Then we find

$$M\Omega^2 4a = G\frac{MM}{(8a)^2} \Rightarrow \Omega^2 = \frac{GM}{4 \times 64a^3} = \frac{GM}{256a^3}$$

and finally $\Omega = \omega$.

57. Consider two planets of equal mass m and radius a separated by a centre-to-centre distance of $8a$. Only gravitational forces act.

(a) Calculate the tidal force at the equator on one of the planets directly, using the formula of the tidal potential (do so at $\lambda = 0°$, i.e. the point in front of the other planet). Express the result in mGal given that $Gm/a^2 = 980\,000$ mGal.

(b) Compare and comment on the reason for the difference between the results of the direct calculation and using the tidal potential.

(c) What relationship must there be between the angular velocities of the planet's spin and of the system's rotation for the centrifugal force due to the planet's spin to be equal to the tidal force at the equator and $\lambda = 0°$?

(a) The exact calculation of the tidal force at a point located at the equatorial plane in front of the other planet is (Fig. 57)

$$f_T = \frac{Gm}{(7a)^2} - \frac{Gm}{(8a)^2} = \frac{Gm}{a^2}\left(\frac{1}{49} - \frac{1}{64}\right) = 46\,875\text{ gu}$$

Using the tidal potential

$$\psi = \frac{Gmr^2}{2(R)^3}\left(3\cos^2\vartheta - 1\right)$$

where R is the centre-to-centre distance between the planets ($8a$), r the radius to the point where the tide is evaluated (a), and ϑ the angle between r and R (at the equator in front of the other planet $\vartheta = 0°$), the radial component of the tidal force can be derived from the potential. At the equator this is the total tidal force

$$f_r^T = \frac{\partial \psi}{\partial r} = \frac{\partial}{\partial r}\left(\frac{Gmr^2}{2(8a)^3}\left(3\cos^2\vartheta - 1\right)\right) = \frac{2Gma}{2(512a^3)}2 = 38\,281\text{ gu}$$

(b) The difference between the value obtained by the exact calculation and by using the tidal potential is 8594 gu, that is, 18%. This is explained because the tidal

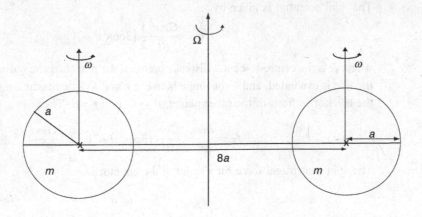

Fig. 57

potential is a first-order approximation corresponding to terms of the order of $(r/R)^2$. For a relatively large value of r/R (1/8) this approximation is not very good.

(c) Now we make the spin centrifugal force equal to the tidal force for a point at the equator of one of the planets which is given by exact calculation

$$f_T = \frac{Gm}{a^2}\left(\frac{1}{49} - \frac{1}{64}\right)$$

If the spin angular velocity is ω_p, the spin centrifugal force at the equator is given by

$$f_C = \omega_p^2 a$$

Equating these two expressions we obtain

$$f_C = f_T \Rightarrow \omega_p^2 a = \frac{Gm}{a^2}\left(\frac{1}{49} - \frac{1}{64}\right) \Rightarrow \omega_p^2 = \frac{Gm}{a^3}\left(\frac{1}{49} - \frac{1}{64}\right) \qquad (57.1)$$

The centrifugal force due the rotation of the system with angular velocity ω_s is equal to the gravitational attraction between the two planets:

$$m\omega_s^2 4a = \frac{Gm}{64a^2} \Rightarrow \omega_s^2 = \frac{Gm^2}{256a^3} \qquad (57.2)$$

From (57.1) and (57.2) we obtain the relation between the two angular velocities:

$$\frac{\omega_p^2}{\omega_s^2} = \frac{\dfrac{Gm}{a^3}\left(\dfrac{1}{49} - \dfrac{1}{64}\right)}{\dfrac{Gm}{256a^3}} = 1.22 \Rightarrow \frac{\omega_p}{\omega_s} = 1.10$$

58. Two planets of equal mass m and radius a are separated by a distance R. The spin angular velocity of each planet is such that the centrifugal force at the equator is equal to the maximum of the tidal force. If the sum of the two forces at the equator cancels the gravitational force, what is the distance R?

The tidal potential is given by

$$\psi = \frac{Gmr^2}{R^3}\frac{1}{2}\left(3\cos^2\vartheta - 1\right)$$

where R is the centre-to-centre distance between the planets, r the radius to the point where the tide is evaluated, and ϑ the angle between r and R. The maximum value is at a point at the equator in front of the other planet, $\vartheta = 0°$ and $r = a$. Then

$$f_T = \frac{\partial\psi}{\partial r} = \frac{Gmr}{R^3}\left(3\cos^2\vartheta - 1\right) = \frac{2Gma}{R^3} \qquad (58.1)$$

The spin centrifugal force for a point at the equator is

$$f_C = \omega^2 a \qquad (58.2)$$

Equating (58.1) and (58.2) we obtain the spin angular velocity,

$$\frac{2Gma}{R^3} = \omega^2 a \Rightarrow \omega^2 = \frac{2Gm}{R^3}$$

If, at the point considered, the sum of the spin centrifugal force and the tidal force cancel the gravitational force of the planet, then

$$F + f_C + f_T = 0$$

where $F = -Gm/a^2$.

The value of R must be

$$\frac{4Gma}{R^3} = \frac{Gm}{a^2} \Rightarrow R = \sqrt[3]{4a}$$

59. Two spherical planets A and B of radii $2a$ and a and masses $5M$ and M spin on their axes with equal angular velocities. They are separated by a centre-to-centre distance of $8a$, and form a system that rotates in the equatorial plane of both planets with an angular velocity that is equal to that of the spin angular velocity of each one.

(a) Determine the total potential for points on planet A.

(b) Determine the expression for the three components of the total gravity, including the tide, for a point on the surface of planet A at longitude $0°$.

(c) If the Love number h on planet A is 0.5, determine the height of the terrestrial tide as a multiple of a at the equator, at local noon with respect to planet B, in the case that the system's rotational angular velocity is the same as that of the spin of the two planets about their axes.

(a) We calculate the centre of gravity of the system, measured from the centre of planet A (Fig. 59a):

$$X = \frac{5M \times 0 + M \times 8a}{5M + M} = \frac{4}{3}a$$

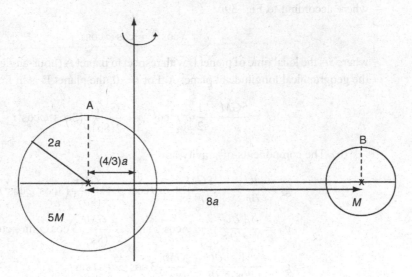

Fig. 59a

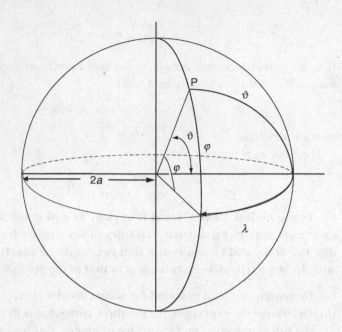

The total potential U at a point on the surface of planet A at latitude φ is given by the sum of the gravitational potential V, plus the spin potential Φ, plus the tidal potential ψ produced by planet B:

$$U = V + \Phi + \psi$$

$$U = \frac{5GM}{r} + \frac{1}{2}\omega^2 r^2 \cos^2\varphi + \frac{GMr^2}{(8a)^3 2}\left(3\cos^2\vartheta - 1\right) \tag{59.1}$$

where according to Fig. 59b

$$\cos\vartheta = \cos\varphi\cos t$$

where t is the local time of planet B with respect to planet A (hour-angle), at a point of $\lambda = 0°$, the geographical longitude at planet A. For $t = 0$, the planet B is in front of the point, so

$$U = \frac{5GM}{r} + \frac{1}{2}\omega^2 r^2 \cos^2\varphi + \frac{GMr^2}{2(8a)^3}\left(3\cos^2\varphi\cos^2 t - 1\right)$$

(b) The components of gravity are

$$g_r = \frac{\partial U}{\partial r} = -\frac{5GM}{r^2} + \omega^2 r\cos^2\varphi + \frac{GMr}{(8a)^3}\left(3\cos^2\varphi\cos^2 t - 1\right)$$

$$g_\theta = -\frac{1}{r}\frac{\partial U}{\partial \varphi} = \omega^2 r\cos\varphi\sin\varphi + \frac{GMr}{(8a)^3}3\cos\varphi\sin\varphi\cos^2 t$$

$$g_\lambda = \frac{1}{r\cos\varphi}\frac{\partial U}{\partial t} = \frac{GMr}{(8a)^3}3\cos\varphi\cos t\sin t$$

At a point on the surface of planet A, $r = 2a$,

$$g_r = -\frac{5GM}{4a^2} + \omega^2 2a\cos^2\varphi + \frac{GM2a}{(8a)^3}\left(3\cos^2\varphi\cos^2 t - 1\right)$$

$$g_\theta = \omega^2 2a\cos\varphi\sin\varphi + \frac{GM2}{8^3 a^2}3\cos\varphi\sin\varphi\cos^2 t$$

$$g_\lambda = \frac{GM2}{8^3 a^2}3\cos\varphi\cos t\sin t$$

(c) At the equator $\varphi = 0°$, at 12 h with respect to B, $t = 180°$, $h = 1/2$, and $\omega = \Omega$. The height of the equilibrium terrestrial tide is given by

$$\varsigma = h\frac{\psi}{g}$$

At the equator of planet A the tidal potential (59.1) is

$$\psi = \frac{2GM(2a)^2}{2(8a)^3} = \frac{GM}{128a}$$

If we approximate g by g_r

$$g_r = -\frac{5GM}{4a^2} + \omega^2 2a + \frac{GM2a}{(8a)^3}2 = -\frac{5GM}{4a^2} + \frac{GM}{128a^2} + \omega^2 2a$$

$$= -\frac{159GM}{128a^2} + 2\omega^2 a$$

and the height of the equilibrium tide is

$$\varsigma = \frac{1}{2}\frac{\dfrac{GM}{128a}}{\left(\dfrac{-159MG}{128a^2} + 2a\omega^2\right)} \tag{59.2}$$

We know that the angular velocity of the rotation of the system is equal to the spin angular velocity of both planets, so the spin angular velocity is given by

$$5M\omega^2\frac{4}{3}a = \frac{G5M^2}{(8a)^2} \Rightarrow \omega^2 = \frac{3GM}{256a^2}$$

By substitution in (59.2)

$$\varsigma = \frac{1}{2}\frac{\dfrac{GM}{128a}}{\left(\dfrac{-159MG}{128a^2} + 2a\dfrac{3GM}{256a^2}\right)} = -\frac{a}{312}$$

60. The Earth is formed by a sphere of radius a and density ρ, and a core of radius $a/2$ and density 2ρ, in the northern hemisphere, centred on the axis of rotation and

tangent to the equatorial plane. The Moon has mass $M/4$ (where $M = (4/3)\pi\rho a^3$), is at a distance (centre-to-centre) of $4a$, and orbits in the equatorial plane. Determine:

(a) The total potential and the components of gravity including the tidal forces.
(b) The total deviation of the vertical from the radial at lunar noon, and the deviation due to the tide at the same hour for latitude 45° N, with $m = 1/8$.

(a) The total potential U is equal to the gravitational potential of the planet V_1 with uniform density plus that of the core V_2 using the differential mass, the spin potential Φ, and the tidal potential ψ. The gravitational potentials are given by (Fig. 60):

$$V_1 = \frac{GM}{r}$$

$$V_2 = \frac{GM'}{q}$$

where the differential mass of the core is

$$M' = \frac{4}{3}\pi(2\rho - \rho)\frac{a^3}{8} = \frac{M}{8} \Rightarrow V_2 = \frac{GM}{8q}$$

and q is the distance to the centre of the core. Its inverse can be approximated by

$$\frac{1}{q} = \frac{1}{r}\left[1 + \frac{a}{2r}\cos\theta + \left(\frac{a}{2r}\right)^2\frac{1}{2}(3\cos^2\theta - 1)\right]$$

The spin potential is

$$\Phi = \frac{1}{2}\omega^2 r^2\sin^2\theta$$

and the total potential is

$$U = \frac{GM}{r}\left[1 + \frac{1}{8}\left(1 + \frac{a}{2r}\cos\theta + \frac{a^2}{(2r)^2}\frac{1}{2}(3\cos^2\theta - 1)\right) + \left(\frac{r}{a}\right)^3\frac{m}{2}\sin^2\theta\right] + \psi$$

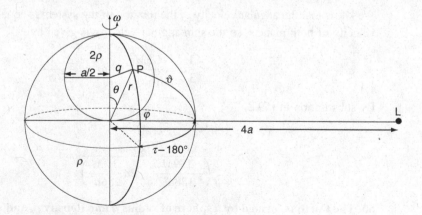

Fig. 60

where $m = \omega^2 a^3/GM$.

The tidal potential ψ due to the Moon is given by

$$\psi = \frac{GM_L r^2}{2R^3}\left(3\cos^2\vartheta - 1\right) \tag{60.1}$$

According to Fig. 60

$$\cos^2\vartheta = \cos^2\varphi\cos^2(\tau - 180)$$

By substitution in (60.1), since R (the centre-to-centre distance between the Earth and the Moon) is $4a$, $M_L = M/4$, and $\varphi = 90° - \theta$, we obtain

$$\psi = \frac{GMr^2}{512a^3}\left(3\cos^2\varphi\cos^2(\tau - 180) - 1\right)$$

The potential U is

$$U = GM\left[\frac{5}{8r} + \frac{a}{16r^2}\sin\varphi + \frac{a^2}{64r^3}\left(3\sin^2\varphi - 1\right)\right.$$
$$\left. + \frac{r^2}{a^3}\frac{m}{2}\cos^2\varphi + \frac{r^2}{512a^3}\left(3\cos^2\varphi\cos^2\tau - 1\right)\right]$$

The components of gravity are found by taking the derivatives of U with respect to r and φ:

$$g_r = GM\left[-\frac{5}{8r^2} - \frac{2a}{16r^3}\sin\varphi - \frac{3a^2}{64r^4}\left(3\sin^2\varphi - 1\right)\right.$$
$$\left. + \frac{r}{a^3}m\cos^2\varphi + \frac{2r}{512a^3}\left(3\cos^2\varphi\cos^2\tau - 1\right)\right]$$

$$g_\theta = -\frac{1}{r}\frac{\partial U}{\partial\varphi}$$
$$= \frac{GM}{r}\left(-\frac{a}{16r^2}\cos\varphi - \frac{a^2}{64r^3}\left(6\cos\varphi\sin\varphi\right)\right.$$
$$\left. + \frac{r^2}{a^3}\frac{m}{2}2\sin\varphi\cos\varphi + \frac{r^2}{512a^3}6\sin\varphi\cos\varphi\cos^2\tau\right)$$

By substituting $r = a$, $\varphi = 45°$, $m = 1/8$, and at 12 h lunar time, $\tau = 180°$, we obtain

$$g_r = -1.128\frac{GM}{a^2}$$

$$g_\theta = -0.023\frac{GM}{a^2}$$

(b) The deviation of the vertical with respect to the radial direction at 12 h lunar time is given by

$$\tan i = \frac{g_\theta}{g_r} = 0.020 \Rightarrow i = 1.2°$$

The part of the deviation due to the lunar tide is given by

$$\tan i' = \frac{g_\theta^M}{g_r} = \frac{1}{g_r}\frac{1}{r}\frac{\partial \psi}{\partial \varphi} = \frac{1}{g_r}\frac{GMr^2}{r512a^3}6\sin\varphi\cos\varphi\cos^2\tau$$

and by substitution of the same values

$$i' = 0.3°$$

The greater part of the deviation is due to the core.

61. Two spherical planets, planet A of radius $2a$ and mass $3m$ and planet B of radius a and mass m, are separated by a centre-to-centre distance of $6a$. The system rotates in the equatorial plane and each planet spins on its axis with the same angular velocity. What, for each planet, is the ratio between the force of gravity and the maximum of the tidal force at the equator in front of the other planet?

The centre of gravity of the system measured from the centre of planet A is (Fig. 61)

$$\frac{0 \times 3m + 6am}{4m} = \frac{3}{2}a$$

The angular velocity of the system is given by

$$\frac{3mGm}{(6a)^2} = 3m\omega^2\frac{3}{2}a \Rightarrow \omega^2 = \frac{Gm}{54a^3}$$

The tidal force (radial component) can be calculated from the tidal potential

$$\psi = \frac{GMr^2}{2R^3}\left(3\cos^2\vartheta - 1\right) \Rightarrow f_r^T = \frac{\partial \psi}{\partial r} = \frac{GMr}{R^3}\left(3\cos^2\vartheta - 1\right)$$

where R is the centre-to-centre distance between the planets. For a point on the equator in front of the other planet, $\vartheta = 0°$, and

$$f_r^T = \frac{2GMr}{(6a)^3}$$

$$\text{Planet A}: f_r^T = \frac{2Gm2a}{216a^3} = 0.018\frac{Gm}{a^2}$$

$$\text{Planet B}: f_r^T = \frac{2G3ma}{216a^3} = 0.028\frac{Gm}{a^2}$$

Gravity without tides is the sum of the gravitational and centrifugal forces:

$$g = \frac{Gm}{r^2} - \omega^2 r$$

$$\text{Planet A}: g = \frac{G3m}{4a^2} - \omega^2 2a = \frac{3Gm}{4a^2} - \frac{Gm}{54a^3}2a = 0.71\frac{Gm}{a^2}$$

$$\text{Planet B}: g = \frac{Gm}{a^2} - \omega^2 a = \frac{Gm}{a^2} - \frac{Gm}{54a^3}a = 0.98\frac{Gm}{a^2}$$

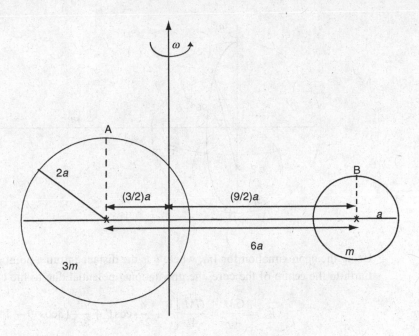

Fig. 61

The ratios between the gravity and tidal forces are:

$$\text{Planet A}: \frac{g}{f_r^T} = \frac{0.71}{0.018} = 39.44$$

$$\text{Planet B}: \frac{g}{f_r^T} = \frac{0.98}{0.028} = 35.00$$

62. The Earth is of radius a and density ρ, with a core of radius $a/2$ and density 3ρ on the axis of rotation in the southern hemisphere tangent to the equatorial plane. The Moon has mass $M/2$ and its centre is at $4a$ from the centre of the Earth ($M = 4/3\pi\rho a^3$).

(a) Write down the total potential.

(b) What is the value of the angular velocity of the Earth if at the point 30° N, 30° E at 06:00 lunar time the radial component of gravity is equal to GM/a^2?

(c) In this case, what is the ratio between the angular velocity of the Earth's rotation and that of the system?

(a) As in previous problems the total potential U is given by

$$U = V_1 + V_2 + \Phi + \psi$$

The differential mass of the core is:

$$M_1 = \frac{4\pi}{3}(3\rho - \rho)\frac{a^3}{8} = \frac{M}{4}$$

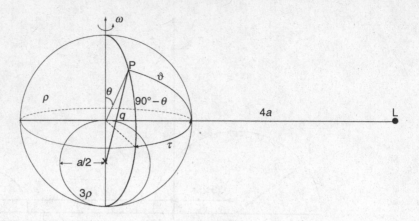

Fig. 62

Using the approximation for $1/q$, where q is the distance from a point on the surface of the Earth to the centre of the core, the gravitational potential due to the core is

$$V_2 = \frac{GM}{4q} = \frac{GM}{4}\left[\frac{1}{r} - \frac{a}{2r^2}\cos\theta + \frac{a^2}{8r^3}\left(3\cos^2\theta - 1\right)\right]$$

The potential due to the spin of the Earth is given by

$$\Phi = \frac{1}{2}\omega^2 r^2 \sin^2\theta = \frac{GM}{a^3}r^2\frac{m}{2}\sin^2\theta$$

where

$$m = \frac{\omega^2 a^3}{GM}$$

and the tidal potential due to the Moon is

$$\psi = \frac{GM}{2}\frac{r^2}{64a^3}\frac{1}{2}\left(3\cos^2\vartheta - 1\right) = \frac{GMr^2}{256a^3}\left(3\sin^2\theta\cos^2\tau - 1\right)$$

where τ is the hour-angle of the Moon (Fig. 62)

Then, the total potential U is

$$U = GM\left[\frac{5}{4r} - \frac{a}{8r^2}\cos\theta + \frac{a^2}{32r^3}\left(3\cos^2\theta - 1\right)\right.$$
$$\left. + \frac{r^3}{a^3}\frac{m}{2}\sin^2\theta + \frac{r^2}{256a^3}\left(3\sin^2\theta\cos^2\tau - 1\right)\right]$$

(b) The radial component of gravity is given by

$$g_r = \frac{\partial U}{\partial r}$$

$$= GM\left[-\frac{5}{4r^2} + \frac{2a}{8r^3}\cos\theta - \frac{3a^2}{32r^4}\left(3\cos^2\theta - 1\right)\right.$$
$$\left. + \frac{rm\sin^2\theta}{a^3} + \frac{2r}{256a^3}\left(3\sin^2\theta\cos^2\tau - 1\right)\right]$$

Substituting $\theta = 60°$ and at 6 h, $\tau = 90$, we have

$$g_r = \frac{GM}{a^2}\left(-\frac{141}{128} + m\frac{3}{4} - \frac{1}{128}\right) = \frac{GM}{a^2}$$

Solving for m we obtain $m = 2.81$, and the spin angular velocity is

$$\omega = \sqrt{\frac{2.81\,GM}{a^3}}\,\text{rad s}^{-1}$$

(c) The centre of gravity r' of the Earth–Moon system, measured from the centre of the Earth, is given by

$$r' = \frac{M_T r_1 + M_L r_2}{M_T + M_L} = \frac{\frac{M}{2}4a}{\left(\frac{5}{4} + \frac{1}{2}\right)M} = \frac{8a}{7}$$

where the mass of the Earth M_T includes that of the core,

$$M_T = M + \frac{M}{4} \quad \text{and } M_L = \frac{M}{2}$$

and $r_1 = 0$ and $r_2 = 4a$.

To calculate the angular velocity of the system, Ω, we put the centripetal force at the Moon equal to the gravitational force between the Earth and the Moon:

$$M_L(r_2 - r')\Omega^2 = G\frac{M_T M_L}{r_2}$$

Substituting and solving for Ω, we obtain

$$G\frac{\frac{5}{8}M^2}{16a^2} = \frac{M}{2}\Omega^2\left(4 - \frac{8}{7}\right)a; \quad \text{so } \Omega = \sqrt{\frac{GM}{a^3}\frac{7}{256}}$$

Then, the ratio between the angular velocities of the spin of the Earth and of the system is:

$$\frac{\omega}{\Omega} = \frac{\sqrt{\dfrac{2.81\,GM}{a^3}}}{\sqrt{\dfrac{7GM}{256a^3}}} = 10.14$$

63. Consider two planets of equal mass m and radius a separated by a centre-to-centre distance of $8a$. The planets revolve around their centre of mass and spin around their own axes. Their spin angular velocity is such that the value of the centrifugal force is equal to the maximum tidal force of the two at the equator.

(a) **Calculate, for a point on the equator of one of the planets and longitude $\lambda = 90°$, the value of the vector g including all the forces acting at that point ($\lambda = 0°$ corresponds to the point on the line joining the two centres of the planets) at $t = 0$.**

(b) **What is the deviation of the vertical from the radial at the point $\varphi = 45°$, $\lambda = 0°$?**

(a) If ω is the spin angular velocity of the two planets, the centrifugal force at the equator $\theta = 90°$, $r = a$, only has radial component:

$$f_r^C = \omega^2 a \tag{63.1}$$

The radial component of the tidal force can be obtained from the tidal potential ψ which, in the first-order approximation, is given by (Fig. 63)

$$\psi = \frac{Gmr^2}{2R^3}\left(3\cos^2\vartheta - 1\right)$$

where, if φ is the latitude and λ the longitude,

$$\cos\vartheta = \cos\varphi\cos(t - \lambda)$$

On the equator $\phi = 0°$, so

$$\psi = \frac{Gmr^2}{2R^3}\left(3\cos^2(t - \lambda) - 1\right)$$

The radial component of the tidal force is

$$f_r^T = \frac{\partial\psi}{\partial r} = \frac{\partial}{\partial r}\left[\frac{Gmr^2}{2R^3}\left(3\cos^2(t - \lambda) - 1\right)\right] = \frac{Gmr}{R^3}\left(3\cos^2(t - \lambda) - 1\right)$$

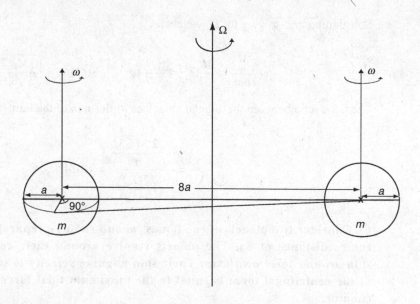

Fig. 63

The maximum value is for $t = \lambda$ and putting $R = 8a$ and $r = a$, we obtain

$$f_r^T = \frac{Gma}{(8a)^3} 2 = \frac{Gm}{256a^2} \tag{63.2}$$

Then, as the centrifugal force is equal to the tidal force, we put (63.1) equal to (63.2) and solve for ω:

$$\omega^2 a = \frac{Gm}{256a^2} \Rightarrow \omega^2 = \frac{Gm}{256a^3} \tag{63.3}$$

We know that $m \, \Omega^2 \, 4a = Gm^2/R^2$, so solving for Ω,

$$\Omega^2 = \frac{Gm}{256a^3}$$

and then using (63.3), we obtain $\omega/\Omega = 1$.

The total potential U is the sum of the gravitational, spin, and tidal potentials, which for $t = 0$, is given by

$$U = V + \Phi + \psi = \frac{Gm}{r} + \frac{1}{2}\omega^2 r^2 \sin^2\theta + \frac{Gmr^2}{2(8a)^3}\left(3\sin^2\theta\cos^2\lambda - 1\right)$$

The components of gravity including the tidal forces are

$$g_r = \frac{\partial U}{\partial r} = -\frac{Gm}{r^2} + \omega^2 r \sin^2\theta + \frac{Gmr}{(8a)^3}\left(3\sin^2\theta\cos^2\lambda - 1\right)$$

$$g_\theta = \frac{1}{r}\frac{\partial U}{\partial \theta} = \omega^2 r \sin\theta\cos\theta + \frac{Gmr}{(8a)^3} 3\sin\theta\cos\theta\cos^2\lambda \tag{63.4}$$

$$g_\lambda = \frac{1}{r\sin\theta}\frac{\partial U}{\partial \lambda} = -\frac{Gmr}{(8a)^3} 3\sin\theta\cos\lambda\sin\lambda$$

By substitution of $r = a$, $\lambda = 90°$, and $\theta = 90°$ we have

$$g_r = -\frac{Gm}{a^2} + \omega^2 a - \frac{Gm}{512a^2}$$

$$g_\theta = 0$$

$$g_\lambda = 0$$

(b) By substitution of $r = a$, $\lambda = 0$, and $\theta = 45°$ in (63.4) we obtain

$$g_r = -\frac{Gm}{a^2} + \omega^2 a \frac{1}{2} + \frac{Gm}{512a^2}\left(\frac{3}{2} - 1\right)$$

$$= -\frac{Gm}{a^2} + \frac{Gm}{256a^2}\frac{1}{2} + \frac{Gm}{512a^2}\frac{1}{2} = -\frac{1021}{1024}\frac{Gm}{a^2}$$

$$g_\theta = \omega^2 a \frac{1}{2} + \frac{Gm}{512a^2}\frac{3}{2} = \frac{Gm}{256a^2}\frac{1}{2} + \frac{Gm}{512a^2}\frac{3}{2} = \frac{5}{1024}\frac{Gm}{a^2}$$

$$g_\lambda = 0$$

The deviation of the vertical with respect to the radial direction is

$$\tan i = \frac{g_\theta}{g_r} = 0.28°$$

64. Two spherical planets of radii 2a and a and masses 8M and M separated by a centre-to-centre distance of 4a spin on their own axes and rotate in the equatorial plane with the same angular velocity.

(a) **Determine all the forces acting at a point on the smaller planet at geocentric coordinates $\varphi = 60°$ N, $\lambda = 0°$ (00:00 h local time corresponds to passage of the other planet through the zero meridian).**

(b) **For this same point, calculate the astronomical latitude and the tidal deviation of the vertical.**

(a) First we determine the centre of gravity of the system, putting the origin at the centre of the small planet (Fig. 64):

$$x = \frac{0 \times M + 4a \times 8M}{9M} = \frac{32a}{9}$$

Because the spin angular velocity of each planet is equal to the angular velocity of the system ($\omega = \Omega$), we can write, for the small planet, putting the gravitational attraction of the two planets equal to the centripetal force:

$$\frac{G8MM}{(4a)^2} = M\omega^2 \frac{32a}{9}$$

and solving for ω,

$$\omega^2 = \frac{9GM}{64a^3}$$

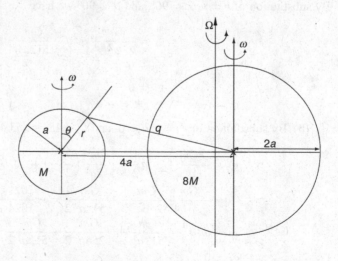

Fig. 64

On the small planet the gravitational force is

$$g_r = -\frac{GM}{r^2}$$

$$g_\theta = 0$$

and the force due to its spin is

$$f_r = \omega^2 r \sin^2 \theta$$

$$f_\theta = \omega^2 r \cos \theta \sin \theta$$

For the point under consideration, $r = a$, $\theta = 30°$, we obtain

$$g_r^{GC} = -\frac{Gm}{a^2} + \frac{9}{64}\frac{GM}{a^2}\frac{1}{4} = -\frac{247GM}{256a^2}$$

$$g_\theta^{GC} = \frac{9}{64}\frac{GM}{a^3}a\frac{\sqrt{3}}{4} = \frac{GM}{a^2}\frac{9\sqrt{3}}{256}$$

To add the tidal force we use the tidal potential in the first-order approximation,

$$\psi = \frac{G8Mr^2}{2R^3}\left(3\cos^2\vartheta - 1\right)$$

where $\cos \vartheta = \sin \theta \cos (\tau - \lambda)$.

The tidal force for the point considered, $r = a$, $\theta = 30°$, $\tau = \lambda = 0°$, and $R = 4a$, is given by

$$f_r^T = \frac{\partial \psi}{\partial r} = -\frac{GM}{a^2}\frac{1}{32}$$

$$f_\theta^T = \frac{1}{r}\frac{\partial \psi}{\partial \theta} = \frac{GM}{a^2}\frac{3\sqrt{3}}{32}$$

The total force acting at the point is the sum of the three forces, gravitational, centrifugal, and tidal:

$$g_r^{total} = -\frac{255GM}{256a^2}$$

$$g_\theta^{total} = \frac{GM}{a^2}\frac{12\sqrt{3}}{256}$$

(b) The astronomical latitude is given by $\varphi_a = \varphi + i$, where i is the deviation of the vertical without considering the tide:

$$\tan i = \frac{g_\theta^{total}}{g_r^{total}} = \frac{9\sqrt{3}}{247} = 0.06 \Rightarrow i = 3.6° \Rightarrow \varphi_a = 60 + 3.6 = 63.6°$$

The maximum deviation of the vertical due to the tide at the point considered is i' given by

$$\tan i' = \frac{f_\theta^{\mathrm{T}}}{g_r^{\mathrm{total}}} = \frac{\dfrac{3\sqrt{3}}{32}}{\dfrac{255}{256}} = 0.163 \Rightarrow i' = 9.3°$$

Gravity observations

65. Determine the values of gravity at the following series of points belonging to a gravimetric survey with a Worden gravimeter, specifying the drift correction for each of them.

Station	Time	Reading
A (base)	08:30	562.5
B	09:21	400.7
C	11:34	437.9
D	13:20	360.1
A	14:20	568.8

The gravity at the base is 980.139 82 Gal, and the gravimeter constant is 0.301 81 mGal/ru (ru: reading unit).

The instrument drift is given by

$$d = \frac{L_{\mathrm{Ae}} - L_{\mathrm{Ab}}}{t_{\mathrm{Ae}} - t_{\mathrm{Ab}}}$$

where L_{Ab} and L_{Ae} are the readings at the base A at the beginning and end of the measurements taken at times t_{Ab} and t_{Ae}, respectively. By substitution we obtain

$$d = \frac{568.8 - 562.5}{14.33 - 8.50} = 1.08\,\mathrm{ru/hour}$$

The corrected reading for station j is given by

$$L_j^{\mathrm{c}} = L_j - d\,(t_j - t_{\mathrm{Ab}})$$

where L_j is the reading taken at time t_j.

For a Worden gravimeter the increment in gravity between two points (Δg) is proportional to the increment in the readings corrected by the instrument drift (ΔL^{c}):

$$\Delta g = K\,\Delta L^{\mathrm{c}}$$

where K is the instrument constant.

Thus, from the readings we obtain the following results.

Station	Corrected reading	Δg (mGal)	g (mGal)
A (base)	562.5		980 139.82
B	399.8	−49.10	980 090.72
C	434.6	10.50	980 101.22
D	354.9	−24.05	980 077.17
A	562.5	62.66	980 139.83

66. Point A is at a geopotential level of 97.437 43 gpu. Point B is at a difference of −15.213 m in height relative to A, and has a value of gravity of 9.712 611 m/s^{-2}. Calculate:

(a) **The value of gravity at point A, if the difference in readings of a Worden gravimeter between B and A is 17.8 ru, and the gravimeter constant is 0.308 21 mGal/ru.**

(b) **The geopotential number, dynamic height, and Helmert height of the point B given that the normal gravity at a point of latitude 45° on the ellipsoid is 980 629.40 mGal.**

(a) Using a Worden gravimeter the increment of gravity between points A and B is given by

$$\Delta g_A^B = K \Delta L_A^B = 5.5 \, \text{mGal}$$

where K is the instrument constant and ΔL is the difference between the readings at points A and B.
The gravity at point A is

$$g_A = g_B + \Delta g_A^B = 971 \, 266.6 \, \text{mGal}$$

(b) The geopotential number at B can be calculated from the value at A in the form

$$C_B = C_A + \left(\frac{g_A + g_B}{2} \right) \Delta h_A^B = 82.661 \, 59 \, \text{gpu}$$

where gravity is given in Gal and increments in height in km, because the geopotential units are, 1 gpu = 1 kGal m = 1 Gal km.
The dynamic height is given by

$$H_D^B = \frac{C_B}{\gamma_{45}} = 84.294 \, \text{m}$$

The Helmert height can be calculated from the dynamic height by

$$H = \frac{C}{g + 0.0424H}$$

where C is given in gpu, g in Gal, and H in km. Solving for H, we obtain

$$H = \frac{-g \pm \sqrt{g^2 + 4 \times 0.0424C}}{2 \times 0.0424}$$

Taking the positive solution because point B is above the geoid ($C_B > 0$) we obtain

$$H_B = 85.107 \, \text{m}$$

67. In a geometric survey with measurements of gravity using a Worden gravimeter, the following values were obtained:

Station	Gravimeter reading (ru)	Time	Gravity (gu)	Height difference (m)
A (base)	1520.23	8 h 50 m	9 793 626.8	
B	1759.15	9 h 15 m	9 794 363.9	−30.410
C	1583.11	9 h 35 m	9 793 820.7	301.863
A	1521.30	9 h 50 m		

Calculate the gravimeter readings corrected for drift, and the gravimeter constant

The instrument drift is given by

$$d = \frac{L_{Ae} - L_{Ab}}{t_{Ae} - t_{Ab}}$$

where L_{Ab} and L_{Ae} are the readings at the base A at the beginning and end of the measurements taken at times t_{Ab} and t_{Ae}, respectively. By substitution we obtain,

$$d = 1.07 \, \text{ru/hour}$$

A reading corrected at station j is given by

$$L_j^c = L_j - d(t_j - t_{Ab})$$

where L_j is the reading at time t_j.

For a Worden gravimeter the increment in gravity between two points (Δg) is proportional to the increment in the readings corrected by the instrument drift (ΔL^c):

$$\Delta g = K \, \Delta L^c$$

where K is the instrument constant. Thus, K can be calculated in the form

$$K = \frac{\Delta g}{\Delta L^c}$$

From each pair of observations we obtain a value of K. Finally we take the arithmetic mean (K_m) from all the values obtained. The results are given in the following table.

Station	Corrected reading (ru)	Gravity (mGal)	K (mGal/ru)
A (base)	1520.23	979 362.68	
B	1758.70	979 436.39	0.3091
C	1582.31	979 382.07	0.3079
A	1520.23		
			$K_m = 0.3085$

68. The following table is obtained from observations with a Lacoste–Romberg gravimeter:

Station	Gravimeter reading	Time
A	3614.351	10:10
B	3650.242	10:25
C	3610.633	10:37
A	3614.414	11:02

The gravimeter scale factor is 1.000 65, and the equivalence between reading units and the relative value of gravity in mGal is given by

Reading	Value in mGal	Interval factor
3600	3846.02	1.071 25
3700	3953.15	1.071 40

Given that the value of gravity at point A is 9.794 6312 m s^{-2}, calculate the values at B and C.

First we correct the readings by the instrument drift:

$$d = \frac{L_{Ae} - L_{Ab}}{t_{Ae} - t_{Ab}}$$

where L_{Ae} and L_{Ab} are the readings at the base A at the end and the beginning of the survey at times t_{Ae} and t_{Ab}. Then

$$d = 0.0727 \, \text{ru/hour}$$

The corrected reading at each station j is given by

$$L_j^c = L_j - d(t_j - t_{Ab})$$

where L_j is the reading at time t_j. The corrected readings are:

$$L_A^c = 3614.351$$

$$L_B^c = 3650.224$$

$$L_C^c = 3610.600$$

These readings are converted into relative gravity values R_j using the conversion table.

The reading at station A is

$$L_A^c = 3600 + 14.351$$

and the relative gravity value is

$$R_A = (3846.02 + (14.351 \times 1.07125)) \times 1.00065 = 3863.90 \, \text{mGal}$$

For stations B and C,

$$R_B = (3846.02 + (50.224 \times 1.07125)) \times 1.00065 = 3902.36 \text{ mGal}$$

$$R_C = (3846.02 + (10.600 \times 1.07125)) \times 1.00065 = 3859.88 \text{ mGal}$$

To convert the relative values into absolute values we need to know both values at one station, in our case in station A:

$$g_A = 9.794\ 6312 \text{ m s}^{-2} = 979\ 463.12 \text{ mGal}$$
$$g_B = g_A - R_A + R_B = 979\ 501.58 \text{ mGal}$$
$$g_C = g_A - R_A + R_C = 979\ 459.10 \text{ mGal}$$

3

Geomagnetism

Main field

69. Assume that the geomagnetic field of the Earth is a geocentric dipole with a North Pole at 80° N, 45° E and a magnetic moment 8×10^{22} A m^2. Calculate for a point with geographical coordinates 45° N, 30° W the components NS, EW, and Z of the Earth's magnetic field, the declination and inclination, and the geomagnetic longitude. Earth's radius: 6370 km and the constant $C = 10^{-7}$ H m^{-1} (this value is used in all problems).

We calculate first the geomagnetic latitude and longitude (ϕ^*, λ^*) from the geographical coordinates (ϕ, λ) of the point and the geographical coordinates of the geomagnetic North Pole (ϕ_B, λ_B) by the equation

$$\sin \varphi^* = \sin \varphi_B \sin \varphi + \cos \varphi_B \cos \varphi \cos(\lambda - \lambda_B)$$

$$\sin \lambda^* = \frac{\sin(\lambda - \lambda_B) \cos \varphi}{\cos \varphi^*}$$

Substituting the values

$$\phi_B = 80° \text{ N}$$
$$\lambda_B = 45° \text{ E}$$
$$\phi = 45° \text{ N}$$
$$\lambda = 30° \text{ W} = 330°$$

we obtain

$$\phi^* = 46.70°$$
$$\lambda^* = -84.82°$$

In the geocentric magnetic dipole model, the vertical (Z^*) and horizontal (H^*) components of the magnetic field can be obtained from

$$Z^* = 2B_0 \sin \phi^*$$

$$H^* = B_0 \cos \phi^*$$

$$B_0 = \frac{Cm}{a^3}$$

(69.1)

In these equations B_0 is the geomagnetic constant, m the magnetic moment of the dipole, a the Earth's radius, and the constant $C = 10^{-7}$ H m^{-1}.

In this case we are given that

$m = 8 \times 10^{22}$ A m^2

$a = 6370$ km $= 6.37 \times 10^6$ m

By substitution in Equations (69.1) we obtain:

$B_0 = 30\ 951$ nT

$Z^* = 45\ 051$ nT

$H^* = 21\ 227$ nT

The geomagnetic declination is given by

$$\sin D^* = \frac{-\cos \phi_B \sin(\lambda - \lambda_B)}{\cos \phi^*}$$

$$D^* = 14.16°$$

The NS (X^*) and EW (Y^*) components are

$$X^* = H^* \cos D^* = 20\ 582\ \text{nT}$$

$$Y^* = H^* \sin D^* = 5193\ \text{nT}$$

Finally, the geomagnetic inclination or dip (I^*) at that point is given by

$$\tan I^* = 2 \tan \phi^* \Rightarrow I^* = 64.77°$$

70. Assume that the geomagnetic field is produced by a geocentric dipole of magnetic moment 8×10^{22} Am^2, with North Pole at 80° N, 70° W, and that the Earth's radius is 6370 km. Calculate for a point with geographical coordinates 60° N, 110° E:

(a) Its geomagnetic coordinates, the components of the Earth's magnetic field (X^*, Y^*, Z^*, H^*), the total field, the declination, and the inclination.

(b) The equation of the line of force passing through it.

(a) For this point the difference in longitude from the Geomagnetic North Pole (GMNP) is 180° (Fig. 70), so both are on the same great circle. Then, the geomagnetic coordinates are obtained from

$$\phi^* = \phi - (90° - \phi_B) = 50°$$
$$\lambda^* = 180°$$

The expressions for the geomagnetic vertical and horizontal components and for the total geomagnetic field are

$$Z^* = 2B_0 \sin \phi^*$$

$$H^* = B_0 \cos \phi^*$$

$$F^* = \sqrt{H^{*2} + Z^{*2}}$$

$$B_0 = \frac{Cm}{a^3}$$

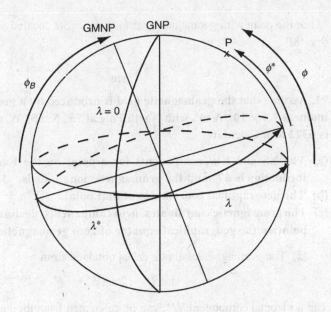

The values of the constants are

$m = 8 \times 10^{22}$ A m^2

$a = 6370$ km $= 6.37 \times 10^6$ m

$C = 10^{-7}$ H m^{-1}

Substituting in the above equations we obtain

$B_0 = 30\,951$ nT

$Z^* = 47\,420$ nT

$H^* = 19\,895$ nT

$F^* = 51\,424$ nT

For this point the geomagnetic declination $D^* = 0$. So the NS (X^*) and EW (Y^*) components are

$$D^* = 0°$$
$$X^* = H^* \cos D^* = H^* = 19\,895 \text{ nT}$$
$$Y^* = H^* \sin D^* = 0$$

The geomagnetic inclination (I^*) is given by

$$\tan I^* = 2 \tan \phi^* \Rightarrow I^* = 67.23°$$

(b) The equation of the line of force passing through a point with geomagnetic co-latitude θ is

$$r = r_0 \sin^2 \theta$$

In this equation r_0 is the distance from the Earth's centre to a point on the line of force with $\theta = 90°$. The distance r_0 is different for each line of force.

For the point with geomagnetic latitude $\phi^* = 50°$ located in the Earth's surface ($r = a$), $\theta = 90° - \phi^* = 40°$, so

$$r_0 = \frac{a}{\sin^2 \theta} = 15\,417\,\text{km}$$

71. Assume that the geomagnetic field is produced by a geocentric dipole of magnetic moment $7.5 \times 10^{22}\,\text{A m}^2$, with North Pole at 75° N, 65° W, and that the Earth's radius is 6372 km. Calculate:

(a) The NS and EW components for a point on the Earth's surface at which the inclination is 67° and the geomagnetic longitude is $-120°$.

(b) The geographical coordinates of that point.

(c) The geomagnetic coordinates, field components, declination, and inclination of the point on the geographical equator of zero geomagnetic longitude.

(a) The geomagnetic latitude ϕ^* is obtained from

$$\tan I^* = 2 \tan \phi^* \Rightarrow \phi^* = 49.7°$$

The horizontal component, H^*, can be calculated from the geomagnetic constant, B_0, and the geomagnetic latitude:

$$B_0 = \frac{Cm}{a^3} = 28\,989\,\text{nT}$$

$$H^* = B_0 \cos \phi^* = 18\,761\,\text{nT}$$

To obtain the NS (X^*) and EW (Y^*) components it is necessary to calculate the declination (D^*) from the spherical triangle with vertices at the Geographical North Pole (GNP), Geomagnetic North Pole (GMNP), and the point P (Fig. 71a). But we need to calculate the geographic latitude firstly by solving the spherical triangle. Applying the cosine law to the angle ($90° - \phi$):

$$\cos(90° - \phi) = \cos(90° - \phi_B) \cos(90° - \phi^*)$$
$$+ \sin(90° - \phi_B) \sin(90° - \phi^*) \cos(180° - \lambda^*)$$
$$\sin \phi = \sin \phi_B \sin \phi^* - \cos \phi_B \cos \phi^* \cos \lambda^* \qquad (71.1)$$

By substitution of the values,

$$\phi = 55.1°$$

To obtain the geomagnetic declination we apply the sine law in the spherical triangle of Fig. 71a:

$$\frac{\sin D^*}{\sin(90° - \phi_B)} = \frac{\sin(180° - \lambda^*)}{\sin(90 - \phi)} \Rightarrow \sin D^* = -\frac{\cos \phi_B \sin \lambda^*}{\cos \phi} \qquad (71.2)$$

and substituting the values we find

$$D^* = 23.1°$$

It is important to note that we have added a minus sign in the last equation in order for the declination be positive toward the east.

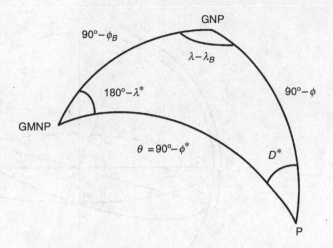

Fig. 71a

The NS (X^*) and EW (Y^*) components are

$$X^* = H^* \cos D^* = 17\,262 \, \text{nT}$$
$$Y^* = H^* \sin D^* = 7\,349 \, \text{nT}$$

(b) The calculated geographical latitude is

$$\phi = 55.1°$$

The geographical longitude is obtained by applying the cosine law to the spherical triangle of Fig. 71a:

$$\cos(90° - \phi^*) = \cos(90° - \phi_B) \cos(90° - \phi)$$
$$+ \sin(90° - \phi_B) \sin(90° - \phi) \cos(\lambda - \lambda_B) \qquad (71.3)$$
$$\cos\theta = \sin\phi^* = \sin\phi_B \sin\phi + \cos\phi_B \cos\phi \cos(\lambda - \lambda_B)$$

so

$$\cos(\lambda - \lambda_B) = \frac{\sin\phi^* - \sin\phi_B \sin\phi}{\cos\phi_B \cos\phi} \qquad (71.4)$$

Substituting the values, gives

$$\lambda - \lambda_B = \pm101.6°$$

To take the inverse cosine in the correct quadrant we bear in mind that $\lambda^* < 0$ implies that the point is to the west of the Geomagnetic North Pole, that is $\lambda - \lambda_B < 0$. So we obtain

$$\lambda = 166.6° \, \text{W}$$

(c) If this point is on the geographical equator ($\phi = 90°$) and has zero geomagnetic longitude ($\lambda^* = 0°$), it is on the same geographical meridian as the Geomagnetic North Pole. Then from Fig. 71b

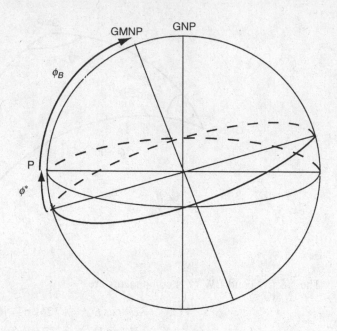

$$\phi^* = 90° - \phi_B = 15.0°$$
$$\lambda^* = 0°$$
$$Z^* = 2B_0 \sin \phi^* = 15\,006\,\text{nT}$$
$$H^* = B_0 \cos \phi^* = 28\,001\,\text{nT}$$
$$D^* = 0°$$
$$X^* = H^*$$
$$Y^* = 0\,\text{nT}$$
$$\tan I^* = 2 \tan \phi^* \Rightarrow I^* = 28.2°$$

72. Assume that the geomagnetic field is that of a dipole with North Pole at 75° N, 0° E. What is the conjugate point of that of geographical coordinates 30° N, 30° E?

First, we calculate the geomagnetic coordinates (ϕ^*, λ^*) (Problem 71; Fig. 71a):

$$\sin \phi^* = \sin \phi_B \sin \phi + \cos \phi_B \cos \phi \cos(\lambda - \lambda_B)$$
$$\sin \lambda^* = \frac{\sin(\lambda - \lambda_B) \cos \phi}{\cos \phi^*} \qquad (72.1)$$

The values of the geographical coordinates are

$$\phi_B = 75° \qquad \lambda_B = 0°$$
$$\phi = 30° \qquad \lambda = 30°$$

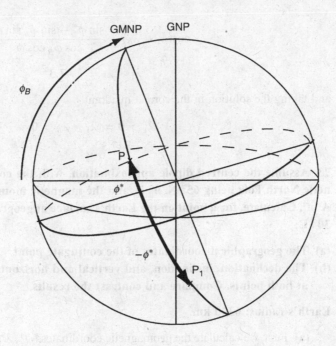

Fig. 72

By substitution in Equations (72.1) we obtain:

$$\phi^* = 42.6°$$

$$\lambda^* = 36.0°$$

A magnetic conjugate point is a point on the Earth's surface that is located on the same line of force and in the opposite hemisphere (Fig. 72, P and P_1). Then, its geomagnetic coordinates (ϕ_1^*, λ_1^*) are

$$\phi_1^* = -\phi^* = -42.6°$$
$$\lambda_1^* = \lambda^* = 36.0°$$

To calculate the geographical coordinates for this point (ϕ_1, λ_1) we use the spherical triangle of Fig. 71a. We calculate the geographical latitude applying the cosine law:

$$\cos(90° - \phi_1) = \cos(90° - \phi_B) \cos(90° - \phi_1^*)$$
$$+ \sin(90° - \phi_B) \sin(90° - \phi_1^*) \cos(180° - \lambda_1^*)$$

$$\sin\phi_1 = \sin\phi_B \sin\phi_1^* - \cos\phi_B \cos\phi_1^* \cos\lambda_1^*$$
$$\phi_1 = -53.9°$$

To calculate the geographical longitude we apply the cosine law again:

$$\cos(90° - \phi_1^*) = \cos(90° - \phi_B) \cos(90° - \phi_1)$$
$$+ \sin(90° - \phi_B) \sin(90° - \phi_1) \cos(\lambda_1 - \lambda_B)$$

$$\cos(\lambda_1 - \lambda_B) = \frac{\sin \phi_1^* - \sin \phi_B \sin \phi}{\cos \phi_B \cos \phi}$$

$$\lambda_1 - \lambda_B = \pm 47.3°$$

and taking the solution in the correct quadrant

$$\lambda_1^* > 0 \Rightarrow \lambda_1 - \lambda_B > 0$$
$$\lambda_1 = 47.3°$$

73. Assume the centred dipole approximation, with the coordinates of the Geomagnetic North Pole being 65° N, 0° E, and the magnetic moment of the dipole 8×10^{22} A m². Calculate, for a point on the Earth's surface at geographical coordinates 30° N, 30° E:

(a) The geographical coordinates of the conjugate point.
(b) The declination, inclination, and vertical and horizontal components of the field at both points. Compare and contrast the results.

Earth's radius: 6370 km.

(a) First, we calculate the geomagnetic coordinates (ϕ^*, λ^*) for point P with geographical coordinates $\phi = 30°$ N, $\lambda = 30°$ E using the equations (Problem 71; Fig. 71a)

$$\sin \phi^* = \sin \phi_B \sin \phi + \cos \phi_B \cos \phi \cos(\lambda - \lambda_B)$$
$$\sin \lambda^* = \frac{\sin(\lambda - \lambda_B) \cos \phi}{\cos \phi^*} \tag{73.1}$$

The geographical coordinates of the Geomagnetic North Pole are

$$\phi_B = 65°, \quad \lambda_B = 0°$$

Substituting in Equations (73.1) results in

$$\phi^* = 50.4°$$
$$\lambda^* = 42.7°$$

The geomagnetic coordinates (ϕ_1^*, λ_1^*) of the magnetic conjugate point P_1 satisfy (Problem 72):

$$\phi_1^* = -\phi^* = -50.4°$$
$$\lambda_1^* = \lambda^* = 42.7°$$

To calculate the geographical coordinates for this point (ϕ_1, λ_1) we use the spherical triangle of Fig. 71a. We calculate the geographical latitude applying the cosine law

$$\cos(90° - \phi_1) = \cos(90° - \phi_B) \cos(90° - \phi_1^*)$$
$$+ \sin(90° - \phi_B) \sin(90° - \phi_1^*) \cos(180° - \lambda_1^*)$$

$$\sin \phi_1 = \sin \phi_B \sin \phi_1^* - \cos \phi_B \cos \phi_1^* \cos \lambda_1^*$$
$$\phi_1 = -63.7°$$

To calculate the geographical longitude we apply the cosine law again and, solving for λ_1,

$$\cos(\lambda - \lambda_B) = \frac{\sin \phi_1^* - \sin \phi_B \sin \phi_1}{\cos \phi_B \cos \phi_1}$$

$$\lambda_1 - \lambda_B = \pm 77.1°$$

$$\lambda_1^* > 0 \Rightarrow \lambda_1 - \lambda_B > 0$$

$$\lambda_1 = 77.1°$$

(b) First we calculate the geomagnetic constant B_0

$$B_0 = \frac{Cm}{a^3} = \frac{8 \times 10^{22}}{(6379)^3 \times 10^9} = 30\,951\,\text{nT}$$

We calculate the declination D^*, inclination I^*, and vertical Z^* and horizontal H^* components of the field at both points. The results are shown in the table, where we notice that except for the declinations, which are very different, all other values are equal for both points except in sign.

P	P_1
$\sin D^* = \dfrac{-\cos \phi_B \sin(\lambda - \lambda_B)}{\cos \phi^*}$	$\sin D_1^* = \dfrac{-\cos \phi_B \sin(\lambda_1 - \lambda_B)}{\cos \phi_1^*}$
$D^* = -19.33°$	$D_1^* = -40.2°$
$\tan I^* = 2 \tan \phi^* \Rightarrow I^* = 67.5°$	$\tan I_1^* = 2 \tan \phi_1^* \Rightarrow I_1^* = -I_1^* = -67.5°$
$Z^* = 2B_0 \sin\phi^* = 47\,662\,\text{nT}$	$Z_1^* = 2B_0 \sin \phi_1^* = -Z^* = -47\,662\,\text{nT}$
$H^* = B_0 \cos\phi^* = 19\,750\,\text{nT}$	$H_1^* = B_0 \cos \phi_1^* = H^* = 19\,750\,\text{nT}$

74. Assume the centred dipole approximation, with the coordinates of the Geomagnetic North Pole 78.5° N, 70.0° W, and the magnetic dipole moment being 8.25×10^{22} A m². Calculate, for a point on the surface with coordinates 60.0° S, 170.0° W:

(a) Its geomagnetic coordinates, declination, inclination, and vertical and horizontal components of the field.

(b) The potential at that point.

(c) The declination and inclination at the point diametrically opposite to it.

Earth's radius: 6370 km.

(a) We calculate the geomagnetic coordinates (ϕ^*, λ^*) using the equations (Problem 71, Fig. 71a)

$$\sin \phi^* = \cos \theta = \sin \phi_B \sin \phi + \cos \phi_B \cos \phi \cos(\lambda - \lambda_B)$$

$$\sin \lambda^* = \frac{\sin(\lambda - \lambda_B) \cos \phi}{\cos \phi^*} \tag{74.1}$$

The values of the geographical coordinates are

$$\phi_B = 78.5° \qquad \lambda_B = -70.0°$$

$$\phi = 60.0° \qquad \lambda = -170.0°$$

Substitution in Equations (74.1) gives

$$\phi^* = -60.0°$$

$$\lambda^* = -80.0°$$

The geomagnetic declination is given by (Problem 71, Fig. 71a)

$$\sin D^* = \frac{-\cos \phi_B \sin(\lambda - \lambda_B)}{\cos \phi^*}$$

$$D^* = 23.1°$$

The inclination (I^*) at that point is given by

$$\tan I^* = 2 \tan \phi^* \Rightarrow I^* = -73.9°$$

The vertical (Z^*) and horizontal (H^*) components of the magnetic field can be obtained from

$$Z^* = 2B_0 \sin \phi^*$$
$$H^* = B_0 \cos \phi^*$$
$$B_0 = \frac{Cm}{a^3}$$

Substituting the values given we obtain:

$$B_0 = 31\,918\,\text{nT}$$
$$Z^* = 2B_0 \sin \phi^* = -55\,279\,\text{nT}$$
$$H^* = B_0 \cos \phi^* = 15\,963\,\text{nT}$$

(b) The potential at a point on the Earth's surface $(r = a)$ at geomagnetic latitude ϕ^* is given by

$$\Phi = \frac{-Cm \cos \theta}{a^2} = \frac{-Cm \sin \phi^*}{a^2} = 176\,\text{T m}$$

(c) We can observe in Fig. 74 that at the point diametrically opposite the geographical and geomagnetic coordinates are

$$\phi_1 = -\phi = 60.0° \text{ N}$$
$$\lambda_1 = \lambda + 180° = 10.0° \text{ E}$$
$$\phi_1^* = -\phi^* = 60.0°$$
$$\lambda_1^* = \lambda^* + 180° = 100.0°$$

The geomagnetic declination at that point satisfies (Fig. 71a)

$$\sin D_1^* = \frac{-\cos \phi_B \sin(\lambda_1 - \lambda_B)}{\cos \phi_1^*} = \frac{-\cos \phi_B \sin(\lambda + 180° - \lambda_B)}{\cos \phi^*}$$

$$= \frac{\cos \phi_B \sin(\lambda - \lambda_B)}{\cos \phi^*} = -\sin D^*$$

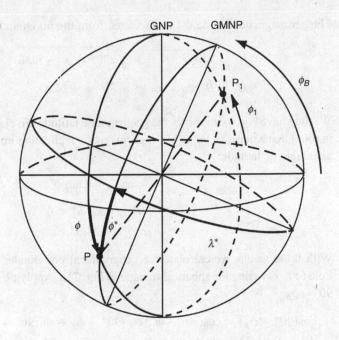

Fig. 74

Then, $D_1^* = -D^*$.

The geomagnetic inclination is given by

$$\tan I_1^* = 2 \tan \phi_1^* = -2 \tan \phi^* \Rightarrow I_1^* = -I^*$$

So, we can notice that the two points P and P_1 (Fig. 74) that are diametrically opposite are not magnetic conjugate points because the geomagnetic longitudes are different by 180°.

75. Consider a point P on the Earth's surface at coordinates 30° S, 10° W at which the NS component of the geomagnetic field is 27 050 nT and the EW component is −5036 nT, with the geomagnetic inclination being negative. Assuming the centred dipole hypothesis with magnetic moment 7.8×10^{22} A m², calculate:

(a) The geographical coordinates of the Geomagnetic North Pole.

(b) The geomagnetic coordinates of P's conjugate point.

(a) We calculate first the geomagnetic constant B_0:

$$B_0 = \frac{Cm}{a^3} = 30\,177 \, \text{nT}$$

The geomagnetic declination D^* is obtained from the NS (X^*) and EW (Y^*) components of the geomagnetic field:

$$\tan D^* = \frac{Y^*}{X^*} = \frac{-5036}{27\,050} \Rightarrow D^* = -10.5°$$

The geomagnetic latitude ϕ^* is calculated from the horizontal component H^*:

$$H^* = \sqrt{X^{*2} + Y^{*2}} = \sqrt{27050^2 + (-5036)^2} = 27\,515\,\text{nT}$$

$$H^* = B_0 \cos \phi^* \Rightarrow \phi^* = \frac{H^*}{B_0} = \pm 24.3°$$

We then have two solutions for the geomagnetic latitude. To choose the correct one we bear in mind that a negative value of the geomagnetic inclination implies a negative value of the geomagnetic latitude:

$$\tan I^* = 2 \tan \phi^*$$

$$I^* < 0 \Rightarrow \phi^* < 0$$

$$\phi^* = -24.3°$$

With these results we calculate the geographical coordinates of the Geomagnetic North Pole (ϕ_B, λ_B) using the spherical triangle in Fig. 71a. Applying the cosine rule for the angle $90° - \phi_B$:

$$\cos(90° - \phi_B) = \cos(90° - \phi^*) \cos(90° - \phi) + \sin(90° - \phi^*) \sin(90° - \phi) \cos D^*$$

$$\sin \phi_B = \sin \phi^* \sin \phi + \cos \phi^* \cos \phi \cos D^*$$

$$\phi_B = 79.0°$$

To calculate the longitude λ_B of the Geomagnetic North Pole, we apply the cosine law for the angle $90° - \phi^*$:

$$\cos(90° - \phi^*) = \cos(90° - \phi_B) \cos(90° - \phi) + \sin(90° - \phi_B) . \sin(90° - \phi) \cos(\lambda - \lambda_B)$$

$$\sin \phi^* = \sin \phi_B \sin \phi + \cos \phi_B \cos \phi \cos(\lambda - \lambda_B)$$

$$\cos(\lambda - \lambda_B) = \frac{\sin \phi^* - \sin \phi_B \sin \phi}{\cos \phi_B \cos \phi} \Rightarrow \lambda - \lambda_B = \pm 61.0°$$

To choose the correct sign for the longitude we notice that the declination is negative and then the point must be to the east of the Geomagnetic North Pole:

$$D^* < 0 \Rightarrow \lambda - \lambda_B > 0$$

$$\lambda_B = -61° - 10° = 71.0°\,\text{W}$$

(b) The geomagnetic coordinates (ϕ_1^*, λ_1^*) of P's conjugate point verify that

$$\phi_1^* = -\phi^* = 24.3°$$

$$\lambda_1^* = \lambda^*$$

We calculate the geomagnetic longitude λ^* by

$$\sin \lambda^* = \frac{\sin(\lambda - \lambda_B) \cos \phi}{\cos \phi^*}$$

$$\lambda^* = 56.2° = \lambda_1^*$$

76. At a point P on the Earth's surface with coordinates 45° N, 30° W, the value of the total geomagnetic field is 49 801 nT, the horizontal component is 21 227 nT, and the EW component is 5171 nT, with the magnetic inclination being positive. Calculate:

(a) The geographical coordinates of the Geomagnetic North Pole.
(b) The value of the geomagnetic potential at P.
(c) The distance from the Earth's centre to the point at which the line of force passing through P intersects the geomagnetic equator.

Earth's radius: 6370 km.

(a) We calculate first the geomagnetic inclination, latitude, and declination by

$$\cos I^* = \frac{H}{F} \Rightarrow I^* = 64.8°$$

$$\tan I^* = 2 \tan \phi^* \Rightarrow \phi^* = 46.7°$$

$$\sin D^* = \frac{Y}{H} \Rightarrow D^* = 14.1°$$

With these results we calculate the geographical coordinates of the Geomagnetic North Pole solving the spherical triangle (Fig. 71a) in the same way as in Problem 71:

$$\sin \phi_B = \sin \phi^* \sin \phi + \cos \phi^* \cos \phi \cos D^*$$

$$\phi_B = 80.0°$$

$$\sin \phi^* = \sin \phi_B \sin \phi + \cos \phi_B \cos \phi \cos(\lambda - \lambda_B)$$

$$\cos(\lambda - \lambda_B) = \frac{\sin \phi^* - \sin \phi_B \sin \phi}{\cos \phi_B \cos \phi}$$

$$\lambda - \lambda_B = \pm 75.2°$$

$$D^* > 0 \Rightarrow \lambda - \lambda_B < 0$$

$$\lambda_B = 45.2° \, \text{E}$$

(b) The geomagnetic potential at point P on the Earth's surface ($r = a = 6370$ km) is given by

$$\Phi = \frac{-Cm \cos \theta}{a^2} = -B_0 a \sin \phi^* \tag{76.1}$$

We calculate the geomagnetic constant B_0 from the horizontal component H^*:

$$H^* = B_0 \cos \phi^* \Rightarrow B_0 = \frac{21\,227}{\cos(46.7°)} = 30\,951 \, \text{nT}$$

Substituting in the potential equation (76.1) we obtain

$$\Phi = -143 \, \text{T m}$$

(c) The equation of the line of force passing through a point with geomagnetic co-latitude θ is

$$r = r_0 \sin^2 \theta$$

In this equation r_0 is the distance from the Earth's centre to the point at which the line of force passing through P intersects the geomagnetic equator. Substituting $r = a = 6370$ km gives

$$r_0 = \frac{a}{\sin^2 \theta} = \frac{a}{\cos^2 \phi^*} = 13\,543 \text{ km}$$

77. Assume the centred dipole approximation, with the coordinates of the Geomagnetic North Pole being 75° N, 65° W, and the magnetic moment of the dipole 7.5×10^{22} A m^2. For a point on the Earth's surface at which the inclination is 67° and the geomagnetic longitude is −120°, calculate:

(a) The NS and EW components.
(b) Its geographical coordinates.

The Earth's radius: 6372 km.

(a) We calculate first the geomagnetic constant B_0, latitude ϕ^*, and horizontal H^* component:

$$B_0 = \frac{Cm}{a^3} = 28\,989 \text{ nT}$$

$$\tan I^* = 2 \tan \phi^* \Rightarrow \phi^* = 49.7°$$

$$H^* = B_0 \cos \phi^* = 18\,761 \text{ nT}$$

To calculate the NS (X^*) and EW (Y^*) components it is necessary to obtain first the geographic latitude (ϕ^*) and the geomagnetic declination (D^*). Applying the cosine rule for the angle $90° - \phi$ (Fig. 71a),

$$\sin \phi = \sin \phi_B \sin \phi^* - \cos \phi_B \cos \phi^* \cos \lambda^*$$

$$\phi = 55.1°$$

The geomagnetic declination is given by

$$\sin D^* = \frac{- \cos \phi_B \sin \lambda^*}{\cos \phi}$$

$$D^* = 23.1°$$

From this value we obtain the NS and EW components:

$$X^* = H \cos D^* = 17\,262 \text{ nT}$$

$$Y^* = H \sin D^* = 7349 \text{ nT}$$

(b) The geographic latitude was already obtained,

$$\phi = 55.1°$$

To calculate the geographical longitude we apply the cosine law for the angle $90° - \phi^*$ (Fig. 71a):

$$\sin \phi^* = \sin \phi_B \sin \phi + \cos \phi_B \cos \phi \cos(\lambda - \lambda_B)$$

$$\cos(\lambda - \lambda_B) = \frac{\sin \phi^* - \sin \phi_B \sin \phi}{\cos \phi_B \cos \phi}$$

$$\lambda - \lambda_B = \pm 101.7°$$

To choose the correct solution we notice that the declination is positive and then the point must be to the west of the Geomagnetic North Pole

$$D^* > 0 \Rightarrow \lambda - \lambda_B < 0$$

$$\lambda = -101.7° - 65° = 166.7°\text{W}$$

78. Assume a spherical Earth of radius 6370 km, with magnetic field produced by a centred dipole whose northern magnetic pole is at 70° N, 60° W. Given that for a point on the surface with coordinates 50° S, 80° W the horizontal component is 24 890 nT, calculate:

(a) The magnetic dipole moment.

(b) The geographical coordinates of the conjugate point.

(a) We calculate first the geomagnetic latitude by (Fig. 71a)

$$\sin \phi^* = \sin \phi_B \sin \phi + \cos \phi_B \cos \phi \cos(\lambda - \lambda_B)$$

$$\phi^* = -30.9°$$

To obtain the magnetic dipole moment m we need the geomagnetic constant B_0, which is related with the horizontal component H by

$$B_0 = \frac{H}{\cos \phi^*} = 29\,007 \, \text{nT}$$

$$B_0 = \frac{Cm}{a^3} \Rightarrow m = \frac{B_0 a^3}{C} = 7.5 \times 10^{22} \, \text{A m}^2$$

(b) Let us obtain first the geomagnetic longitude by (Problem 71, Fig. 71a)

$$\sin \lambda^* = \frac{\sin(\lambda - \lambda_B) \cos \phi}{\cos \phi^*}$$

$$\lambda^* = -14.8°$$

The geomagnetic coordinates of the conjugate point (ϕ_1^*, λ_1^*) are (Fig. 78)

$$\phi_1^* = -\phi^* = 30.9°$$

$$\lambda_1^* = \lambda^* = -14.8°$$

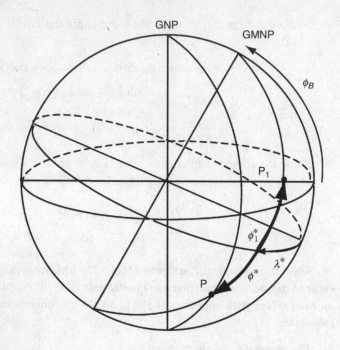

Fig. 78

Solving again the spherical triangle (Fig. 71a) we calculate the geographical coordinates (ϕ_1, λ_1):

$$\sin \phi_1 = \sin \phi_B \sin \phi_1{}^* - \cos \phi_B \cos \phi_1{}^* \cos \lambda_1{}^*$$

$$\phi_1 = 11.4°$$

$$\cos(\lambda - \lambda_B) = \frac{\sin \phi_1{}^* - \sin \phi_B \sin \phi_1}{\cos \phi_B \cos \phi_1}$$

$$\lambda_1 - \lambda_B = \pm 13.0°$$

$$\lambda^* < 0 \Rightarrow \lambda_1 - \lambda_B < 0$$

$$\lambda_1 = 73.0°$$

79. If the Earth's geomagnetic field is produced by a centred dipole, tilted 15° away from the axis of rotation, of magnetic moment 7.6×10^{22} A m^2, and the Geomagnetic North Pole is at longitude 65° W, calculate:

(a) **The geomagnetic constant in nT.**
(b) **The geographical coordinates of a point on the Earth's surface at which the declination is $D^* = 14° \ 15.5'$ and the inclination is $I^* = -65° \ 23.5'$. Discuss the possible solutions.**
(c) **The geographical and geomagnetic longitude of the agonic line.**

Assume a spherical Earth of radius 6370 km.

(a) We calculate the geomagnetic constant from the magnetic moment m, the Earth's radius a, and the constant $C = 10^{-7}\,\mathrm{H\,m^{-1}}$, by

$$B_0 = \frac{Cm}{a^3} = 29\,403\,\mathrm{nT}$$

(b) Let us obtain first the geomagnetic latitude from the inclination I^* by

$$\tan I^* = 2\tan\phi^* \Rightarrow \phi^* = -47.5°$$

If the dipole is tilted $15°$ away from the axis of rotation the latitude of the Geomagnetic North Pole will be

$$\phi_B = 90° - 15° = 75.0°$$

With these results we calculate the geographical latitude ϕ solving the spherical triangle (Fig. 71a). Applying the cosine rule,

$$\cos(90 - \phi_B) = \cos(90 - \phi^*)\cos(90 - \phi) + \sin(90 - \phi^*)\sin(90 - \phi)\cos D^*$$
$$\sin\phi_B = \sin\phi^*\sin\phi + \cos\phi^*\cos\phi\cos D^* \tag{79.1}$$

To obtain the geographical latitude from this equation we can carry out a change of variables, introducing two new variables $(m,\,N)$ such that

$$\sin\phi^* = m\cos N$$
$$\cos\phi^*\cos D^* = m\sin N \tag{79.2}$$

From these equations we can calculate P and N:

$$\tan N = \frac{\cos\phi^*\cos D^*}{\sin\phi^*} = \frac{\cos D^*}{\tan\phi^*} \Rightarrow N = -41.6°$$

$$P = \frac{\cos N}{\sin\phi^*} = 1.02$$

Substituting Equations (79.2) in Equations (79.1) we obtain

$$\sin\phi_B = P\cos N\sin\phi + P\cos N\cos\phi = P\sin(\phi + N)$$
$$\sin(\phi + N) = \frac{\sin\phi_B}{m} = \frac{\sin\phi_B\cos N}{\sin\phi^*}$$
$$\phi + N = -78.4° \Rightarrow \phi = -36.8°$$

But another solution is also possible:

$$\phi + N = 180° - (-78.4°) = 258.4° \Rightarrow \phi = -60.0°$$

The two solutions are correct and we don't have any additional information to choose one or the other.

(c) The agonic line is the line where the declination is zero and this implies that the point is on the great circle that contains the Geographic North Pole and the Geomagnetic North Pole. So the geomagnetic longitude λ^* is zero or $180°$:

$$\lambda^* = 0° \Rightarrow \lambda = \lambda_B = 65°\,\mathrm{W}$$
$$\lambda^* = 180° \Rightarrow \lambda = \lambda_B + 180 = 115°\,\mathrm{E}$$

80. The Earth's magnetic field is produced by two dipoles of equal moment ($M = Cm = 9.43 \times 10^9$ nT m^3) and polarity, forming angles of 30° and 45° with the axis of rotation, and contained in the plane corresponding to the 0° meridian. Find the potential of the total field and the coordinates of the resulting magnetic North pole, taking the Earth's radius to be 6000 km.

The total potential at a point is the sum of the potentials of the two dipoles. If M is the magnetic moment ($M = Cm$), r is the distance from the dipole's centre, and θ_1 and θ_2 are the geomagnetic co-latitude relative to each dipole (Fig. 80), the total potential Φ is given by

$$\Phi = \Phi_1 + \Phi_2 = \frac{-M\cos\theta_1}{r^2} - \frac{M\cos\theta_2}{r^2} = \frac{-M(\cos\theta_1 + \cos\theta_2)}{r^2} \qquad (80.1)$$

We calculate the angles θ_1 and θ_2 (Fig. 71a) by

$$\cos\theta_1 = \sin\phi_{B1}\sin\phi + \cos\phi_{B1}\cos\phi\cos(\lambda - \lambda_{B1})$$
$$\cos\theta_2 = \sin\phi_{B2}\sin\phi + \cos\phi_{B2}\cos\phi\cos(\lambda - \lambda_{B2}) \qquad (80.2)$$

The geographical coordinates of the two Geomagnetic North Poles are given by

$$\phi_{B1} = 90° - 30° = 60°; \lambda_{B1} = 0°$$
$$\phi_{B2} = 90° - 45° = 45°; \lambda_{B2} = 180°$$

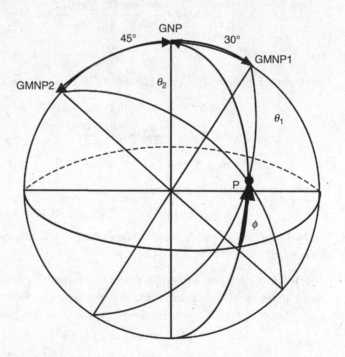

Fig. 80

Substituting these values in Equations (80.2):

$$\cos \theta_1 = \sin \phi_{B1} \sin \phi + \cos \phi_{B1} \cos \phi \cos \lambda$$
$$\cos \theta_2 = \sin \phi_{B2} \sin \phi - \cos \phi_{B2} \cos \phi \cos \lambda$$

Adding the two equations gives

$$\cos \theta_1 + \cos \theta_2 = (\sin \phi_{B1} + \sin \phi_{B2}) \sin \phi + (\cos \phi_{B1} - \cos \phi_{B2}) \cos \phi \cos \lambda$$

and substituting in the equation of the potential (80.1)

$$\Phi = \frac{-M[(\sin \phi_{B1} + \sin \phi_{B2}) \sin \phi + (\cos \phi_{B1} - \cos \phi_{B2}) \cos \phi \cos \lambda]}{r^2}$$

$$\Phi = \frac{-M[(\sqrt{3} + \sqrt{2}) \sin \phi + (1 - \sqrt{2}) \cos \phi \cos \lambda]}{2r^2}$$

If we call ϑ the geographic co-latitude, $\vartheta = 90° - \phi$, then

$$\Phi = \frac{-M[(\sqrt{3} + \sqrt{2}) \cos \vartheta + (1 - \sqrt{2}) \sin \vartheta \cos \lambda]}{2r^2}$$

The resulting magnetic North Pole, the point where the inclination $I = 90°$, due to the combined effect of the two dipoles is given by

$$\tan I = \frac{Z}{H}$$

and therefore at the magnetic Pole, $H = 0$.

We derive the component H by taking the gradient of the potential Φ

$$X = -B_\vartheta = \frac{1}{r} \frac{\partial \Phi}{\partial \vartheta} = -\frac{M[(-(\sqrt{3} + \sqrt{2}) \sin \vartheta + (1 - \sqrt{2}) \cos \vartheta \cos \lambda)]}{2r^3}$$

$$Y = B_\lambda = \frac{-1}{r \sin \vartheta} \frac{\partial \Phi}{\partial \lambda} = \frac{M(1 - \sqrt{2}) \sin \lambda}{2r^3}$$

$$H = \sqrt{X^2 + Y^2}$$

Since the magnetic North Pole is contained in the plane corresponding to the 0° geographical meridian, then its longitude is either 0° or 180°.

If the longitude is 0°

$$\lambda = 0° \Rightarrow H = \frac{-M}{2r^3} \left[-\left(\sqrt{3} + \sqrt{2}\right) \sin \vartheta + \left(1 - \sqrt{2}\right) \cos \vartheta \right] = 0$$
$$\vartheta = -8° = 172°$$

But this result doesn't correspond to the north hemisphere. Then we must take the geographical longitude 180°:

$$\lambda = 180° \Rightarrow H = \frac{-M}{2r^3} \left[-\left(\sqrt{3} + \sqrt{2}\right) \sin \vartheta - \left(1 - \sqrt{2}\right) \cos \vartheta \right] = 0$$
$$\vartheta = 8° \Rightarrow \phi_B = 82°$$

This is the correct result and the coordinates of the magnetic North Pole are

$$\phi_B = 82°, \lambda_B = 180°$$

81. The Earth's magnetic field is produced by one dipole in the direction of the axis of rotation (negative pole in the northern hemisphere) and another with the same moment in the equatorial plane which rotates with differential angular velocity ω with respect to the points on the surface of the Earth (consider that the Earth doesn't rotate). Its negative pole passes through the 45° E meridian at time $t = 0$ and completes a rotation with respect to that point in 24 hours. Consider a point of geographical coordinates 45° N, 45° E.

(a) Calculate the magnetic field components (B_r, B_θ, B_λ) at that point.

(b) Illustrate graphically how each of them varies with local time.

(a)　The total potential at a point on the surface of the Earth is the sum of the potentials of the two dipoles (Problem 80, Equation 80.1):

$$\Phi = \Phi_1 + \Phi_2 = \frac{-M \cos\theta_1}{r^2} - \frac{M \cos\theta_2}{r^2} = \frac{-M(\cos\theta_1 + \cos\theta_2)}{r^2}$$

Dipole 1 is in the direction of the axis of rotation and so the geomagnetic co-latitude of the point with respect to this dipole (Fig. 81a) is equal to the geographical co-latitude,

$$\theta_1 = 90° - \phi$$

$$\cos\theta_1 = \sin\phi$$

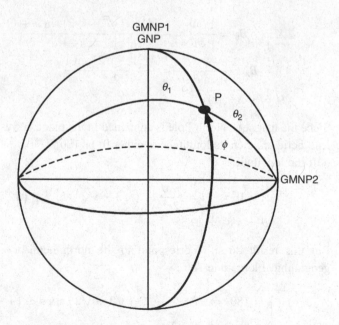

Dipole 2 is on the equatorial plane ($\phi_{B2} = 0$) and rotates with respect to the points of the surface. Owing to this rotation its geographical longitude λ_{B2} changes with time t in the form

$$\lambda_{B2} = \omega t + 45°$$

where ω is the angular velocity, $\omega = 360°/T$, T being the rotation period of 24 h.
   The co-latitude θ_2 is

$$\cos \theta_2 = \sin \phi_{B2} \sin \phi + \cos \phi_{B2} \cos \phi \cos(\lambda - \lambda_{B2})$$

Substituting the geographical coordinates of the negative geomagnetic equatorial Pole (ϕ_{B2}, λ_{B2}):

$$\cos \theta_2 = \cos \varphi \cos(\lambda - \omega t - 45°)$$

Substituting in the equation for the potential

$$\Phi = \Phi_1 + \Phi_2 = \frac{-M[\sin \phi + \cos \phi \cos(\lambda - \omega t - 45°)]}{r^2}$$

If we consider the geographical co-latitude $\vartheta = 90° - \phi$, the potential is given by

$$\Phi = \frac{-M[\cos \vartheta + \sin \vartheta \cos(\lambda - \omega t - 45°)]}{r^2}$$

We obtain the magnetic field components (B_r, B_θ, B_λ) at the point (ϑ, λ) by taking the gradient in spherical coordinates of the potential Φ:

$$B_r = -\frac{\partial \Phi}{\partial r} = \frac{-2M[\cos \vartheta + \sin \vartheta \cos(\lambda - \omega t - 45°)]}{r^3}$$

$$B_\vartheta = -\frac{1}{r}\frac{\partial \Phi}{\partial \vartheta} = \frac{M[-\sin \vartheta + \cos \vartheta \cos(\lambda - \omega t - 45°)]}{r^3}$$

$$B_\lambda = -\frac{1}{r \sin \vartheta}\frac{\partial \Phi}{\partial \lambda} = -\frac{M \sin(\lambda - \omega t - 45°)}{r^3}$$

Substituting the values $\vartheta = 45°$, $\lambda = 45°$, $B_0 = M/a^3$, gives

$$B_r = -\frac{-2B_0}{\sqrt{2}}(1 + \cos \omega t)$$

$$B_\vartheta = -\frac{B_0}{\sqrt{2}}(1 - \cos \omega t)$$

$$B_\lambda = -B_0 \sin \omega t$$

(b) The variation of each component with local time is shown in Fig. 81b

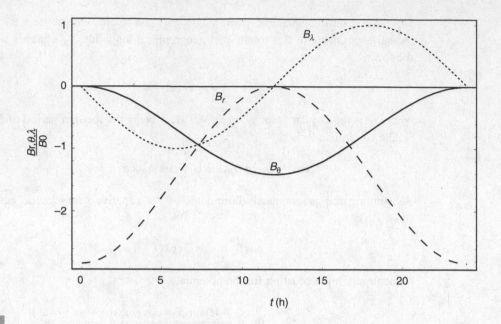

Magnetic anomalies

82. Calculate the magnetic anomaly created by a magnetic dipole buried at depth d, arbitrarily oriented, at an angle to the vertical of α. The negative pole is upwards.

Consider a point P with coordinates (x, z), where x is measured along the horizontal from the projection of the centre of the dipole and z is the vertical from the reference level (the Earth's surface). The position vector r forms an angle β to the vertical (Fig. 82). The anomalous magnetic potential created by the dipole for this point is

$$\Delta\Phi = \frac{-Cm\cos(\alpha + \beta)}{r^2} = \frac{-Cm(\sin\beta\cos\alpha + \cos\beta\sin\alpha)}{r^2} \quad (82.1)$$

where (Fig. 82)

$$\cos\beta = \frac{z+d}{\sqrt{x^2 + (z+d)^2}}$$

$$\sin\beta = \frac{x}{\sqrt{x^2 + (z+d)^2}}$$

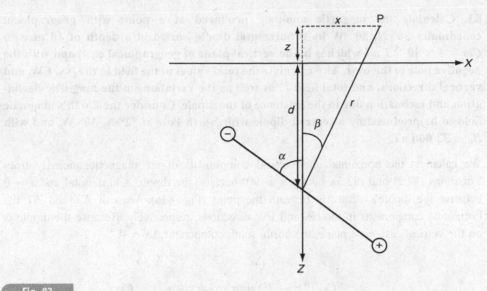

Fig. 82

Substituting in Equation (82.1) we obtain

$$\Delta\Phi = \frac{-Cm[(z+d)\cos\alpha - x\sin\alpha]}{\left[x^2 + (z+d)^2\right]^{3/2}}$$

To calculate the magnetic anomaly ΔB:

$$\Delta B = -\nabla(\Delta\Phi)$$

The vertical component of the magnetic field anomaly, taking the z-coordinate positive downward, is

$$\Delta Z = \frac{\partial(\Delta\Phi)}{\partial z} = \frac{-Cm\left[\left(x^2 + (z+d)^2\right)\cos\alpha - 3(z+d)[(z+d)\cos\alpha - x\sin\alpha]\right]}{\left[x^2 + (z+d)^2\right]^{5/2}}$$

For points on the Earth's surface ($z = 0$)

$$\Delta Z = \frac{-Cm[(x^2 - 2d^2)\cos\alpha + 3dx\sin\alpha]}{[x^2 + d^2]^{5/2}} \tag{82.2}$$

The component of the magnetic anomaly in an arbitrary horizontal direction x for the Earth's surface points ($z = 0$) is given by

$$\Delta X = -\frac{\partial(\Delta\Phi)}{\partial x} = \frac{Cm[(2x^2 - d^2)\sin\alpha - 3dx\cos\alpha]}{[x^2 + d^2]^{5/2}} \tag{82.3}$$

83. Calculate the magnetic anomaly produced at a point with geographical coordinates 38° N, 30° W by a horizontal dipole buried at a depth of 10 m with $Cm = 5 \times 10^{-5}$ T m³ which is in the vertical plane of geographical east, and with the negative pole to the west. Also calculate the total values of the field in the NS, EW, and vertical directions, and total field F, as well as the variations in the magnetic declination and inclination due to the existence of the dipole. Consider the Earth's magnetic field to be produced by a centred dipole with North Pole at 72° N, 30° W, and with $B_0 = 32\,000$ nT.

We calculate the horizontal and vertical components of the magnetic anomaly from Equations (82.2) and (82.3), taking $\alpha = 90°$ because the dipole is horizontal and $x = 0$ because the dipole's centre is beneath the point (Fig. 83a). We call ΔX and ΔY the horizontal components in the NS and EW directions, respectively. Because the dipole is on the vertical east–west plane, the north–south component $\Delta X = 0$,

$$\Delta Z = 0$$

$$\Delta Y = \frac{Cm[(2x^2 - d^2)\sin\alpha - 3dx\cos\alpha]}{[x^2 + d^2]^{5/2}} = -\frac{Cm}{d^3}$$

Substituting

$$Cm = 5 \times 10^{-5}\,\text{T m}^3$$
$$d = 10\,\text{m}$$

we obtain

$$\Delta Y = -50\,\text{nT}$$
$$|\Delta\boldsymbol{B}| = \Delta Y = 50\,\text{nT}$$
$$\Delta X = 0$$

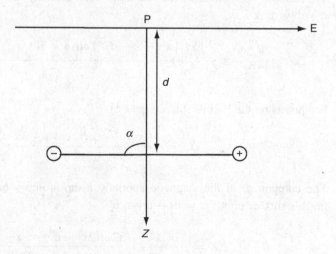

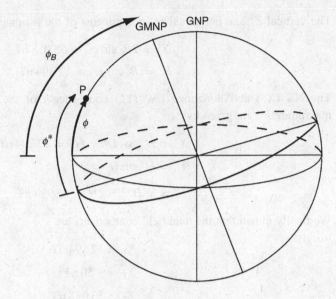

Fig. 83b

To calculate the components of the magnetic anomaly in the direction of the Earth's magnetic field, F, and their horizontal component, H, we need to determine the magnetic declination and inclination at the point:

$$\Delta H = \Delta X \cos D^* + \Delta Y \sin D^*$$
$$\Delta F = \Delta H \cos I^* + \Delta Z \sin I^* \tag{83.1}$$

Because the point has the same longitude as the Geomagnetic North Pole (Fig. 83b),

$$\phi^* = 90° - (\phi_B - \phi) = 56.0°$$
$$D^* = 0°$$

The inclination is given by

$$\tan I^* = 2 \tan \phi^* \Rightarrow I^* = 71.4°$$

Substituting these values in Equations (83.1) we obtain

$$\Delta H = 0$$
$$\Delta F = 0$$

The total value of the field in the NS, EW, and vertical directions, and total field F are

$$X_T = X^* + \Delta X$$
$$Y_T = Y^* + \Delta Y$$
$$Z_T = Z^* + \Delta Z$$
$$F_T = F^* + \Delta F$$

The vertical Z^* and horizontal H^* components of the geomagnetic field are given by

$$Z^* = 2B_0 \sin \phi^* = 53\,058\,\text{nT}$$
$$H^* = B_0 \cos \phi^* = 17\,894\,\text{nT}$$

The NS (X^*) and horizontal EW (Y^*) components of the geomagnetic field and its magnitude F^* are given by

$$X^* = H^* \cos D^* = H^* = 17\,894\,\text{nT}$$
$$Y^* = H^* \sin D^* = 0$$
$$F^* = \sqrt{H^{*2} + Z^{*2}} = 55\,994\,\text{nT}$$

We finally obtain that the total field components are

$$X_\text{T} = 17\,894\,\text{nT}$$
$$Y_\text{T} = -50\,\text{nT}$$
$$Z_\text{T} = 53\,058\,\text{nT}$$
$$F_\text{T} = 55\,994\,\text{nT}$$

The variations in magnetic declination and inclination due to the presence of the buried dipole are

$$\tan D' = \frac{Y_\text{T}}{X_\text{T}} \Rightarrow D' = -0.02°$$
$$D^* - D' = 0.02°$$
$$\tan I' = \frac{Z_\text{T}}{H_\text{T}} \Rightarrow I' \approx 71.4° = I^*$$

84. Buried at a point with magnetic latitude 30° N and a depth of 50 m is a horizontal magnetic dipole with $Cm = 10^7$ nT m^3 with the positive pole to the geographical north.

(a) **Calculate ΔF if $B_0 = 30\,000$ nT and the declination at that point is 15°. Find the ratio $\Delta F / F$.**
(b) **How far from the dipole's centre along the north–south line will the dipole field strength be in the same direction as that of the Earth (take $D^* = 0°$).**

(a) The component of the magnetic anomaly ΔF in the direction of the Earth's magnetic field (the total field anomaly) is given by

$$\Delta F = \Delta H \cos I^* + \Delta Z \sin I^* \tag{84.1}$$

We first calculate the components in the geographical directions of the magnetic anomaly produced by the buried dipole using Equations (82.2) and (82.3) of Problem 82, substituting $\alpha = 90°$ because the dipole is horizontal, and $x = 0$ because the dipole's centre is beneath the point. In this problem $\Delta Y = 0$ because the dipole is on the geographical north–south vertical plane (Fig. 84). Then

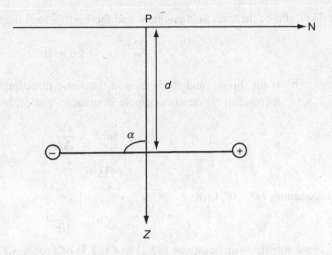

Fig. 84

$$\Delta X = -\frac{Cm}{d^3}$$

$$\Delta Y = 0$$

$$\Delta Z = 0$$

Substituting the values

$$Cm = 10^7 \, \text{nT m}^3$$

$$d = 50 \, \text{m}$$

we obtain

$$\Delta X = -80 \, \text{nT}$$

Substituting $D^* = 15°$, the component of the magnetic anomaly in the direction of the horizontal component H of the Earth's magnetic field is

$$\Delta H = \Delta X \cos D^* = -77 \, \text{nT}$$

At a point of magnetic latitude $\phi^* = 30°$ the magnetic inclination is

$$\tan I^* = 2 \tan \phi^* \Rightarrow I^* = 49.1°$$

Substituting in Equation (84.1), the total field anomaly is

$$\Delta F = -50 \, \text{nT}$$

To calculate the geomagnetic field F^* we first obtain the components H^* and Z^*:

$$Z^* = 2B_0 \sin \phi^* = 30\,000 \, \text{nT}$$

$$H^* = B_0 \cos \phi^* = 25\,981 \, \text{nT}$$

$$F^* = \sqrt{H^{*2} + Z^{*2}} = 39\,686 \, \text{nT}$$

The ratio of the total field anomaly and the Earth's total magnetic field is

$$\frac{\Delta F}{F} = -1.26 \times 10^{-3}$$

(b) If the dipole field strength is in the same direction as that of the Earth then the inclination I' due to the dipole is equal to that of the Earth's field I^*, where

$$\tan I' = \frac{\Delta Z}{\Delta H}$$

$$\tan I^* = \frac{Z^*}{H^*}$$

Assuming $D^* = 0°$, then

$$\Delta H = \Delta E$$

If we substitute in Equations (82.2) and (82.3) of Problem 82, the angle $\alpha = 90°$ because the dipole is horizontal, we obtain

$$\Delta Z = \frac{3Cmdx}{\left[x^2 + d^2\right]^{5/2}}$$

$$\Delta X = \frac{Cm(2x^2 - d^2)}{\left[x^2 + d^2\right]^{5/2}}$$

We have changed the sign of the vertical component because the negative pole is toward the south.

Applying the condition, $\tan I' = \tan I^*$, we obtain

$$\frac{\Delta Z}{\Delta H} = \frac{\Delta Z}{\Delta X} = \frac{Z^*}{H^*}$$

$$\frac{\dfrac{3Cmdx}{\left(x^2 + d^2\right)^{5/2}}}{\dfrac{Cm\left(2x^2 - d^2\right)}{\left(x^2 + d^2\right)^{5/2}}} = \frac{3dx}{2x^2 - d^2} = \frac{Z^*}{H^*}$$

$$2Z^*x^2 - 3dH^*x - Z^*d^2 = 0$$

Substituting the values
   $d = 50$ m
   $Z^* = 30\ 000$ nT
   $H^* = 25\ 981$ nT
and solving the equation, we obtain

$$x_1 = 80\,\text{m}$$

$$x_2 = -15\,\text{m}$$

We have two solutions: a point 80 m to the north from the surface projection of the dipole's centre and another 15 m to the south.

85. Located at a point with geocentric geographical coordinates 45° N, 30° W, at a depth of 100 m, is a dipole of magnetic moment $Cm= 1$ T m^3, tilted 45° from the horizontal to true north, with the negative pole to the north and downwards. At this point on the surface, the following magnetic field values were observed (in nT):

   $F = 55\ 101; \quad H = 12\ 413; \quad \Delta F = -1268; \quad \Delta H = 547.$

Determine:

(a) At the indicated point, the main field components X^*, Y^*, Z^*.

(b) At the indicated point, the deviation of the compass with respect to geomagnetic north.

(c) The geocentric geographical coordinates of the North Pole of the Earth's dipole.

Precision 1 nT.

(a) We calculate first the magnetic anomaly produced by the dipole, applying Equations (82.2) and (82.3) of Problem 82, substituting $\alpha = 225°$ and $x = 0$. The horizontal component is in the NS direction (ΔX) (Fig. 85a)

$$\Delta Z = \frac{2Cm\cos\alpha}{d^3} = -1414\,\text{nT}$$

$$\Delta X = \frac{-Cm\sin\alpha}{d^3} = 707\,\text{nT} \tag{85.1}$$

$$\Delta Y = 0$$

To calculate the declination we use the equation

$$\Delta H = \Delta X\cos D^* \Rightarrow \cos D^* = \frac{\Delta H}{\Delta X}$$

$$D^* = 39.3°$$

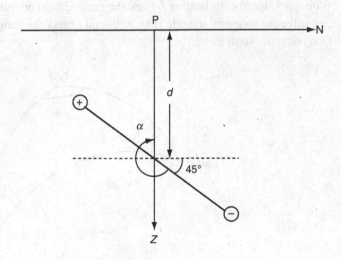

To obtain the Earth's main field we eliminate the buried dipole contribution from the observed values:

$$F^* = F - \Delta F = 56\,369\,\text{nT}$$
$$H^* = H - \Delta H = 11\,866\,\text{nT}$$

$$Z^* = \sqrt{(F^*)^2 - (H^*)^2} = 55\,106\,\text{nT}$$

and the X^* and Y^* components

$$X^* = H^* \cos D^* = 9182\,\text{nT}$$
$$Y^* = H^* \sin D^* = 7516\,\text{nT}$$

(b) The observed declination is given by

$$\tan D' = \frac{Y}{X} = \frac{Y^* + \Delta Y}{X^* + \Delta X} \Rightarrow D' = 37.2°$$

The deviation of the compass due to the buried dipole with respect to geomagnetic north is

$$D' - D^* = -2.1°$$

(c) We calculate first the geomagnetic latitude of the point from the vertical and horizontal components:

$$Z^* = 2B_0 \sin \phi^*$$
$$H^* = B_0 \cos \phi^*$$

$$\tan \phi^* = \frac{Z^*}{2H^*}$$
$$\phi^* = 66.7°$$

With this value, the declination D^* and the geographical coordinates of the point (ϕ, λ), we can solve the spherical triangle (Fig. 85b) and obtain the geographical coordinates of the Geomagnetic North Pole:

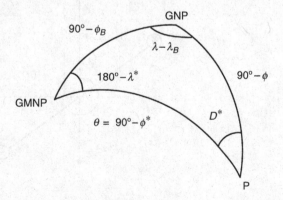

Fig. 85b

$$\cos(90 - \phi_B) = \cos(90 - \phi^*) \cos(90 - \phi) + \sin(90 - \phi^*) \sin(90 - \phi) \cos D^*$$

$$\sin \phi_B = \sin \phi^* \sin \phi + \cos \phi^* \cos \phi \cos D^*$$

$$\phi_B = 60.0°$$

$$\cos(90 - \phi^*) = \cos(90 - \phi_B) \cos(90 - \phi) + \sin(90 - \phi_B) \sin(90 - \phi) \cos(\lambda - \lambda_B)$$

$$\sin \phi^* = \sin \phi_B \sin \phi + \cos \phi_B \cos \phi \cos(\lambda - \lambda_B)$$

$$\cos(\lambda - \lambda_B) = \frac{\sin \phi^* - \sin \phi_B \sin \phi}{\cos \phi_B \cos \phi}$$

$$\lambda - \lambda_B = \pm 30.0°$$

The correct solution is the negative one because a positive value of the declination implies that the point is to the west of the Geomagnetic North Pole:

$$D^* > 0 \Rightarrow \lambda - \lambda_B < 0$$

$$\lambda_B = 0°$$

86. Located at a point with geographical coordinates 45° N, 30° W, at a depth of 100 m, is a dipole of magnetic moment $Cm = 1$ T m^3, inclined 45° to the vertical towards the south, with the positive pole upwards, and in the geographical north–south vertical plane. The Earth's dipole has its north pole at 60° N, 0° E and $B_0 = 30\ 000$ nT. Calculate:

(a) The values of Z, H, F at the given point.

(b) Where does the compass point to at that same point?

(a) We calculate first the geomagnetic latitude corresponding to the point by

$$\sin \phi^* = \sin \phi_B \sin \phi + \cos \phi_B \cos \phi \cos(\lambda - \lambda_B)$$

$$\phi^* = 66.7°$$

From this value we obtain the geomagnetic components Z^*, H^* and the total main field F^*:

$$Z^* = 2B_0 \sin \phi^* = 55\ 107 \, \text{nT}$$

$$H^* = B_0 \cos \phi^* = 11\ 866 \, \text{nT}$$

$$F^* = \sqrt{(H^*)^2 + (Z^*)^2} = 56\ 370 \, \text{nT}$$

The geomagnetic declination D^* is given by

$$\sin D^* = \frac{-\cos \phi_B \sin(\lambda - \lambda_B)}{\cos \phi^*}$$

$$D^* = 39.2°$$

and the geomagnetic inclination I^* by

$$\tan I^* = 2 \tan \phi^* \Rightarrow I^* = 77.8°$$

The magnetic anomaly created by the dipole buried at depth d is given by Equations (85.1) of Problem 85. Substituting $Cm = 1$ T m^3, $d = 100$ m, and $\alpha = 45°$, we obtain

$$\Delta Z = \frac{2Cm \cos \alpha}{d^3} = -1414 \, \text{nT}$$

$$\Delta X = \frac{-Cm \sin \alpha}{d^3} = 707 \, \text{nT}$$

$$\Delta Y = 0$$

The field anomalies ΔH and ΔF are given by

$$\Delta H = \Delta X \cos D^* = 548 \, \text{nT}$$

$$\Delta F = \Delta H \cos I^* + \Delta Z \sin I^* = -1266 \, \text{nT}$$

Finally, the observed values are

$$Z = Z^* + \Delta Z = 53\,693 \, \text{nT}$$

$$F = F^* + \Delta F = 55\,104 \, \text{nT}$$

$$H = H^* + \Delta H = 12\,414 \, \text{nT}$$

(b) To calculate in what direction the compass points we need the value of the observed declination D' including the effects of the geomagnetic field and the buried dipole:

$$\tan D' = \frac{Y}{X} = \frac{Y^* + \Delta Y}{X^* + \Delta X}$$

$$Y^* = H^* \sin D = 7500 \, \text{nT}$$

$$X^* = H^* \cos D = 9195 \, \text{nT}$$

$$D' = 37.1°$$

87. Located at a point on the Earth with geographical coordinates 45° N, 30° E, at a depth of 100 m, is a dipole of magnetic moment $Cm = 10^7$ nT m^3, tilted 45° to the vertical towards the south, with the positive pole downwards, and contained in the plane of true north. The Earth's field is produced by a centred dipole tilted 30° from the axis of rotation in the plane of the 0° meridian, with $B_0 = 30\,000$ nT. Calculate the total values of F, Z, and H observed at the point of the surface above the centre of the buried dipole.

We first calculate the geographical coordinates of the Geomagnetic North Pole and the geomagnetic latitude

$$\phi_B = 90° - 30° = 60°$$
$$\lambda_B = 0°$$

$$\sin \phi^* = \sin \phi_B \sin \phi + \cos \phi_B \cos \phi \cos(\lambda - \lambda_B)$$
$$\phi^* = 66.7°$$

The geomagnetic field components Z^*, H^* and the total main field F^* are given by

$$Z^* = 2B_0 \sin \phi^* = 55\,107\,\text{nT}$$
$$H^* = B_0 \cos \phi^* = 11\,866\,\text{nT}$$
$$F^* = \sqrt{(H^*)^2 + (Z^*)^2} = 56\,370\,\text{nT}$$

We calculate the geomagnetic declination D^* by

$$\sin D^* = \frac{-\cos \phi_B \sin(\lambda - \lambda_B)}{\cos \phi^*}$$
$$D^* = -39.2°$$

The inclination I^* is given by

$$\tan I^* = 2 \tan \phi^* \Rightarrow I^* = 77.8°$$

We obtain the magnetic anomaly produced by the buried dipole applying Equations (82.2) and (82.3) of Problem 82, substituting $\alpha = 45°$ and $x = 0$. The horizontal component is in the NS direction (ΔX):

$$\Delta Z = \frac{2Cm \cos \alpha}{d^3} = 14\,\text{nT}$$
$$\Delta X = \frac{-Cm \sin \alpha}{d^3} = -7\,\text{nT}$$
$$\Delta Y = 0$$

The field anomalies ΔH and ΔF are given by

$$\Delta H = \Delta X \cos D^* = -5\,\text{nT}$$
$$\Delta F = \Delta H \cos I^* + \Delta Z \sin I^* = 13\,\text{nT}$$

Finally the observed values are

$$Z = Z^* + \Delta Z = 55\,121\,\text{nT}$$
$$F = F^* + \Delta F = 56\,383\,\text{nT}$$
$$H = H^* + \Delta H = 11\,861\,\text{nT}$$

88. Buried at a point with geographical latitude 20° N and the same longitude as the geomagnetic pole, at a depth of 200 m, is a sphere of 50 m radius of material with magnetic susceptibility 0.01. The Earth's field is produced by a centred dipole tilted 10° from the axis of rotation and magnetic moment $M = 10^{30}$ γ cm^3 (Earth's radius: 6000 km). Calculate:

(a) **The anomaly produced by induced magnetization in the sphere at a point on the Earth's surface above the centre of the sphere. Give the vertical and horizontal components in units of nT.**

(b) **The total anomaly for a point on the Earth's surface 100 m south of the above point.**

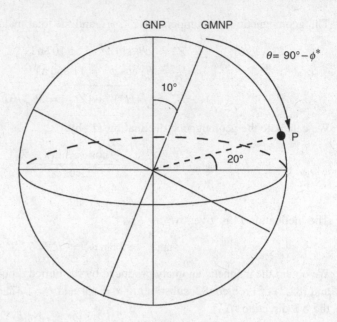

(a) We first calculate the geomagnetic co-latitude (θ) and latitude (ϕ^*) of the point, knowing that it is in the same meridian as the Geomagnetic North Pole (Fig. 88a):

$$\theta = 90° - \phi^* = 90° - 10° - 20° = 60°$$
$$\phi^* = 30°$$

The geomagnetic field is given by

$$B_0 = \frac{M}{a^3} = 4630 \, \text{nT}$$
$$Z^* = 2B_0 \sin \phi^* = 4630 \, \text{nT}$$
$$H^* = B_0 \cos \phi^* = 4009 \, \text{nT}$$
$$F^* = \sqrt{(H^*)^2 + (Z^*)^2} = 6124 \, \text{nT}$$

The inclination is given by

$$\tan I^* = 2 \tan \phi^* \Rightarrow I^* = 49°$$

The magnetic anomaly created by a sphere is the same as the anomaly created by a magnetic dipole oriented in the same direction as the geomagnetic field, that is, tilted $90° - I^*$ to the vertical and with the negative pole upwards (Fig. 88b). So we use Equations (82.2) and (82.3) taking $\alpha = 41°$ and $x = 0$, but we change the sign of Equation (82.3) because the negative pole is toward the north. The horizontal component is $\Delta X = \Delta H$ because the dipole is in the magnetic north-vertical plane and $\Delta Y = 0$:

$$\Delta Z = \frac{2Cm \cos \alpha}{d^3}$$
$$\Delta H = \frac{Cm \sin \alpha}{d^3}$$

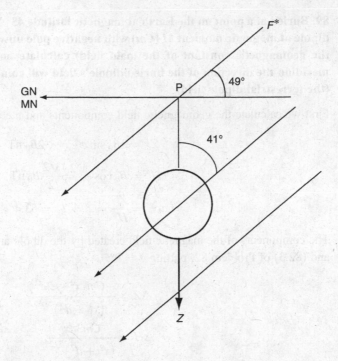

Fig. 88b

To calculate Cm we use the magnetic susceptibility χ and the volume V of the sphere:

$$Cm = \chi F^* V = 3.2 \times 10^7 \, \text{nT m}^3$$

Substituting this value in the equation of the components of the magnetic anomaly we obtain

$$\Delta Z = \frac{2Cm \cos \alpha}{d^3} = \frac{2 \times 3.2 \times 10^7 \times \cos 41°}{8 \times 10^6} = 6 \, \text{nT}$$

$$\Delta X = \frac{Cm \sin \alpha}{d^3} = \frac{3.2 \times 10^7 \times \sin 41°}{8 \times 10^6} = 3 \, \text{nT}$$

(b) The anomaly created by the sphere at a point at a distance $x = 100$ m to the south of the above point is given by

$$\Delta Z = \frac{-Cm \left[(x^2 - 2d^2) \cos \alpha + 3dx \sin \alpha\right]}{\left[x^2 + d^2\right]^{5/2}} = 0.8 \, \text{nT}$$

$$\Delta H = \frac{-Cm \left[(2x^2 - d^2) \sin \alpha - 3dx \cos \alpha\right]}{\left[x^2 + d^2\right]^{5/2}} = 3.3 \, \text{nT}$$

The total magnetic field anomaly is therefore

$$\Delta F = \Delta H \cos I^* + \Delta Z \sin I^* = 3.4 \, \text{nT}$$

89. Buried at a point on the Earth at magnetic latitude 45° N, at a depth *d*, is a vertical dipole of magnetic moment *M (Cm)* with negative pole upwards. If *M/d³* = 10*B₀* (*B₀* is the geomagnetic constant of the main field) calculate how far along the magnetic meridian the direction of the buried dipole's field will coincide with that of the Earth (the terrestrial dipole field).

First we calculate the geomagnetic field components and the inclination by

$$Z^* = 2B_0 \sin \phi^* = \sqrt{2}B_0 \, \text{nT}$$

$$H^* = B_0 \cos \phi^* = \frac{\sqrt{2}}{2}B_0 \, \text{nT}$$

$$\tan I^* = \frac{Z^*}{H^*} = 2 \Rightarrow I^* = 63.4°$$

The components of the magnetic field created by the dipole are given by Equations (82.2) and (82.3) of Problem 82, putting $\alpha = 0°$:

$$\Delta Z = \frac{-Cm\left(x^2 - 2d^2\right)}{\left(x^2 + d^2\right)^{5/2}}$$

$$\Delta H = \frac{Cm3dx}{\left(x^2 + d^2\right)^{5/2}}$$

If the buried dipole's field coincides with that of the Earth the magnetic inclinations due to both have to be equal and so

$$\frac{\Delta Z}{\Delta H} = \tan I^* = 2$$

$$x^2 + 6xd - 2d^2 = 0 \Rightarrow x = d\left(-3 \pm \sqrt{28}\right)$$

Of the two solutions, $x = 2.3d$ and $x = -8.3d$, only the positive corresponds to the equal direction of the two fields.

External magnetic field

90. The Earth's magnetic field is produced by two dipoles of equal moment and polarity that are at an angle of 60° to each other, with the bisector being the axis of rotation. The dipoles are contained in the plane of the 0° geographical meridian.

(a) **Calculate the potential on points of the Earth's surface as a function of geographical coordinates ϕ and λ.**
(b) **At what points on the surface are the magnetic poles located?**
(c) **What form would the external field have in order to annul the internal field at the magnetic equator?**

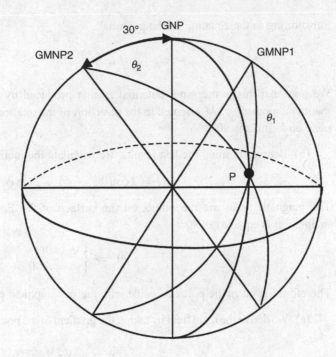

Fig. 90

(a) The total potential at a point is the sum of the potentials of the two dipoles (see Equation 80.1 of Problem 80):

$$\Phi = \Phi_1 + \Phi_2 = \frac{-M\cos\theta_1}{r^2} - \frac{M\cos\theta_2}{r^2} = \frac{-M(\cos\theta_1 + \cos\theta_2)}{r^2}$$

We calculate the geomagnetic co-latitudes by Equation (71.3) of Problem 71:

$$\cos\theta_1 = \sin\phi_{B1}\sin\phi + \cos\phi_{B1}\cos\phi\cos(\lambda - \lambda_{B1})$$
$$\cos\theta_2 = \sin\phi_{B2}\sin\phi + \cos\phi_{B2}\cos\phi\cos(\lambda - \lambda_{B2})$$

(90.1)

The geographical coordinates of the North Pole of each dipole are given by (Fig. 90)

$$\phi_{B1} = \phi_{B2} = 90° - 30° = 60°$$
$$\lambda_{B1} = 0°$$
$$\lambda_{B2} = 180°$$

Substituting these values in Equation (90.1):

$$\cos\theta_1 = \frac{\sqrt{3}}{2}\sin\phi + \frac{1}{2}\cos\phi\cos\lambda$$
$$\cos\theta_2 = \frac{\sqrt{3}}{2}\sin\phi - \frac{1}{2}\cos\phi\cos\lambda$$

Adding the two equations:

$$\cos\theta_1 + \cos\theta_2 = \sqrt{3}\sin\phi$$

Substituting in the equation of the potential:

$$\Phi = \frac{-\sqrt{3}M \sin\phi}{r^2}$$

We note that this is the same potential as that produced by only a centred dipole with magnetic moment $\sqrt{3}M$ oriented in the direction of the rotation axis, because the geomagnetic co-latitude is $90° - \phi$.

(b) Bearing in mind the last results, we calculate the inclination by

$$\tan I = 2 \cot(90° - \phi) = 2 \tan\phi \tag{90.2}$$

The magnetic poles are the points on the surface of the Earth where the value of the inclination is equal to $\pm 90°$:

$$I = \pm 90° \Rightarrow \begin{cases} \phi = 90° \\ \phi = -90° \end{cases}$$

Therefore the magnetic poles coincide with the geographical poles.

(c) We derive the main field by taking the gradient of the potential Φ. The components are

$$B_r = -\frac{\partial \Phi}{\partial r} = \frac{-2\sqrt{3}M \sin\phi}{r^3}$$

$$B_\phi = \frac{1}{r}\frac{\partial \Phi}{\partial \phi} = \frac{-\sqrt{3}M \cos\phi}{r^3}$$

$$B_\lambda = -\frac{1}{r\cos\phi}\frac{\partial \Phi}{\partial \lambda} = 0$$

At the magnetic equator the inclination is null ($I = 0$) and according to Equation (90.2) the latitude is null too ($\phi = 0$). Substituting in the last equations:

$$B_r = 0$$

$$B_\phi = \frac{-\sqrt{3}M}{r^3}$$

$$B_\lambda = 0$$

The external magnetic field to annul out the internal field is therefore

$$B_e = \left(0, \frac{\sqrt{3}M}{r^3}, 0\right)$$

91. The Earth's magnetic field is formed by a centred dipole with northern geomagnetic pole at 60° N, 0° E and $B_0 = 32\,000$ nT and a uniform external field from the Sun of 10 000 nT parallel to the equatorial plane.

(a) **For a point at coordinates 60° N, 60° W, calculate the components X, Y, Z of the total field, and the values of D and I.**

(b) **How do D and Z of the total field vary throughout the day with local time t?**

(a) We calculate the geomagnetic main field components Z^*, H^* obtaining the geomagnetic latitude given by (71.3):

$$\sin \phi^* = \sin \phi_B \sin \phi + \cos \phi_B \cos \phi \cos(\lambda - \lambda_B)$$
$$\phi^* = 61°$$

$$Z^* = 2B_0 \sin \phi^* = 55\,976\,\text{nT}$$
$$H^* = B_0 \cos \phi^* = 15\,514\,\text{nT}$$

To calculate the NS (X^*) and EW (Y^*) components we need the geomagnetic declination D^* (71.2):

$$\sin D^* = \frac{-\cos \phi_B \sin(\lambda - \lambda_B)}{\cos \phi^*}$$
$$D^* = -63°$$
$$X^* = H^* \cos D^* = 7043\,\text{nT}$$
$$Y^* = H^* \sin D^* = -13\,823\,\text{nT}$$

The external field is parallel to the equatorial plane and has a diurnal period ($\omega = 2\pi/24$) because it comes from the Sun. We assume that at local time $t = 0$ the Sun is at the point's meridian. If we denote by N the modulus of the external field ($N = 10\,000$ nT) and bearing in mind Fig. 91a (representation of the plane parallel to the equator that contains the point) we have at time $t \neq 0$

$$Y = B_\lambda^e = N \sin \omega t$$

The radial and tangential components (Fig. 91b) are

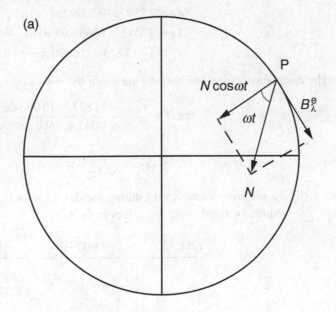

(a)

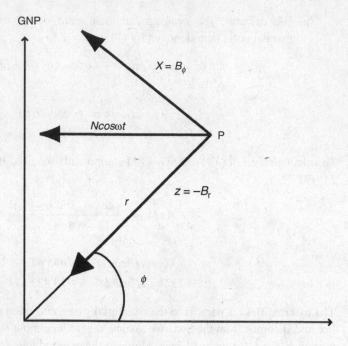

$$Z = -B_r = N \cos \omega t \cos \phi$$
$$X = B_\phi = N \cos \omega t \sin \phi$$

The total magnetic field is the sum of the two contributions:

$$Z_T = (55\,976 + 5000 \cos \omega t)\,\text{nT}$$
$$X_T = (7043 + 8660 \cos \omega t)\,\text{nT}$$
$$Y_T = (-13\,823 + 10\,000 \sin \omega t)\,\text{nT}$$

The declination D and inclination I are given by

$$\tan D = \frac{Y_T}{X_T} = \frac{-13\,823 + 10\,000 \sin \omega t}{7043 + 8660 \cos \omega t}$$

$$\tan I = \frac{Z_T}{H_T} = \frac{Z_T}{\sqrt{X_T^2 + Y_T^2}}$$

(b) To see how D and Z vary during the day with local time t we substitute several values for t, obtaining the values in the table:

t (h)	Z (nT)	D (°)
0	60976	−41
6	55976	−28
12	50976	83
18	55976	−73

92. The Earth's magnetic field is formed by a centred dipole of moment m and north pole 60° N, 0° E, and a uniform external field of magnitude $N = B_0/4$ (B_0 is the geomagnetic constant of the internal field) parallel to the axis of rotation. Determine:

(a) The total potential at any point.
(b) The coordinates of the boreal magnetic pole.
(c) The magnetic declination at the point 45° N, 45° E.
(d) The angle along the meridian between that point and the magnetic equator.

(a) The total potential (Φ) is the sum of two contributions: the main (internal) field (Φ^i) and the external field (Φ^e):

$$\Phi = \Phi^i + \Phi^e$$

The main field is formed by a centred dipole of moment m so the potential is given by

$$\Phi^i = \frac{-Cm\cos\theta}{r^2}$$

We calculate the geomagnetic co-latitude $\theta = 90 - \phi^*$ at a point with geographical coordinates (ϕ, λ) by (71.3):

$$\cos\theta = \sin\phi_B\sin\phi + \cos\phi_B\cos\phi\cos(\lambda - \lambda_B) = \frac{\sqrt{3}}{2}\sin\phi + \frac{1}{2}\cos\phi\cos\lambda$$

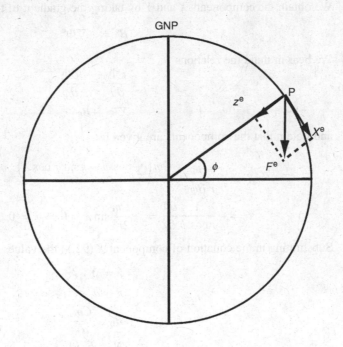

Fig. 92a

The external field is parallel to the axis of rotation; therefore its components are in the vertical and NS directions (Fig. 92a). If we call $\vartheta = 90° - \phi$ the geographical co-latitude, the components are given by

$$B_r^e = -N \cos \vartheta = -N \sin \phi$$
$$B_\vartheta^e = N \sin \vartheta = N \cos \phi$$
$$Z^e = -B_r^e = N \sin \phi$$
$$X^e = -B_\vartheta^e = -N \cos \phi$$

Bearing in mind that $\boldsymbol{B}^e = -\nabla \Phi^e$ the potential for the external field is

$$\Phi^e = Nr \cos \vartheta = Nr \sin \phi$$

Therefore the total potential is given by

$$\Phi = \frac{-Cm\left(\frac{\sqrt{3}}{2}\sin \phi + \frac{1}{2}\cos \phi \cos \lambda\right)}{r^2} + Nr \sin \phi$$

(b) At the magnetic boreal pole the inclination is $I = 90°$ and the horizontal field components H, X, Y are given by

$$\tan I = \frac{Z}{H} \Rightarrow H = 0$$

$$H = \sqrt{X^2 + Y^2} \Rightarrow \begin{cases} X = 0 \\ Y = 0 \end{cases}$$

We obtain the components X and Y by taking the gradient of the potential

$$\boldsymbol{B}^e = -\nabla \Phi^e$$

We bear in mind the relations

$$\frac{\partial \Phi}{\partial \theta} = -\frac{\partial \Phi}{\partial \phi}$$
$$X = -B_\theta$$

and obtain that the components are given by

$$X = -\frac{1}{r}\frac{\partial \Phi}{\partial \phi} = \frac{Cm\left(\frac{\sqrt{3}}{2}\cos \phi - \frac{1}{2}\sin \phi \cos \lambda\right)}{r^3} - N \cos \phi = 0$$

$$Y = -\frac{1}{r \cos \phi}\frac{\partial \Phi}{\partial \lambda} = -\frac{Cm}{2r^3}\sin \lambda = 0 \Rightarrow \lambda = 0$$

(92.1)

Substituting in the equation of component X (92.1) the values

$$\lambda = 0$$
$$r = a$$
$$B_0 = \frac{Cm}{a^3}$$
$$N = B_0/4$$

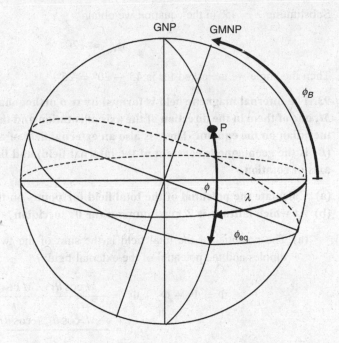

Fig. 92b

we obtain the geographical latitude of the boreal magnetic pole

$$\tan \phi_{BM} = \frac{2\sqrt{3} - 1}{2} \Rightarrow \phi_{BM} = 51°$$

Therefore the geographical coordinates of the magnetic boreal pole are 51° N, 0°. Notice that this is different from the geomagnetic pole.

(c) The declination is given by

$$\tan D = \frac{Y}{X} = \frac{-\dfrac{B_0}{2}\sin \lambda}{B_0\left(\dfrac{\sqrt{3}}{2}\cos \phi - \dfrac{1}{2}\sin \phi \cos \lambda\right) - \dfrac{B_0}{4}\cos \phi}$$

$$= \frac{-\sin \lambda}{\left(\sqrt{3} - \dfrac{1}{2}\right)\cos \phi - \sin \phi \cos \lambda}$$

Substituting the geographical coordinates of the point (45° N, 45° E) we obtain

$$D = -62°$$

(d) We call ϕ_{eq} the angle between the geographical and magnetic equators at longitude $\lambda = 45°$ (Fig. 92b). Then the angle to calculate will be $\phi_{eq} + 45°$.

To calculate ϕ_{eq} we take into account that at the magnetic equator the vertical component is $Z = 0$:

$$Z = -B_r = \frac{\partial \Phi}{\partial r} = \frac{2CM}{r^3}\left(\frac{\sqrt{3}}{2}\sin \phi_{eq} + \frac{1}{2}\cos \phi_{eq}\cos \lambda\right) + N\sin \phi_{eq} = 0$$

$$\tan \phi_{eq} = \frac{-4\cos \lambda}{4\sqrt{3} + 1}$$

Substituting $\lambda = 45°$ in the equation we obtain

$$\phi_{eq} = -20°$$

Then the angle we are asked for is $45° - 20° = 25°$.

93. The internal magnetic field is formed by two orthogonal dipoles of equal moment M, one of them in the direction of the axis of rotation and the other contained in the $0°$ meridian on the equator. There is also an external field of constant intensity $N = B_0/4$ (B_0 is the geomagnetic constant of the internal field) and lines of force parallel to the axis of rotation.

(a) Calculate the potential of the total field for points on the surface.
(b) At which latitude is Z maximum on the $0°$ meridian?

(a) The potential of the total field is the sum of the two potentials of the internal dipoles and the potential of the external field:

$$\Phi = \Phi_1 + \Phi_2 + \Phi^e = \frac{-M \cos \theta_1}{r^2} - \frac{M \cos \theta_2}{r^2} + \Phi^e$$
$$= \frac{-M(\cos \theta_1 + \cos \theta_2)}{r^2} + \Phi^e$$

In this equation r is the distance from the dipole's centre (the Earth's centre), θ_1 and θ_2 are the co-latitudes relative to each dipole, and $M = Cm$.

Dipole 1 is on the direction of the axis of rotation and so the geomagnetic co-latitude of the point with respect to this dipole is

$$\theta_1 = 90° - \phi$$

$$\cos \theta_1 = \sin \phi$$

Dipole 2 is in the equatorial plane ($\phi_{B2} = 0$) and contained in the $0°$ meridian so the geomagnetic co-latitude θ_2 is given by (71.3)

$$\cos \theta_2 = \sin \phi_{B2} \sin \phi + \cos \phi_{B2} \cos \phi \cos(\lambda - \lambda_{B2})$$

$$\phi_{B2} = 0°$$
$$\lambda_{B2} = 0°$$

$$\cos \theta_2 = \cos \phi \cos \lambda$$

The equation for the potential of the external field is the same as that of Problem 92:

$$\Phi^e = +Nr \sin \phi$$

Therefore the potential of the total field is given by

$$\Phi = \frac{-M(\sin \phi + \cos \phi \cos \lambda)}{r^2} + Nr \sin \phi$$

(b) We obtain the component Z by taking the vertical component of the gradient of the potential $(\boldsymbol{B} = -\nabla\Phi)$

$$Z = -B_r = \frac{\partial \Phi}{\partial r} = \frac{2M(\sin\phi + \cos\phi\cos\lambda)}{r^3} + N\sin\phi$$

To calculate the maximum of Z on the $0°$ meridian we substitute $\lambda = 0$ and apply the condition that the first derivate with respect to the latitude is null:

$$\frac{\partial Z}{\partial \phi} = \frac{2M}{r^3}(\cos\phi - \sin\phi) + N\cos\phi = 0$$

At the Earth's surface $r = a$ and we know that

$$B_0 = \frac{M}{a^3}$$

Substituting this constant and solving the equation we determine the latitude at which the Z component is maximum:

$$2B_0(\cos\phi - \sin\phi) + \frac{B_0}{4}\cos\phi = 0 \Rightarrow \phi = 48°$$

94. The internal field has its northern geomagnetic pole at the coordinates $60°$ N, $0°$ E, and $B_0 = 30\,000$ nT. At a point with coordinates $30°$ N, $45°$ W, one observes an increase of $7.7°$ in the value of the declination from $00:00$ h to $09:00$ h. There is known to be an external field parallel to the Earth's axis of rotation in the direction from N to S which is null at $00:00$ h and maximum at $12:00$ h local times. Calculate:

(a) The components of the internal and external fields.
(b) The difference in the inclination at $00:00$ h and $09:00$ h.
(c) The maximum value of the declination during the day.

(a) To calculate the geomagnetic main field intensity components we obtain first the geomagnetic latitude (Equation 71.3):

$$\sin\phi^* = \sin\phi_B\sin\phi + \cos\phi_B\cos\phi\cos(\lambda - \lambda_B)$$
$$\phi^* = 47.7°$$

The declination and inclination are given by (71.2):

$$\sin D^* = \frac{-\cos\phi_B\sin(\lambda - \lambda_B)}{\cos\phi^*} \Rightarrow D^* = 31.7°$$
$$\tan I^* = 2\tan\phi^* \Rightarrow I^* = 65.5°$$

So the geomagnetic main field components are

$$Z^* = 2B_0\sin\phi^* = 44\,378\,\text{nT}$$
$$H^* = B_0\cos\phi^* = 20\,190\,\text{nT}$$
$$X^* = H^*\cos D^* = 17\,178\,\text{nT}$$
$$Y^* = H^*\sin D^* = 10\,609\,\text{nT}$$

The external field is parallel to the axis of rotation so its components are in the vertical and NS directions. This field is null at 00:00 h and maximum at 12:00 h local time (period $T = 24$ h). Its components are given by

$$Z^e = \frac{N}{2}(1 - \cos \omega t) \sin \phi$$

$$X^e = -\frac{N}{2}(1 - \cos \omega t) \cos \phi$$

$$Y^e = 0$$

$$H^e = X^e$$

$$\omega = \frac{2\pi}{T} = \frac{2\pi}{24}$$

(b) To calculate the difference in the inclination we obtain first the value of N, bearing in mind the time variation of the declination. The observed declination as a function of time is given by

$$\tan D = \frac{Y}{X} = \frac{Y^* + Y^e}{X^* + X^e} = \frac{Y^*}{X^* - \frac{N}{2}(1 - \cos \omega t) \cos \phi}$$

For $t = 0$ h:

$$\tan D_1 = \frac{Y^*}{X^*} = \tan D^* \Rightarrow D_1 = D^* = 31.7°$$

Since we know the change in declination between 0 h and 9 h, we find the declination at 9 h, D_2:

$$D_2 - D_1 = 7.7° \Rightarrow D_2 = 39.4°$$

We know that at $t = 9$ h

$$\tan D_2 = \frac{Y^*}{X^* - \frac{N}{2}(1 - \cos \omega t) \cos \phi} = 0.82$$

Solving for N we obtain

$$N = \frac{2}{(1 - \cos \omega t) \cos \phi}\left(X^* - \frac{Y^*}{\tan D_2}\right) = 5766 \, \text{nT}$$

The magnetic inclination is given by

$$\tan I = \frac{Z}{H} = \frac{Z^* + Z^e}{H^* + H^e} = \frac{Z^* + \frac{N}{2}(1 - \cos \omega t) \sin \phi}{H^* - \frac{N}{2}(1 - \cos \omega t) \cos \phi}$$

At $t = 0$ h:

$$I_1 = I^* = 65.5°$$

At $t = 9$ h:

$$I_2 = 71.2°$$

The difference in the inclination is therefore

$$I_2 - I_1 = 5.7°$$

(c) The declination is given by

$$\tan D = \frac{Y^*}{X^* - \dfrac{N}{2}(1 - \cos \omega t)\cos \phi} \qquad (94.1)$$

The maximum value is at 12 h because at that time the external field has the maximum value. Substituting $\omega t = \pi$ and the values obtained for X^*, Y^*, N, and ϕ:

$$D_{max} = 41.0°$$

95. At a point on the Earth with coordinates 45° N, 45° E, measurements are made of the magnetic field components at 00:00 h and 12:00 h in nT with a 2 nT precision:

0 h	$X = 20\ 732$	$Y = 2500$	$Z = 57\ 768$
12 h	$X = 24\ 267$	$Y = 2500$	$Z = 54\ 232$

It is known that the modulus of the magnetic field intensity has a harmonic diurnal variation, and that the geomagnetic pole is on the zero meridian. Calculate:

(a) The moment and coordinates of the main field dipole.

(b) Expressions for the potential and components of the external field.

(Earth's radius $a = 6400$ km, $\mu_0 = 4\pi \times 10^{-7}$ kg m s^{-2} A^{-2}).

(a) To calculate the moment and coordinates of the main field dipole we need to obtain the geomagnetic main field intensity components. The observed values are equal to the sum of the geomagnetic main field (X^*, Y^*, Z^*) and the external field (X^e, Y^e, Z^e):

$$X = X^* + X^e$$
$$Y = Y^* + Y^e$$
$$Z = Z^* + Z^e$$

The geomagnetic main field is constant but the external field changes with time. So if we denote by (X_0, Y_0, Z_0) and (X_{12}, Y_{12}, Z_{12}) the observed values at 0 h and at 12 h respectively then the differences are due to the variations of the external field:

$$X_{12} - X_0 = 3535\,\text{nT} = X_{12}^e - X_0^e$$
$$Y_{12} - Y_0 = 0\,\text{nT} = Y_{12}^e - Y_0^e$$
$$Z_{12} - Z_0 = -3536\,\text{nT} = Z_{12}^e - Z_0^e$$

We notice that the Y component doesn't vary, which implies that the component Y^e is zero, and so the external field is parallel to the axis of rotation. We also notice that the NS component increases in the time interval between 0 h and 12 h, while the vertical component diminishes, which implies that the polarity of the external field is inverted with respect to that of the main field.

The modulus of the magnetic field intensity has a harmonic diurnal variation and increases with time. Therefore the components of the external field are (Fig. 95a)

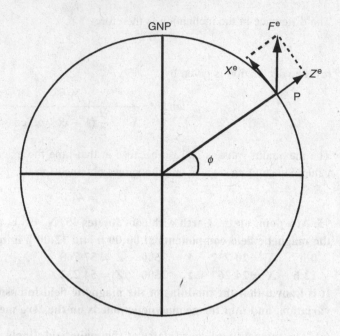

Fig. 95a

$$X^e = N(1 - \cos \omega t) \cos \phi$$
$$Z^e = -N(1 - \cos \omega t) \sin \phi$$
$$\omega = \frac{2\pi}{T} = \frac{2\pi}{24h}$$

We notice that at time $t = 0$ the external field is null and so

$$X^e_{12} = 3535 \, \text{nT}$$
$$Y^e_{12} = 0 \, \text{nT}$$
$$Z^e_{12} = -3536 \, \text{nT}$$

Therefore we calculate the main field components by

$$X^* = X_{12} - X^e_{12} = 20\,732 \, \text{nT}$$
$$Y^* = Y_{12} - Y^e_{12} = 2500 \, \text{nT}$$
$$Z^* = Z_{12} - Z^e_{12} = 57\,768 \, \text{nT}$$

$$H^* = \sqrt{(X^*)^2 + (Y^*)^2} = 20\,882 \, \text{nT}$$

If we consider the centred magnetic dipole model, the vertical and horizontal field components are given by the equations

$$Z^* = 2B_0 \sin \phi^*$$
$$H^* = B_0 \cos \phi^*$$

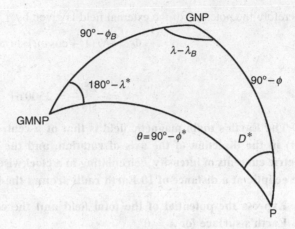

From these equations we obtain the geomagnetic latitude (ϕ^*) and the geomagnetic constant B_0:

$$\tan \phi^* = \frac{Z^*}{2H^*} \Rightarrow \phi^* = 54.1°$$

$$B_0 = \frac{Z^*}{2 \sin \phi^*} = 35\,657\,\text{nT}$$

From B_0 we calculate the magnetic moment:

$$B_0 = \frac{\mu_0}{4\pi} \frac{m}{a^3} \Rightarrow m = B_0 10^7 a^3 = 9.3 \times 10^{22}\,\text{A m}^2$$

The longitude of the Geomagnetic North Pole is $\lambda_B = 0$ and we calculate the latitude ϕ_B from the spherical triangle (Fig. 95b), but obtaining first the declination from the X^* and Y^* components:

$$\tan D^* = \frac{Y^*}{X^*} \Rightarrow D^* = 6.9°$$

Applying the cosine rule:

$$\cos(90 - \phi_B) = \cos(90 - \phi^*) \cos(90 - \phi) + \sin(90 - \phi^*) \sin(90 - \phi) \cos D^*$$
$$\sin \phi_B = \sin \phi^* \sin \phi + \cos \phi^* \cos \phi \cos D^*$$
$$\phi_B = 79.9°$$

(b) We obtain the radial and transverse components of the external field from the vertical and NS components:

$$B_r^e = -Z^e = -N(1 - \cos \omega t) \sin \phi = \frac{-\partial \Phi^e}{\partial r}$$

$$B_\phi^e = X^e = N(1 - \cos \omega t) \cos \phi = \frac{1}{r} \frac{\partial \Phi^e}{\partial \phi}$$

Therefore the potential of the external field is given by

$$\Phi^e = Nr(1 - \cos \omega t) \sin \phi$$

where

$$N = \frac{Z_{12}^e}{2 \sin \phi} = 2500 \, \text{nT}.$$

96. The Earth's main magnetic field is that of a centred dipole of moment M ($M = Cm$) in the direction of the axis of rotation, and the external field is produced by electric currents of intensity J circulating in a clockwise sense in a ring in the plane of the ecliptic at a distance of 10 Earth radii around the Earth.

(a) Express the potential of the total field and the components B_r and B_θ on the Earth's surface for $\lambda = 0°$.

(b) If the inclination of the ecliptic is $30°$, and the external field strength is $N = M / 4R^3$, what is the latitude of the northern magnetic pole?

(a) The potential of the total field is the sum of the potentials of the dipole and of the external field:

$$\Phi^T = \Phi + \Phi^e = \frac{-M \cos \theta}{r^2} + \Phi^e$$

where r is the distance from the dipole's centre (Earth's centre) and θ is the geomagnetic co-latitude. The dipole is in the direction of the axis of rotation and so the geomagnetic co-latitude is equal to the geographic co-latitude:

$$\theta = 90° - \phi$$

$$\cos \theta = \sin \phi$$

To calculate the potential of the external field we know that it is produced by electric currents of intensity J circulating in a clockwise sense at a distance of 10 Earth radii around the Earth. These electric currents produce, at remote points, a magnetic dipolar field whose modulus is $\mu_0 J/2R$, J being the current intensity and R the radius of the circular currents; the dipole is oriented in the direction of the ecliptic's axes with the negative pole in the southern hemisphere, because the currents are clockwise. So the potential of the external field is given by

$$\Phi^e = \frac{-Cm_e \cos \theta_2}{r^2}$$

In this equation $m_e = J\pi 100a^2$, a is the Earth's radius, and θ_2 is the angle between the axes of the circular currents and the direction of the point from the negative pole (Fig. 96) Therefore

$$\theta_2 = 180° - (\theta + \varepsilon)$$

$$\cos \theta_2 = -\cos(\theta + \varepsilon)$$

where ε is the angle between the axes of the circular currents and the axes of rotation of the Earth.

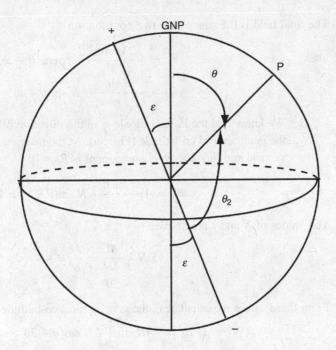

Fig. 96

The potential of the external field is

$$\Phi^e = \frac{Cm_e \cos(\theta + \varepsilon)}{r^2}$$

and the total potential is given by

$$\Phi^T = -\frac{M \cos \theta}{r^2} + \frac{Cm_e \cos(\theta + \varepsilon)}{r^2}$$

We calculate the components of the main field intensity by taking the gradient of the potential:

$$B_r = -\frac{\partial \Phi}{\partial r} = \frac{2M \cos \theta}{r^3}$$

$$B_\theta = -\frac{1}{r}\frac{\partial \Phi}{\partial \theta} = -\frac{M \sin \theta}{r^3}$$

The magnitude of the external field is given by

$$B^e = \frac{\mu_0}{4}\frac{J}{10a} = N$$

The radial and tangential components are

$$B_r^e = N \cos(\theta + \varepsilon)$$
$$B_\theta^e = N \sin(\theta + \varepsilon)$$

The total field is the sum of the two contributions:

$$B_r = -\frac{2M\cos\theta}{r^3} + N\cos(\theta + \varepsilon)$$

$$B_\theta = -\frac{M\sin\theta}{r^3} + N\sin(\theta + \varepsilon)$$

(b) We know that the Earth's dipole is in the direction of the axis of rotation and θ is the geographical co-latitude (Fig. 96). At the magnetic North Pole the total field is vertical and the tangential component is $B_\theta = 0$:

$$B_\theta = -\frac{M\sin\theta}{r^3} + N\sin(\theta + \varepsilon) = 0$$

The values of N and ε are known:

$$N = \frac{M}{4r^3} \Rightarrow \frac{M}{r^3} = 4N$$

$$\varepsilon = 30°$$

From these values we calculate θ, the geographical co-latitude of the magnetic North Pole,

$$B_\theta = -4N\sin\theta + N\sin(\theta + 30°) = 0$$

$$-4\sin\theta + \frac{\sqrt{3}}{2}\sin\theta + \frac{1}{2}\cos\theta = 0 \Rightarrow \theta = 9°$$

97. Two spherical planets of radius a and separated by a centre-to-centre distance of $4a$ orbit around each other and spin in the equatorial plane. Each has a magnetic field produced by a centred dipole in the direction of the axis of rotation, with the positive pole in the northern hemisphere and $B_0 = 10\,000$ nT. Determine the components X, Y, Z, D, and I of the total magnetic field at the North Pole of one of the planets (precision 1 nT).

The total magnetic field in each planet is the sum of its main field and the external field created by the other planet. To determine the main field of either of the planets we need the geomagnetic latitude, which is positive toward the negative pole, in this case, the South Pole. Therefore the geomagnetic latitude and the components of the main field at the North Pole are

$$\phi^* = -90°$$
$$Z^* = 2B_0\sin\phi^* = -20\,000\,\text{nT}$$
$$H^* = B_0\cos\phi^* = 0$$
$$X^* = Y^* = 0$$

The external field at one of the planets is created by the main field of the other planet and corresponds to that of a magnetic dipole. Its components are (Fig. 97)

$$B_r^{\text{e}} = \frac{2Cm\cos\theta}{r^3}$$

$$B_\theta^{\text{e}} = \frac{Cm\sin\theta}{r^3}$$

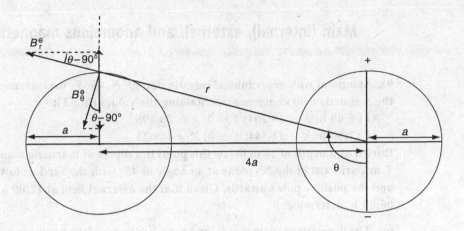

Fig. 97

At the North Pole $r = \sqrt{17}a = 4.12a$ and we calculate the geomagnetic co-latitude θ by (Fig. 97)

$$\tan(\theta - 90°) = \frac{a}{4a} = \frac{1}{4}$$
$$\theta = 14° + 90° = 104°$$

Substituting these values in the equations for the radial and tangential components with respect to the planet producing the external field:

$$B_r^e = \frac{2Cm\cos\theta}{r^3} = \frac{Cm\cos\theta}{35a^3} = \frac{B_0\cos\theta}{35} = -69\,\text{nT}$$
$$B_\theta^e = \frac{Cm\sin\theta}{r^3} = \frac{Cm\sin\theta}{70a^3} = \frac{B_0\sin\theta}{70} = 139\,\text{nT}$$

From this value we calculate the vertical and horizontal components (Fig. 97):

$$Z^e = B_r^e\cos(180° - \theta) + B_\theta^e\cos(\theta - 90°) = -B_r^e\cos\theta + B_\theta^e\sin\theta = 118\,\text{nT}$$
$$H^e = B_r^e\sin(180° - \theta) + B_\theta^e\sin(\theta - 90°) = -B_r^e\sin\theta - B_\theta^e\cos\theta = -33\,\text{nT}$$
$$X^e = H^e = -33\,\text{nT}$$
$$Y^e = 0$$

The components of the total field are finally

$$Z = Z^* + Z^e = -19882\,\text{nT}$$
$$H = H^* + H^e = -33\,\text{nT}$$
$$X = X^* + X^e = -33\,\text{nT}$$
$$Y = Y^* + Y^e = 0$$
$$\tan D = \frac{Y}{X} \Rightarrow D = 0°$$
$$\tan I = \frac{Z}{H} \Rightarrow I = 89.9°$$

Main (internal), external, and anomalous magnetic fields

98. At a point with geographical coordinates 40° N, 45° E, measurements are made of the magnetic field components, obtaining the values (in nT):

At 06:00 h: $X = 19\,204$; $Y = 0$; $Z = 38\,195$

At 12:00 h: $X = 11\,544$; $Y = 0$; $Z = 44\,623$

Buried at a depth of 20 m below this point is a dipole of magnetic moment $Cm = 0.01$ T m^3, oriented in the NS plane at an angle of 45° with the vertical towards the south, and the positive pole upwards. Given that the external field at 12:00 h is twice that at 06:00 h, determine:

(a) The geomagnetic constant B_0 and the coordinates of the northern geomagnetic pole.

(b) The magnitude and direction of the external field. How does the magnitude of the external field vary with time?

(a) The observed magnetic field is composed of three parts: the geomagnetic main (internal) field, the anomalous field (magnetic anomaly) created by the buried dipole, and the external field.

We determine first the magnetic anomaly produced by the buried dipole, applying Equations (82.2) and (82.3) in Problem 82, substituting $\alpha = 225°$ and $x = 0$. The horizontal component is in the NS direction (ΔX), because the buried dipole is in the NS-vertical plane (Fig. 98a):

$$\Delta Z = \frac{2Cm\cos\alpha}{d^3} = -1768\,\text{nT}$$

$$\Delta X = \frac{-Cm\sin\alpha}{d^3} = 884\,\text{nT}$$

$$\Delta Y = 0$$

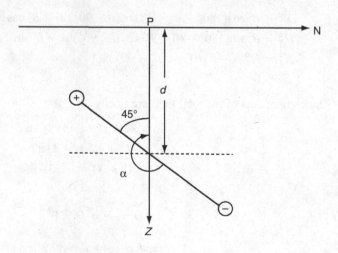

Fig. 98a

The components of the total magnetic field at 06:00 h are given by

$$X_1 = X^* + \Delta X + X^e = 19\,204\,\text{nT}$$
$$Y_1 = Y^* + \Delta Y + Y^e = 0$$
$$Z_1 = Z^* + \Delta Z + Z^e = 38\,195\,\text{nT}$$

At 12:00 h given that the external field is twice that at 06:00 h:

$$X_2 = X^* + \Delta X + 2X^e = 11\,544\,\text{nT}$$
$$Y_2 = Y^* + \Delta Y + 2Y^e = 0$$
$$Z_2 = Z^* + \Delta Z + 2Z^e = 44\,623\,\text{nT}$$

If we subtract both sets of equations and obtain

$$X_2 - X_1 = X^e = -7660\,\text{nT}$$
$$Y_2 - Y_1 = Y^e = 0$$
$$Z_2 - Z_1 = Z^e = 6428\,\text{nT}$$

Now we can calculate the elements of the main field

$$X^* = X_1 - \Delta X - X^e = 25\,980\,\text{nT}$$

$$Y^* = Y_1 - \Delta Y - Y^e = 0$$

$$Z^* = Z_1 - \Delta Z - Z^e = 33\,535\,\text{nT}$$

$$H^* = \sqrt{(X^*)^2 + (Y^*)^2} = X^* = 25\,980\,\text{nT}$$

$$\tan D^* = \frac{Y^*}{H^*} \Rightarrow D^* = 0°$$

We calculate the geomagnetic latitude of the point ϕ^* and the geomagnetic constant B_0 from the vertical and horizontal geomagnetic main field components by

$$Z^* = 2B_0 \sin \phi^*$$
$$H^* = B_0 \cos \phi^*$$

$$\tan \phi^* = \frac{Z^*}{2H^*} \Rightarrow \phi^* = 32.8°$$

$$B_0 = \frac{Z^*}{2 \sin \phi^*} = 30\,953\,\text{nT}$$

We obtain the coordinates of the Geomagnetic North Pole by (Fig. 98b)

$$D^* = 0 \Rightarrow \lambda_B = 180 + \lambda = 225° \text{ E} = 135° \text{ W}$$
$$90° - \phi_B = \phi - \phi^* \Rightarrow \phi_B = 82.8°$$

(b) The magnitude of the external field at 06:00 h is

$$B_e^6 = \sqrt{X_e^2 + Y_e^2 + Z_e^2} = 10\,000\,\text{nT}$$

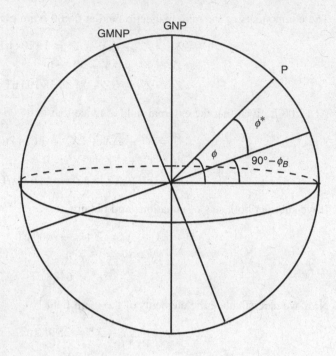

At 12:00 h the magnitude is

$$B_e^{12} = 2B_e^6 = 20\,000\,\text{nT}$$

The direction of the external field is in the NS-vertical plane because the EW component is null, forming with the horizontal an angle I_e (Fig. 98c). This direction is the same at 06:00 h and at 12:00 h. We calculate the angle I_e by

$$\tan I_e = \frac{Z_e}{-X_e}$$

$$I_e = 40°$$

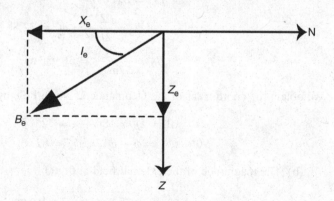

Because at

$$06{:}00\,\text{h}\ (t = \pi/2) \rightarrow N = 10\,000$$

$$12{:}00\,\text{h}(t = \pi) \rightarrow N = 20\,000$$

the variation of the magnitude of the external field with time is given by

$$N = 10\,000(1 - \cos t)$$

99. Buried at a depth of 100 m at a point with geographical coordinates 45° N, 45° W is a dipole anomaly of $Cm = 0.1\ \text{T m}^3$, inclined 45° from the horizontal northwards in the vertical plane with the negative pole downwards. Measurements gave the following results (in nT):

    09:00 h $X = 27\,759$; $Y = 0$; $Z = 30\,141$
    12:00 h $X = 28\,052$; $Y = 0$; $Z = 30\,141$

Find:

(a) The coordinates of the magnetic dipole's North Pole.
(b) The value of B_0.
(c) An expression for the variation S_q knowing that it is zero at 00:00 h and maximum at 12:00 h.

(a) As in Problem 98, the observed field is the result of three parts: the main (internal) field, the buried dipole field, and the external field. To calculate the coordinates of the magnetic dipole's North Pole we need to obtain the components of the geomagnetic main field from the components of the total field. With this aim we begin by calculating the magnetic anomaly created by the buried dipole, applying Equations (82.2) and (82.3), and substituting $\alpha = 225°$ and $x = 0$. The horizontal component is in the NS direction (ΔX) given that the dipole is on the NS-vertical plane (Fig. 99a).

$$\Delta Z = \frac{-2Cm\cos\alpha}{d^3} = 141\,\text{nT}$$

$$\Delta X = \frac{-Cm\sin\alpha}{d^3} = -71\,\text{nT}$$

$$\Delta Y = 0$$

The total observed field at 09:00 h is

$$X_1 = X^* + \Delta X + X_1^e$$
$$Y_1 = Y^* + \Delta Y + Y_1^e = 0 \Rightarrow Y^* = -Y_1^e$$
$$Z_1 = Z^* + \Delta Z + Z_1^e$$

The total field at 12:00 h is

$$X_2 = X^* + \Delta X + X_2^e$$
$$Y_2 = Y^* + \Delta Y + Y_2^e$$
$$Z_2 = Z^* + \Delta Z + Z_2^e$$

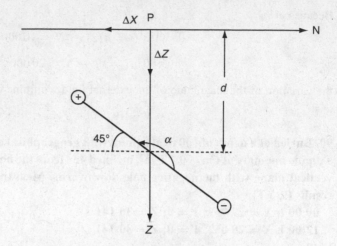

Subtracting both sets of equations we obtain

$$X_2 - X_1 = X_2^e - X_1^e = 293 \, \text{nT}$$
$$Y_2 - Y_1 = Y_2^e - Y_1^e = 0 \Rightarrow Y_2^e = Y_1^e = -Y^* \qquad (99.1)$$
$$Z_2 - Z_1 = Z_2^e - Z_1^e = 0 \Rightarrow Z_2^e = Z_1^e$$

We assume that the time variation of the observations is due to the diurnal S_q variation which is zero at 00:00 h and maximum at 12:00 h. Therefore the only possible values for the components Y^e and Z^e are zero because these components have the same values at 09:00 h and at 12:00 h:

$$Y_2^e = Y_1^e = -Y^* = 0$$
$$Z_2^e = Z_1^e = 0$$

Then, the intensity of the external field is given by

$$X^e = N(1 - \cos \omega t)$$

$$\omega = \frac{2\pi}{24}$$

The NS components of this field at 09:00 h (X_1) and at 12:00 h (X_2) are

$$X_1^e = N\left(1 - \cos\frac{3\pi}{4}\right) = N\left(1 + \frac{\sqrt{2}}{2}\right)$$
$$X_2^e = N(1 - \cos \pi) = 2N$$

Subtracting the two values and using Equation (99.1) we obtain

$$X_2^e - X_1^e = \left(1 - \frac{\sqrt{2}}{2}\right)N = \left(1 - \frac{\sqrt{2}}{2}\right)\frac{X_2^e}{2} = 293 \, \text{nT} \Rightarrow X_2^e = 2001 \, \text{nT}$$
$$X_1^e = 1708 \, \text{nT}$$

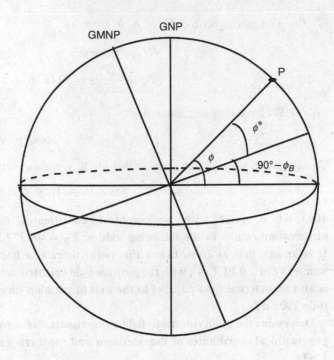

Fig. 99b

The components of the geomagnetic main field intensity are given by

$$X^* = X_1 - \Delta X - X_1^e = 26\,122\,\text{nT}$$

$$Y^* = Y_1 - \Delta Y - Y_1^e = 0$$

$$Z^* = Z_1 - \Delta Z - Z^e = 30\,000\,\text{nT}$$

$$H^* = \sqrt{(X^*)^2 + (Y^*)^2} = X^* = 26\,122\,\text{nT}$$

$$\tan D^* = \frac{Y^*}{H^*} \Rightarrow D^* = 0°$$

The geomagnetic latitude of the point ϕ^* is determined from the vertical and horizontal geomagnetic main field components:

$$Z^* = 2B_0 \sin \phi^*$$

$$H^* = B_0 \cos \phi^*$$

$$\tan \phi^* = \frac{Z^*}{2H^*} \Rightarrow \phi^* = 29.9°$$

We obtain the coordinates of the Geomagnetic North Pole by (Fig. 99b)

$$D^* = 0 \Rightarrow \lambda_B = 180 + \lambda = 135° \,\text{E}$$

$$90° - \phi_B = \phi - \phi^* \Rightarrow \phi_B = 74.9°$$

(b) The geomagnetic constant B_0 is given by

$$H^* = B_0 \cos \phi^*$$

$$B_0 = \frac{H^*}{\cos \phi^*} = 30\,133\,\text{nT}$$

(c) We have obtained above that

$$F^e = X^e = N(1 - \cos \omega t)$$

To calculate N we take into account that the S_q variation is maximum at 12:00 h:

$$X_2^e = 2N \Rightarrow N = 1000\,\text{nT}$$

100. At a point with geographical coordinates 60° N, 45° E, a magnetic observation results in the following values: $F_T = 48\,277$ nT, $D_T = 2.9°$, $I_T = 63.7°$. It is known that at 20 m below this point there is a horizontal magnetic dipole of moment $Cm = 0.01$ T m³, with the positive pole oriented in the direction N 60° E. There is also an external field parallel to the axis of rotation directed southwards of magnitude 1000 nT.

Determine the main (internal) field components, the constant B_0, and the geocentric geographical coordinates of the northern and southern geomagnetic poles (precision 1 nT).

We first calculate the components of the total field intensity from F_T, D_T, and I_T (Fig. 100a) by

$$H_T = F_T \cos I_T = 21\,390\,\text{nT}$$

$$Z_T = F_T \sin I_T = 43\,280\,\text{nT}$$

$$X_T = H_T \cos D_T = 21\,363\,\text{nT}$$

$$Y_T = H_T \sin D_T = 1082\,\text{nT}$$

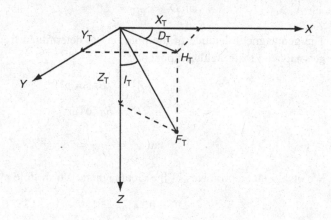

Fig. 100a

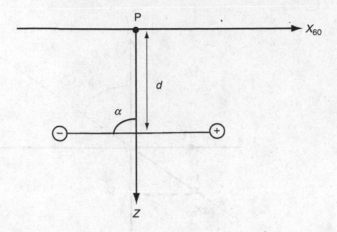

The observed field is the result of three parts: the main field (X^*, Y^*, Z^*), the buried dipole field (ΔX, ΔY, ΔZ), and the external field (X^e, Y^e, Z^e):

$$X_T = X^* + \Delta X + X^e$$
$$Y_T = Y^* + \Delta Y + Y^e$$
$$Z_T = Z^* + \Delta Z + Z^e$$

We determine the magnetic anomaly created by the buried dipole from Equations (82.2) and (82.3) substituting $\alpha = 90°$ and $x = 0$. If we call X_{60} the direction N 60° E the horizontal component of this anomaly is ΔX_{60} (Fig. 100b):

$$\Delta Z = \frac{2Cm \cos \alpha}{d^3} = 0 \, \text{nT}$$

$$\Delta X_{60} = \frac{-Cm \sin \alpha}{d^3} = -1250 \, \text{nT}$$

The NS and EW components will be given by (Fig. 100c)

$$\Delta X = \Delta X_{60} \cos 60 = -625 \, \text{nT}$$
$$\Delta Y = \Delta X_{60} \sin 60 = -1083 \, \text{nT}$$

The external field is parallel to the Earth's axis of rotation in the southwards direction so its components are in the vertical and NS direction and are given by (Fig. 100d)

$$Z^e = 1000 \sin \phi = 866 \, \text{nT}$$
$$X^e = -1000 \cos \phi = 500 \, \text{nT}$$

We calculate the main field components from these values:

$$X^* = X_T - \Delta X - X^e = 21\,238 \, \text{nT}$$
$$Y^* = Y_T - \Delta Y - Y^e = 1082 \, \text{nT}$$
$$Z^* = Z_T - \Delta Z - Z^e = 42\,414 \, \text{nT}$$

$$H^* = \sqrt{(X^*)^2 + (Y^*)^2} = 21\,266 \, \text{nT}$$

$$\tan D^* = \frac{Y^*}{X^*} \Rightarrow D^* = 2.9°$$

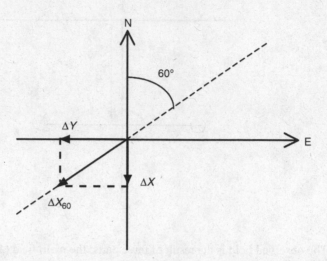

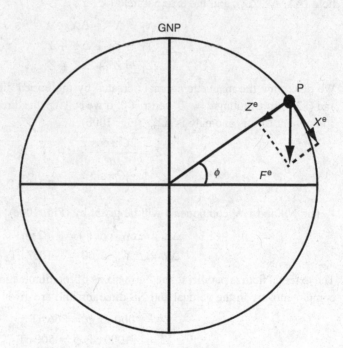

The geomagnetic latitude of the point ϕ^* and the geomagnetic constant B_0 are found from the vertical and horizontal geomagnetic main field components by

$$Z^* = 2B_0 \sin \phi^*$$
$$H^* = B_0 \cos \phi^*$$
$$\tan \phi^* = \frac{Z^*}{2H^*} \Rightarrow \phi^* = 44.9°$$
$$B_0 = \frac{Z^*}{2 \sin \phi^*} = 30\,953\,\text{nT} = 30\,044\,\text{nT}$$

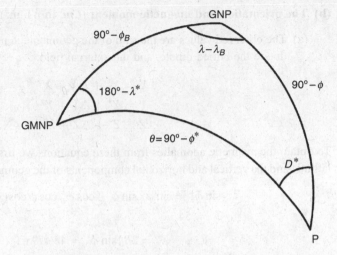

Fig. 100e

To calculate the geographical coordinates of the Geomagnetic North Pole we use the corresponding spherical triangle (Fig. 100e). We obtain the latitude ϕ_B by applying the cosine law for the angle $90° - \phi_B$:

$$\cos(90 - \phi_B) = \cos(90 - \phi^*)\cos(90 - \phi) + \sin(90 - \phi^*)\sin(90 - \phi)\cos D^*$$
$$\sin \phi_B = \sin \phi^* \sin \phi + \cos \phi^* \cos \phi \cos D^*$$
$$\phi_B = 75.0°$$

We obtain the longitude λ_B by applying the cosine rule for the angle $90° - \phi^*$:

$$\cos(90 - \phi^*) = \cos(90 - \phi_B)\cos(90 - \phi) + \sin(90 - \phi_B)\sin(90 - \phi)\cos(\lambda - \lambda_B)$$
$$\sin \phi^* = \sin \phi_B \sin \phi + \cos \phi_B \cos \phi \cos(\lambda - \lambda_B)$$

$$\cos(\lambda - \lambda_B) = \frac{\sin \phi^* - \sin \phi_B \sin \phi}{\cos \phi_B \cos \phi}$$

$$\lambda - \lambda_B = \pm 180.0°$$

$$\lambda_B = 135.0° \, \text{W}$$

Therefore the coordinates of the Geomagnetic North Pole are

$$\phi_\mathrm{B} = 75° \, \text{N} \qquad \lambda_\mathrm{B} = 135° \, \text{W}$$

The coordinates of the Geomagnetic South Pole (the antipodal point) are:

$$\phi_\mathrm{A} = -\phi_B = 75° \, \text{S} \qquad \lambda_\mathrm{A} = 180° + \lambda_B = 45° \, \text{E}$$

101. At a point with geographical coordinates 30° N, 30° E, the observed geomagnetic field components are (in nT): $X = 15\,364$, $Y = -7660$, $Z = 48\,980$. The northern geomagnetic pole is at 60° N, 0° E, and $B_0 = 30\,000$ nT. There is also a constant external magnetic field normal to the equatorial plane, with a southwards direction, of 1000 nT intensity. Buried 10 m below the observation point is a magnetic dipole. Calculate:

(a) The magnetic anomalies ΔX, ΔY, ΔZ, ΔH, ΔF.

(b) The orientation and magnetic moment (Cm, in nT m^3) of the buried dipole.

(a) The observed values are the sum of the geomagnetic main field, the magnetic field due to the buried dipole, and the external field:

$$X = X^* + \Delta X + X^e$$
$$Y = Y^* + \Delta Y + Y^e$$
$$Z = Z^* + \Delta Z + Z^e$$

To obtain the magnetic anomalies from these equations we first calculate the geomagnetic latitude and the vertical and horizontal components of the geomagnetic main field by (71.3):

$$\sin \phi^* = \sin \phi_B \sin \phi + \cos \phi_B \cos \phi \cos(\lambda - \lambda_B)$$
$$\phi^* = 54°$$

$$Z^* = 2B_0 \sin \phi^* = 48\,479\,\text{nT}$$
$$H^* = B_0 \cos \phi^* = 17\,676\,\text{nT}$$

The declination and inclination are given by

$$\sin D^* = \frac{-\cos \phi_B \sin(\lambda - \lambda_B)}{\cos \phi^*} \Rightarrow D^* = -25°$$
$$\tan I^* = 2 \tan \phi^* \Rightarrow I^* = 70°$$

The NS and EW components are

$$X^* = H^* \cos D^* = 16\,007\,\text{nT}$$
$$Y^* = H^* \sin D^* = -7498\,\text{nT}$$

The external field is parallel to the axis of rotation directed southwards so its components are in the vertical and NS direction and are given by (Fig. 101a)

$$Z^e = 1000 \sin \phi = 500\,\text{nT}$$
$$X^e = -1000 \cos \phi = -866\,\text{nT}$$
$$Y^e = 0$$

From these values we calculate the magnetic anomalies ΔX, ΔY, ΔZ, ΔH, ΔF:

$$\Delta X = X - X^* - X^e = 223\,\text{nT}$$
$$\Delta Y = Y - Y^* - Y^e = -162\,\text{nT}$$
$$\Delta Z = Z - Z^* - Z^e = 1\,\text{nT}$$
$$\Delta H = \Delta X \cos D^* + \Delta Y \sin D^* = 271\,\text{nT}$$
$$\Delta F = \Delta H \cos I^* + \Delta Z \sin I^* = 94\,\text{nT}$$

(b) We call β the angle between the geographical north and the buried dipole directions. Then using Equations (82.2) and (82.3) we obtain

$$\Delta Z = \frac{2Cm \cos \alpha}{d^3} \tag{101.1}$$

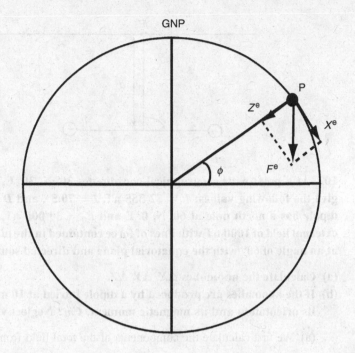

$$\Delta X = \frac{-Cm \sin \alpha}{d^3} \cos \beta \qquad (101.2)$$

$$\Delta Y = \frac{-Cm \sin \alpha}{d^3} \sin \beta \qquad (101.3)$$

To solve this system of three equations in three unknowns (Cm, α, β) we divide Equation (101.3) by (101.2):

$$\tan \beta = \frac{\Delta Y}{\Delta X} \Rightarrow \beta = -36° + 180° = 144°$$

This value of the angle β implies that the dipole is oriented in the N 144° E direction.

To calculate the angle α between the buried dipole and the vertical we divide Equation (101.2) by (101.1):

$$\frac{\Delta X}{\Delta Z} = -\frac{1}{2} \tan \alpha \cos \beta$$

$$\tan \alpha = -\frac{2}{\cos \beta} \frac{\Delta X}{\Delta Z} \Rightarrow \alpha \approx 90°$$

Therefore the dipole is practically horizontal (Fig. 101b) and this explains the small value of the vertical component ΔZ.

Finally we calculate the magnetic moment of the buried dipole from Equation (101.2):

$$Cm = -\frac{d^3}{\sin \alpha \cos \beta} \Delta X = 2.8 \times 10^5 \, \text{nT m}^3$$

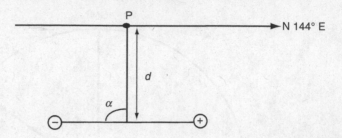

102. At a point with geographical coordinates 30° N, 30° E, magnetic measurements give the following values: $F = 52\,355$ nT, $I = 70.5°$, and $D = -26.0°$. The terrestrial dipole has a north pole at 60° N, 0° E and $B_0 = 30\,000$ nT. There is also a constant external field of 1000 nT with lines of force contained in the plane of the 30° E meridian at an angle of 60° with the equatorial plane and directed southwards.

(a) Calculate the anomalies ΔX, ΔY, ΔZ.

(b) If the anomalies are produced by a dipole buried at 10 m below the point, what is its orientation and its magnetic moment, Cm? Neglect values less than 10 nT.

(a) We first calculate the components of the total field from F, D, and I by

$$H = F \cos I = 17\,476\,\text{nT}$$
$$Z = F \sin I = 49\,352\,\text{nT}$$
$$X = H \cos D = 15\,707\,\text{nT}$$
$$Y = H \sin D = -7661\,\text{nT}$$

The magnetic anomalies are found by subtracting from the observed values the main and external field contributions:

$$\Delta X = X - X^* - X^{\text{e}}$$
$$\Delta Y = Y - Y^* - Y^{\text{e}}$$
$$\Delta Z = Z - Z^* - Z^{\text{e}}$$

First we determine the geomagnetic latitude and declination using (71.3) and (71.2):

$$\sin \phi^* = \sin \phi_B \sin \phi + \cos \phi_B \cos \phi \cos(\lambda - \lambda_B)$$
$$\phi^* = 53.9°$$

$$\sin D^* = \frac{-\cos \phi_B \sin(\lambda - \lambda_B)}{\cos \phi^*} \Rightarrow D^* = -25.1°$$

The vertical and horizontal components are given by

$$Z^* = 2B_0 \sin \phi^* = 48\,479\,\text{nT}$$
$$H^* = B_0 \cos \phi^* = 17\,676\,\text{nT}$$
$$X^* = H^* \cos D^* = 16\,007\,\text{nT}$$
$$Y^* = H^* \sin D^* = -7498\,\text{nT}$$

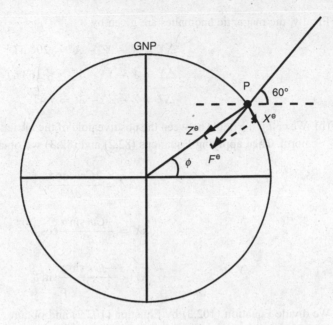

Fig. 102a

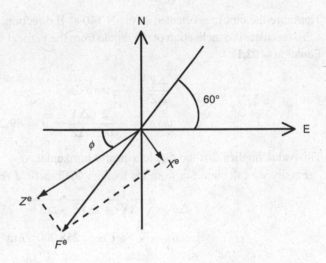

Fig. 102b

The external field is on the plane containing the vertical and NS directions (Figs 102a and 102b):

$$Z^e = 1000\cos(60° - \phi) = 866\,\text{nT}$$
$$X^e = -1000\sin(60° - \phi) = -500\,\text{nT}$$
$$Y^e = 0$$

Finally, the magnetic anomalies are given by

$$\Delta X = X - X^* - X^e = 200\,\text{nT}$$

$$\Delta Y = Y - Y^* - Y^e = -163\,\text{nT}$$

$$\Delta Z = Z - Z^* - Z^e = 7\,\text{nT}$$

(b) We call β the angle between the positive pole of the buried dipole and the geographical north. Then applying Equations (82.2) and (82.3) we obtain

$$\Delta Z = \frac{2Cm\cos\alpha}{d^3} \qquad\qquad (102.1)$$

$$\Delta X = \frac{-Cm\sin\alpha}{d^3}\cos\beta \qquad\qquad (102.2)$$

$$\Delta Y = \frac{-Cm\sin\alpha}{d^3}\sin\beta \qquad\qquad (102.3)$$

We divide Equation (102.3) by Equation (102.2) and obtain

$$\tan\beta = \frac{\Delta Y}{\Delta X} \Rightarrow \beta = -39.2° + 180° = 140.8°$$

Therefore the dipole is oriented in the N 140.8° E direction.

To calculate the inclination of the dipole from the vertical we divide Equation (102.2) by Equation (102.1):

$$\frac{\Delta X}{\Delta Z} = -\frac{1}{2}\tan\alpha\cos\beta$$

$$\tan\alpha = -\frac{2}{\cos\beta}\frac{\Delta X}{\Delta Z} \Rightarrow \alpha = 89.2°$$

This value implies that the dipole is nearly horizontal.

Finally we calculate Cm from the total anomalous field ΔB

$$\Delta B = \sqrt{\Delta X^2 + \Delta Y^2 + \Delta Z^2} = 258\,\text{nT}$$

$$\Delta B = \frac{Cm}{d^3} \Rightarrow Cm = 258\,000\,\text{nT m}^3$$

103. The Earth's magnetic field is formed by a centred dipole with a geomagnetic pole at 60° N, 0° E, and $B_0 = 30\,000$ nT, and an external field of 10 000 nT parallel to the equatorial plane and to the zero meridian.

(a) Calculate the components X, Y, Z of the total field at a point P with geographical coordinates 60° N, 30° W.

(b) If at 30 m in the direction of the compass needle from P there is a vertical dipole of moment $Cm = 4000$ nT m^3 buried 40 m deep, what would be the anomaly ΔZ produced at P?

(a) The components of the total field are the sum of the geomagnetic main field and the external field:

$$X = X^* + X^e$$
$$Y = Y^* + Y^e$$
$$Z = Z^* + Z^e$$

Let us first calculate the geomagnetic latitude, declination, and inclination using (71.3) and (71.2):

$$\sin \phi^* = \sin \phi_B \sin \phi + \cos \phi_B \cos \phi \cos(\lambda - \lambda_B)$$
$$\phi^* = 75.1°$$

$$\sin D^* = \frac{- \cos \phi_B \sin(\lambda - \lambda_B)}{\cos \phi^*} \Rightarrow D^* = 76.5°$$
$$\tan I^* = 2 \tan \phi^* \Rightarrow I^* = 82.4°$$

The vertical and horizontal components are given by

$$Z^* = 2B_0 \sin \phi^* = 57\,983\,\text{nT}$$
$$H^* = B_0 \cos \phi^* = 7714\,\text{nT}$$
$$X^* = H^* \cos D^* = 1801\,\text{nT}$$
$$Y^* = H^* \sin D^* = 7501\,\text{nT}$$

The external field is parallel to the equatorial plane and to the zero meridian. If its magnitude is $N = 10\,000$ nT, its components, from Fig. 103a (plane through the point parallel to the equator) and Fig. 103b (plane through the geographical meridian of the point), are given by

$$Z^e = -B_r^e = N \cos \lambda \cos \varphi = 4330\,\text{nT}$$
$$X^e = B_\phi^e = N \cos \lambda \sin \varphi = 7500\,\text{nT}$$
$$Y^e = B_\lambda^e = N \sin \lambda = 5000\,\text{nT}$$

Therefore the components of the total field are

$$Z = 62\,313\,\text{nT}$$
$$X = 9301\,\text{nT}$$
$$Y = 12\,501\,\text{nT}$$

(b) The buried vertical dipole is in the direction of the compass, that is, in the direction of the magnetic north (Fig. 103c). We calculate the anomaly ΔZ produced at P using Equations (82.2) and (82.3) substituting $\alpha = 0°$ and $x = 30$:

$$\Delta Z = \frac{-Cm[(x^2 - 2d^2) \cos \alpha + 3dx \sin \alpha]}{[x^2 + d^2]^{5/2}}$$
$$\Delta Z = \frac{-Cm(x^2 - 2d^2)}{[x^2 + d^2]^{5/2}} = 0.029\,\text{nT}$$

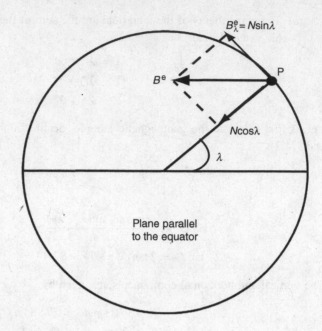

Fig. 103a

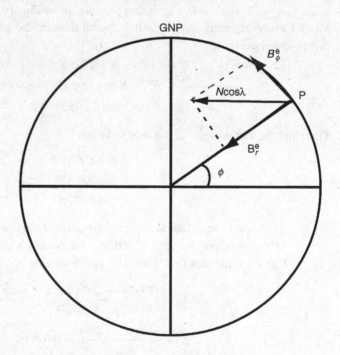

Fig. 103b

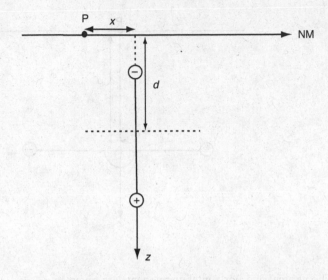

104. The Geomagnetic North Pole is at 60° N, 150° W, with $B_0 = 30\ 000$ nT, and there is an external magnetic field of intensity 3000 nT parallel to the axis of rotation pointing away from the North Pole. Buried 10 m below a point with coordinates 30° N, 30° E there is a horizontal dipole with $Cm = 40\ 000$ nT m^3 and the negative pole pointing in the direction N 45° E.

(a) What are the components X, Y, Z of the total field?
(b) Calculate the total field anomaly ΔF.
(c) What is the angle between the direction of the compass and geographic north?

(a) The components of the total field are the sum of the geomagnetic main field, the magnetic field due to the buried dipole, and the external field:

$$X = X^* + \Delta X + X^{\mathrm{e}}$$
$$Y = Y^* + \Delta Y + Y^{\mathrm{e}}$$
$$Z = Z^* + \Delta Z + Z^{\mathrm{e}}$$

We determine first the geomagnetic latitude, declination, and inclination using (71.3) and (71.2):

$$\sin \phi^* = \sin \phi_B \sin \phi + \cos \phi_B \cos \phi \cos(\lambda - \lambda_B)$$
$$\phi^* = 0°$$

$$\sin D^* = \frac{-\cos \phi_B \sin(\lambda - \lambda_B)}{\cos \phi^*} \Rightarrow D^* = 0°$$

$$\tan I^* = 2 \tan \phi^* \Rightarrow I^* = 0°$$

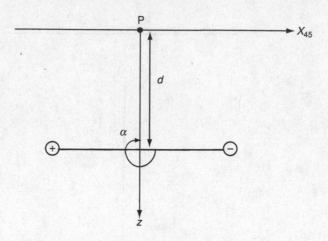

Fig. 104

The vertical and horizontal components are given by

$$Z^* = 2B_0 \sin \phi^* = 0\,\mathrm{nT}$$
$$H^* = B_0 \cos \phi^* = 30\,000\,\mathrm{nT}$$
$$X^* = H^* \cos D^* = 30\,000\,\mathrm{nT}$$
$$Y^* = H^* \sin D^* = 0\,\mathrm{nT}$$

We calculate the magnetic anomaly created by the buried dipole from Equations (82.2) and (82.3), substituting $\alpha = 270°$ and $x = 0$. The vertical component ΔZ and the horizontal component ΔX_{45} of the anomaly in the direction N 45° E are (Fig. 104)

$$\Delta Z = \frac{2Cm \cos \alpha}{d^3} = 0\,\mathrm{nT}$$

$$\Delta X_{45} = \frac{-Cm \sin \alpha}{d^3} = 40\,\mathrm{nT}$$

The NS and EW components are given by

$$\Delta X = 40 \sin 45° = 28.3\,\mathrm{nT}$$
$$\Delta Y = 40 \cos 45° = 28.3\,\mathrm{nT}$$

The external field is parallel to the axis of rotation directed southwards, so its components are in the vertical and NS direction and are given by

$$N_r = Z^e = N \cos 60° = 1500\,\mathrm{nT}$$
$$N_\vartheta = -X^e = N \sin 60° = -2598\,\mathrm{nT}$$

The components of the observed total magnetic field are

$$X = 27\,430\,\mathrm{nT}$$
$$Y = 28\,\mathrm{nT}$$
$$Z = 1500\,\mathrm{nT}$$

(b) The total field anomaly ΔF is given by

$$\Delta F = \Delta X = 28\,\text{nT}$$

(c) The direction of the compass is affected by the three fields. The angle D between the direction of the compass and the geographic north is obtained from the horizontal components of the total observed field

$$\tan D = \frac{Y}{X} = \frac{28}{27430} \Rightarrow D = 0.06°$$

105. The geomagnetic field is that of a dipole in the direction of the axis of rotation and $B_0 = 30\,000$ nT. There is also a constant external field of 2500 nT normal to the equatorial plane in the direction of the South Pole.

(a) **Calculate the value of the inclination observed at a point P with coordinates 45° N, 45° E given that, at 10 m below it, there is a vertical dipole with the negative pole upwards and moment $Cm = 40\,000$ nT m^3.**

(b) **For a point 20 m north of P, calculate the observed inclination and declination, and the total field anomaly ΔF.**

(a) The components of the total observed field are the sum of the geomagnetic main field, the magnetic field due to the buried dipole, and the external field:

$$X = X^* + \Delta X + X^e$$
$$Y = Y^* + \Delta Y + Y^e$$
$$Z = Z^* + \Delta Z + Z^e$$

The magnetic dipole is oriented in the direction of the axis of rotation and therefore

$$\phi^* = \phi = 45°$$
$$D^* = 0°$$

Then the components of the geomagnetic main field are

$$Z^* = 2B_0 \sin \phi^* = 42\,426\,\text{nT}$$

$$H^* = B_0 \cos \phi^* = 21\,213\,\text{nT}$$

$$X^* = H^* \cos D^* = 21\,213\,\text{nT}$$

$$Y^* = H^* \sin D^* = 0$$

The magnetic anomaly created by the dipole is obtained from Equations (82.2) and (82.3) substituting $\alpha = 0°$ and $x = 0$ (Fig. 105a):

$$\Delta Z = \frac{2Cm \cos \alpha}{d^3} = 80\,\text{nT}$$

$$\Delta X = \frac{-Cm \sin \alpha}{d^3} = 0$$

$$\Delta Y = 0$$

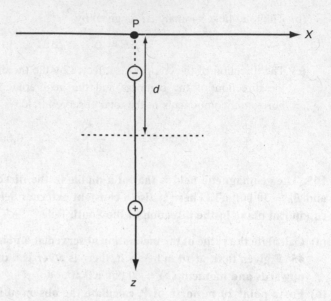

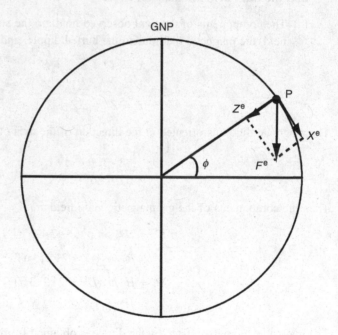

The external field is parallel to the axis of rotation directed southwards so its components are in the vertical and NS direction (Fig. 105b) and are given by

$$Z^e = 2500 \sin \phi = 1768 \, \text{nT}$$
$$X^e = -2500 \cos \phi = -1768 \, \text{nT}$$
$$Y^e = 0$$

Therefore the components of the observed field are

$$X = X^* + \Delta X + X^e = 19\,445\,\text{nT}$$
$$Y = Y^* + \Delta Y + Y^e = 0$$
$$H = X$$
$$Z = Z^* + \Delta Z + Z^e = 44\,274\,\text{nT}$$

From these values we calculate the observed inclination

$$\tan I = \frac{Z}{H} \Rightarrow I = 66.3°$$

(b) For a point Q located 20 m to the north of P we can assume that the main and external fields have the same value as at point P and only the magnetic anomaly created by the buried dipole is different. We calculate this anomaly from Equations (82.2) and (82.3) substituting $\alpha = 0°$ and $x = 20$:

$$\Delta Z = \frac{-Cm[(x^2 - 2d^2)\cos\alpha + 3dx\sin\alpha]}{[x^2 + d^2]^{5/2}} = \frac{-Cm(x^2 - 2d^2)}{[x^2 + d^2]^{5/2}} = -1\,\text{nT}$$

$$\Delta X = \frac{Cm[(2x^2 - d^2)\sin\alpha - 3dx\cos\alpha]}{[x^2 + d^2]^{5/2}} = \frac{-Cm3dx}{[x^2 + d^2]^{5/2}} = -4\,\text{nT}$$

$$\Delta Y = 0$$

Therefore, the components of the observed field at that point are

$$X = X^* + \Delta X + X^e = 19\,441\,\text{nT}$$
$$Y = Y^* + \Delta Y + Y^e = 0$$
$$H = X$$
$$Z = Z^* + \Delta Z + Z^e = 44\,193\,\text{nT}$$

From these values we calculate the observed inclination and declination by the expressions

$$\tan I = \frac{Z}{H} \Rightarrow I = 66.2°$$
$$\tan D = \frac{Y}{X} \Rightarrow D = 0$$

The total field anomaly ΔF is given by

$$\Delta F = \Delta X \cos I + \Delta Z \sin I = -4\cos 66° - 1\sin 66° = -3\,\text{nT}$$

106. Consider a point with coordinates 30° N, 30° E under which is buried at a depth of 100 m a horizontal dipole of moment $\mu_0\,m/4\pi = 1\,\text{T m}^3$, with the positive pole in the direction N 60° E. The terrestrial field is formed by a centred dipole in the direction of the axis of rotation and a constant external field of 10 000 nT from the Sun, $B_0 = 30\,000\,\text{nT}$.

(a) Calculate F, D, and I at that point on December 21 at 12 noon.

(b) How do D and I vary throughout the year?

(a) The components of the total field intensity are the sum of the geomagnetic main field, the magnetic field due to the buried dipole, and the external field:

$$X = X^* + \Delta X + X^e$$

$$Y = Y^* + \Delta Y + Y^e$$

$$Z = Z^* + \Delta Z + Z^e$$

We first calculate the geomagnetic latitude and the declination and inclination. The magnetic dipole is oriented in the direction of the axis of rotation and therefore

$$\phi^* = \phi = 30°$$
$$D^* = 0°$$

Then the components of the geomagnetic main field are

$$Z^* = 2B_0 \sin \phi^* = 30\,000\,\text{nT}$$

$$H^* = B_0 \cos \phi^* = 25\,981\,\text{nT}$$

$$X^* = H^* \cos D^* = 25\,981\,\text{nT}$$

$$Y^* = H^* \sin D^* = 0$$

We calculate the magnetic anomaly produced by the buried dipole from Equations (82.2) and (82.3) substituting $\alpha = 90°$ and $x = 0$. We call X_{60} the direction N 60° E and ΔX_{60} the horizontal component of the anomaly (Fig. 106a):

$$\Delta Z = \frac{2Cm \cos \alpha}{d^3} = 0\,\text{nT}$$

$$\Delta X_{60} = \frac{-Cm \sin \alpha}{d^3} = -1000\,\text{nT}$$

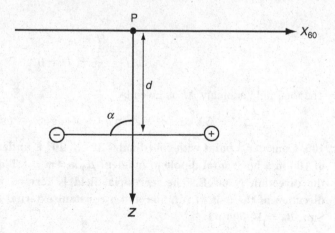

Fig. 106a

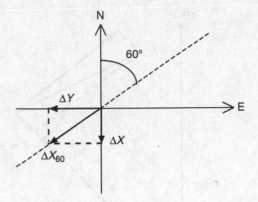

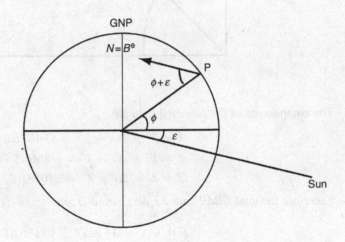

The NS and EW components are given by (Fig. 106b)

$$\Delta X = \Delta X_{60} \cos 60 = -500 \, \text{nT}$$
$$\Delta Y = \Delta X_{60} \sin 60 = -866 \, \text{nT}$$

The external field has a diurnal period ($\omega = 2\pi / 24\,\text{h}$) and we know that the Sun is at the meridian point at 12:00 h (solar time). This field changes through the year because the Sun moves on the ecliptic plane (apparent motion) which is tilted with respect to the equatorial plane by an angle $\varepsilon = 23°$. Therefore the solar declination (δ), the angle from the Sun to the celestial equator, changes through the year. On December 21 (winter solstice) this angle is $\delta = -\varepsilon = -23°$ (Fig. 106c). If we call N the magnitude of the external field ($N = 10\,000\text{nT}$) its components on December 21 at 12:00 are (Fig. 106d)

$$Z^{\text{e}} = -B_r = N \cos(\phi + \varepsilon) = 6018 \, \text{nT}$$
$$X^{\text{e}} = B_\phi = N \sin(\phi + \varepsilon) = 7986 \, \text{nT}$$
$$Y^{\text{e}} = 0$$

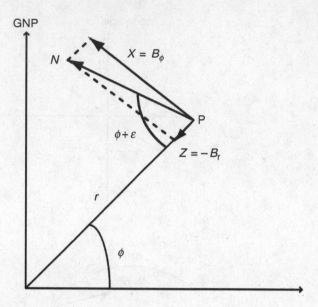

Fig. 106d

The components of the observed field are

$$X = X^* + \Delta X + X^e = 33\,467\,\text{nT}$$
$$Y = Y^* + \Delta Y + Y^e = -866\,\text{nT}$$
$$Z = Z^* + \Delta Z + Z^e = 36\,018\,\text{nT}$$

Therefore the total field F and the declination D are

$$F = \sqrt{X^2 + Y^2 + Z^2} = 49\,166\,\text{nT}$$
$$\tan D = \frac{Y}{X} \Rightarrow D = -1.5°$$
$$\sin I = \frac{Z}{F} \Rightarrow I = 47.1°$$

(b) Any other day at 12:00 h

$$X = X^* + \Delta X + X^e = 25\,481\,\text{nT} + N\sin(\phi - \delta)$$
$$Y = Y^* + \Delta Y + Y^e = -866\,\text{nT}$$
$$Z = Z^* + \Delta Z + Z^e = 30\,000\,\text{nT} + N\cos(\phi - \delta)$$

107. The internal field of the Earth corresponds to a centred dipole with the negative pole in the northern hemisphere at coordinates 80° N, 130° W and $B_0 = 30\,000$ nT. There is a uniform external field from the Sun of 1000 nT. Buried at 500 m depth under a point P with geocentric coordinates 40° N, 50° E there is a positive magnetic pole of strength $CP = 0.5$ T m². Calculate:

(a) The total field components X, Y, and Z, and the magnetic and geomagnetic declination at P on March 21 at 12:00 h.

(b) The same parameters for a point 200 m north of P, assuming that neither the internal nor the external fields change (precision 1 nT).

  (a) The components of the intensity of the total field are the sum of the geomagnetic main field, the external field, and the magnetic field due to the buried pole.

To calculate the main field we determine first the geomagnetic latitude and the declination by (71.3) and (71.2):

$$\sin \phi^* = \sin \phi_B \sin \phi + \cos \phi_B \cos \phi \cos(\lambda - \lambda_B)$$
$$\phi^* = 30°$$

$$\sin D^* = \frac{-\cos \phi_B \sin(\lambda - \lambda_B)}{\cos \phi^*} \Rightarrow D^* = 0°$$

From these values we obtain the vertical and horizontal components of the geomagnetic main field:

$$Z^* = 2B_0 \sin \phi^* = 30\,000 \,\text{nT}$$
$$H^* = B_0 \cos \phi^* = 25\,981 \,\text{nT}$$
$$X^* = H^* \cos D^* = 25\,981 \,\text{nT}$$
$$Y^* = H^* \sin D^* = 0$$

To calculate the external field we notice that it comes from the Sun which on March 21 (spring equinox) is on the equatorial plane so that the external field is parallel to this plane. In addition this field changes during the day as a function of local time t with a diurnal period ($\omega = 2\pi/24$ h). At $t = 12$ h, the external field is maximum given that at this time the Sun is at the meridian point (Fig. 107a). Calling N its magnitude ($N = 1000$ nT), the components of the external field are given by

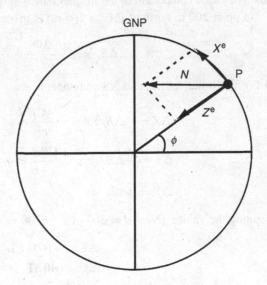

Fig. 107a

$$Z^e = -B_r = N\cos\phi = 766\,\text{nT}$$
$$X^e = B_\phi = N\sin\phi = 643\,\text{nT}$$
$$Y^e = B^e_\lambda = 0$$

The magnetic field anomaly created by the buried pole is derived from its potential $\Delta\Phi$, given by

$$\Delta\Phi = \frac{CP}{r}$$

Applying the gradient, we obtain

$$\Delta\boldsymbol{B} = -\Delta\Phi$$

but the only component is the vertical:

$$\Delta Z = -\Delta B_r = \frac{\partial(\Delta\Phi)}{\partial r} = -\frac{CP}{r^2}$$

Substituting $r = d = 500$ m in this equation we obtain

$$\Delta Z = -2000\,\text{nT}$$

Therefore the components of the total field and the observed declination are given by

$$X = X^* + X^e + \Delta X = 26\,624\,\text{nT}$$
$$Y = Y^* + Y^e + \Delta Y = 0$$
$$Z = Z^* + Z^e + \Delta Z = 28\,766\,\text{nT}$$

$$\tan D = \frac{Y}{X} \Rightarrow D = 0°$$

(b) The radial component of the magnetic field anomaly created by the buried pole for a point 200 m north of P ($x = 200$ m) is given by

$$\Delta B_r = -\frac{\partial(\Delta\Phi)}{\partial r} = \frac{CP}{r^2}$$

From Fig. 107b the vertical and NS components are

$$\Delta Z = -\Delta B_r \cos\alpha = \frac{-CP}{r^2}\frac{d}{r} = \frac{-CPd}{(x^2+d^2)^{3/2}}$$

$$\Delta X = -\Delta B_r \sin\alpha = \frac{CP}{r^2}\frac{x}{r} = \frac{CPx}{(x^2+d^2)^{3/2}}$$

$$\Delta Y = 0$$

Substituting the values given ($d = 500$ m, $x = 200$ m, $CP = 0.5$ Tm^2), we obtain

$$\Delta Z = -1601\,\text{nT}$$
$$\Delta X = 640\,\text{nT}$$
$$\Delta Y = 0$$

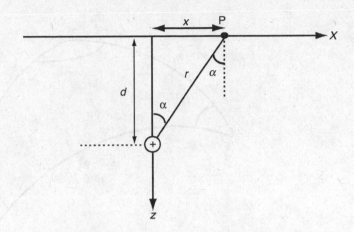

The components of the observed total field and the declination are the sum of the three contributions:

$$X = X^* + X^e + \Delta X = 27\,264\,\text{nT}$$

$$Y = Y^* + Y^e + \Delta Y = 0$$

$$Z = Z^* + Z^e + \Delta Z = 29\,165\,\text{nT}$$

$$\tan D = \frac{Y}{X} \Rightarrow D = 0°$$

Paleomagnetism

108. At a point with geographical coordinates 60° N, 60° W a 1 cm³ sample was taken of a rock with remanent magnetism, age 10 000 years, specific susceptibility 0.01 cm^{-3}. The magnetization components of the rock were:

$X = 40$, $Y = -30$, $Z = 50$ nT (N, E, nadir).

The current field is $B_0 = 30\,000$ nT and the geomagnetic pole coincides with the geographical pole. Calculate:

(a) The coordinates of the virtual geomagnetic pole which corresponds to the sample.

(b) The magnetic moment of the terrestrial dipole 10 000 years ago.

(c) The secular variation of F, D, and I in nT and minutes per year assuming that the variation since that time has been constant.

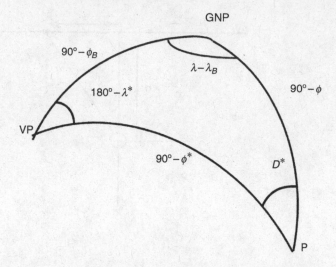

Fig. 108

(a) First we determine the declination D and the geomagnetic co-latitude θ, corresponding to the virtual pole, from the magnetization components of the rock X, Y, and Z:

$$\tan D = \frac{Y}{X} \Rightarrow D = -36.9°$$

$$H = \sqrt{X^2 + Y^2} = 50\,\text{nT}$$

$$\tan I = \frac{Z}{H} \Rightarrow I = 45°$$

$$\tan I = 2 \cot \theta \Rightarrow \theta = 63.4°, \phi^*_{\text{virtual}} = 26.6°$$

Since at present the geomagnetic pole coincides with the geographical pole, the geographical latitude of the point coincides with the present geomagnetic latitude:

$$\phi = \phi^*_{\text{present}} = 60°\text{N}$$

To determine the coordinates of the virtual Geomagnetic North Pole (VP), corresponding to the magnetization of the rock, we solve the spherical triangle of Fig. 108 for φ_B and λ_B using the obtained values of θ and D. The latitude ϕ_B applying the cosine rule is given by

$$\cos(90 - \phi_B) = \cos \theta \cos(90 - \phi) + \sin \theta \sin(90 - \phi) \cos D$$
$$\sin \phi_B = \cos \theta \sin \phi + \sin \theta \cos \phi \cos D$$
$$\phi_B = 48.2$$

To obtain the longitude λ_B we again apply the cosine rule:

$$\cos \theta = \sin(90 - \phi^*) = \cos(90 - \phi_B) \cos(90 - \phi)$$
$$+ \sin(90 - \phi_B) \sin(90 - \phi) \cos(\lambda - \lambda_B)$$
$$\cos \theta = \sin \phi_B \sin \phi + \cos \phi_B \cos \phi \cos(\lambda - \lambda_B)$$

$$\cos(\lambda - \lambda_B) = \frac{\cos\theta - \sin\phi_B \sin\phi}{\cos\phi_B \cos\phi}$$

$$\lambda - \lambda_B = \pm 126.4°$$

To choose between the positive and negative solution we bear in mind that the declination is negative and so the point is to the east of the virtual magnetic North Pole:

$$D < 0 \Rightarrow \lambda - \lambda_B > 0$$

$$\lambda_B = 173.6° \, \text{E}$$

(b) To obtain the magnetic moment we first calculate the constant B_0. The susceptibility χ relates the magnetization and the magnetic field. If we call F the magnitude of the paleomagnetic field and F' the remanent magnetization, the relation between them is

$$F' = \chi F \tag{108.1}$$

$$\chi = 0.01$$

We calculate F' from its components

$$F' = \sqrt{X^2 + Y^2 + Z^2} = 71\,\text{nT}$$

The field F of the virtual pole is given by

$$F = B_0 \sqrt{1 + 3\cos^2\theta}$$

Substituting in Equation (108.1) we obtain

$$F' = \chi B_0 \sqrt{1 + 3\cos^2\theta}$$

$$B_0 = \frac{F'}{\chi\sqrt{1 + 3\cos^2\theta}} = 5610\,\text{nT}$$

From this value we calculate the magnetic moment of the virtual pole taking $a = 6370$ km for the Earth's radius

$$B_0 = \frac{Cm}{a^3} \Rightarrow m = 1.45 \times 10^{22}\,\text{A}\,\text{m}^2$$

(c) The magnetic field, the declination, and inclination 10 000 years ago were

$$F = \frac{F'}{\chi} = 7100\,\text{nT}$$

$$D = -36.9°$$

$$I = 45°$$

At present the values of these parameters are

$$F^p = B_0^a \sqrt{1 + 3\sin^2\phi^*_{\text{present}}} = 54\,083\,\text{nT}$$

$$D^p = 0°$$

$$\tan I^p = 2\tan\phi^* \Rightarrow I^p = 73.9°$$

The secular variation of F, D, and I, for this period of time, is

$$\frac{F^p - F}{10\,000} = 4.7\,\text{nT/yr}$$

$$\frac{D^p - D}{10\,000} = 0.22'/\text{yr}$$

$$\frac{I^p - I}{10\,000} = 0.17'/\text{yr}$$

109. The following table gives the demagnetization data for a sample that was subjected to stepwise thermal demagnetization of its natural remanent magnetization (NRM).

Demagnetization temperature (°C)	Declination (D, °E)	Inclination (I, °)	NRM Intensity (J, mA/m^{-1})
20	32	33	0.056
100	36	22	0.056
200	38	12	0.057
300	39	4	0.058
400	41	−5	0.058
500	41	−5	0.050
600	41	−5	0.016
650	41	−5	0.009
700	300	55	0.000

Calculate the direction of each stable component identified by the demagnetization curve.

First, construct a vector component diagram of the demagnetization data. Decompose each observation into its north (X), east (Y), and vertical (Z) components:

$$H = J \cos I$$
$$X = H \cos D$$
$$Y = H \sin D$$
$$Z = J \sin I$$

In Fig. 109 plotting X versus Y gives the projection of the demagnetization vector onto the horizontal plane, while plotting X versus Z gives the projection onto the vertical plane.

A stable component of NRM is represented by collinear points on the vector component diagrams, so that two stable components can be identified in the range 20–300 °C and in the range 400–700 °C.

The declination of a stable component is determined by measuring or by calculating the angle between the north axis and the trajectory of the stable component in the horizontal plane.

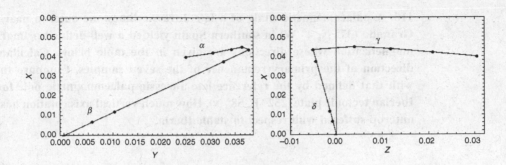

Fig. 109

For the 20–300 °C component:

$$\alpha = \tan^{-1}\left(\frac{X_{300} - X_{20}}{Y_{300} - Y_{20}}\right) = \tan^{-1}\left(\frac{0.0051}{0.0115}\right) = 23.9°$$

$$D = 180 + 23.9 = 203.9°$$

For the 400–700 °C component:

$$D = \beta = \tan^{-1}\left(\frac{Y_{400}}{X_{400}}\right) = \tan^{-1}\left(\frac{0.0379}{0.0436}\right) = 41.0°$$

Note: this value can be obtained directly from the declination of the observations between 400 and 650 °C.

The apparent inclination, I_{ap}, of a stable component is determined by measuring or by calculating the angle between the north axis and the trajectory of the stable component in the vertical plane. I_{ap} is related to the true inclination, I, by:

$$\tan I = \tan I_{ap}|\cos D|$$

For the 20–300 °C component:

$$I_{ap} = \gamma = \tan^{-1}\left(\frac{Z_{20} - Z_{300}}{X_{300} - X_{20}}\right) = \tan^{-1}\left(\frac{0.0265}{0.0051}\right) = 79.1°$$

$$I = \tan^{-1}(\tan(79.1)|\cos(203.9)|) = 78.1°$$

For the 400–700 °C component:

$$I_{ap} = \delta = \tan^{-1}\left(\frac{Z_{400}}{X_{400}}\right) = \tan^{-1}\left(\frac{-0.0051}{0.0436}\right) = -6.7°$$

$$I = \tan^{-1}(\tan(-6.7)|\cos(41)|) = -5.0°$$

Note: this value can be obtained directly from the inclination of the observations between 400 and 650 °C.

Therefore the stable component isolated in the range 20–300 °C has $D = 203.9°$ and $I = 78.1°$ and the stable component isolated in the range 400–700 °C has $D = 41.0°$ and $I = -5.0°$.

110. A palaeomagnetic study of a late Jurassic limestone outcrop near Alhama de Granada (37° N, 4° W) in southern Spain yielded a well-defined primary remanent magnetization whose directions are given in the table below. Calculate the mean direction of the primary remanence of the seven samples. Compare this direction with that defined by the reference late Jurassic palaeomagnetic pole for the stable Iberian tectonic plate (252° E, 58° N). How much vertical axis rotation has the studied outcrop suffered with respect to stable Iberia?

Declination (D, °E)	Inclination (I, °)
30	43
28	39
34	44
25	45
32	38
35	44
26	40

Use unit vector addition to calculate the mean direction of the primary remanence. Calculate the direction cosines of each direction, the resultant total field vector, F, and then the mean direction using:

$$X = \cos I \cos D$$
$$Y = \cos I \sin D$$
$$Z = \sin I$$

$$F = \sqrt{\left(\sum_{i=1}^{N} X_i\right)^2 + \left(\sum_{i=1}^{N} Y_i\right)^2 + \left(\sum_{i=1}^{N} Z_i\right)^2} = 6.98542$$

$$X_{mean} = \frac{\sum_{i=1}^{N} X_i}{F} = 0.6446,$$

$$Y_{mean} = \frac{\sum_{i=1}^{N} Y_i}{F} = 0.37186,$$

$$Z_{mean} = \frac{\sum_{i=1}^{N} Z_i}{F} = 0.66799$$

$$D_{mean} = \tan^{-1}\left(\frac{Y_{mean}}{X_{mean}}\right) = 30.0°$$

$$I_{mean} = \sin^{-1}(Z_{mean}) = 41.9°$$

The mean direction of the primary remanence has a declination of 30.0° and an inclination of 41.9°.

Next, calculate the expected field direction at the site using the reference palaeomagnetic pole. The first step is to determine $\theta = 90 - \phi^*$, (Equation 71.3), from the pole (ϕ_p, λ_p) to the site (ϕ_s, λ_s) using spherical triangles:

$$\sin\phi^* = \sin\varphi_p \sin\varphi_s + \cos\varphi_p \cos\varphi_s \cos(\lambda_s - \lambda_p) \Rightarrow \phi^* = 24.1°$$

The expected inclination can then be calculated using:

$$\tan I_{exp} = 2\tan\phi^* \Rightarrow I_{exp} = 41.8°$$

The expected declination can be calculated by (71.2):

$$\sin D_{exp} = \frac{-\cos\varphi_p \sin(\lambda_s - \lambda_p)}{\cos\phi^*} \Rightarrow D_{exp} = 34.5° \text{ W} = 325.5°$$

rotation about the vertical axis should give rise to a difference between the observed and expected declinations, defined as positive for an observed declination clockwise from the expected declination.

Therefore the outcrop has suffered 64.5° of clockwise rotation with respect to stable Iberia.

4

Seismology

Elasticity

111. Determine the principal stresses and principal axes of the stress tensor:

$$\begin{pmatrix} 2 & -1 & 1 \\ -1 & 0 & 1 \\ 1 & 1 & 2 \end{pmatrix}$$

Find the invariants I_1, I_2, I_3, the deviator tensor, its eigenvalues, and the invariants J_2 and J_3.

To calculate the principal stresses (σ_1, σ_2, σ_3) and principal axes (v_i^1, v_i^2, v_i^3), we calculate the eigenvalues and eigenvectors of the matrix. They are found through the equation

$$(\tau_{ij} - \sigma\delta_{ij})v_i = 0 \tag{111.1}$$

The eigenvalues are the roots of the cubic equation for σ resulting from putting the determinant of the matrix in (111.1) equal to zero:

$$\begin{vmatrix} 2-\sigma & -1 & 1 \\ -1 & -\sigma & 1 \\ 1 & 1 & 2-\sigma \end{vmatrix} = 0$$

$$\Rightarrow (2-\sigma)(-\sigma)(2-\sigma) - 1 - 1 + \sigma - (2-\sigma) - (2-\sigma) = 0$$

$$\sigma^3 - 4\sigma^2 + \sigma + 6 = 0$$

The three roots of the equation are the principal stresses

$$\sigma_1 = -1$$
$$\sigma_2 = 2$$
$$\sigma_3 = 3$$

From these values we obtain $\sigma_0 = \frac{1}{3}(\sigma_1 + \sigma_2 + \sigma_3) = \frac{4}{3}$.

The invariants of the matrix are the coefficients of the characteristic equation

$$\sigma^3 - I_1\sigma^2 + I_2\sigma - I_3 = 0$$

which, in terms of the roots of the equation, are

$$I_1 = 4 = \sigma_1 + \sigma_2 + \sigma_3$$

$$I_2 = 1 = \sigma_1\sigma_2 + \sigma_1\sigma_3 + \sigma_2\sigma_3$$

$$I_3 = -6 = \sigma_1\sigma_2\sigma_3$$

The principal axes of stress are the eigenvectors v_i associated with the three eigenvalues.

For $\sigma_1 = -1$

$$\begin{pmatrix} 3 & -1 & 1 \\ -1 & 1 & 1 \\ 1 & 1 & 3 \end{pmatrix} \begin{pmatrix} v_1 \\ v_2 \\ v_3 \end{pmatrix} = 0 \Rightarrow \left(v_1^1, v_2^1, v_3^1\right) = (-1, -2, 1)$$

For $\sigma_2 = 2$

$$\begin{pmatrix} 0 & -1 & 1 \\ -1 & -2 & 1 \\ 1 & 1 & 0 \end{pmatrix} \begin{pmatrix} v_1 \\ v_2 \\ v_3 \end{pmatrix} = 0 \Rightarrow \left(v_1^2, v_2^2, v_3^2\right) = (-1, 1, 1)$$

For $\sigma_3 = 3$

$$\begin{pmatrix} -1 & -1 & 1 \\ -1 & -3 & 1 \\ 1 & 1 & -1 \end{pmatrix} \begin{pmatrix} v_1 \\ v_2 \\ v_3 \end{pmatrix} = 0 \Rightarrow \left(v_1^3, v_2^3, v_3^3\right) = (1, 0, 1)$$

The deviatoric stress tensor is defined as

$$\tau'_{ij} = \tau_{ij} - \sigma_0 \delta_{ij}$$

where in our problem, $\sigma_0 = 4/3$.

The three components of the principal diagonal of the deviatoric tensor are

$$\tau'_{11} = \frac{2}{3}$$

$$\tau'_{22} = -\frac{4}{3}$$

$$\tau'_{33} = \frac{2}{3}$$

To calculate its eigenvalues we proceed as we did before:

$$\begin{pmatrix} \frac{2}{3} - s & -1 & 1 \\ -1 & -\frac{4}{3} - s & 1 \\ 1 & 1 & \frac{2}{3} - s \end{pmatrix} = 0 \Rightarrow s^3 - \frac{13}{3}s + \frac{38}{27} = 0$$

Comparing with the characteristic equation

$$\sigma^3 - J_1\sigma^2 + J_2\sigma - J_3 = 0$$

the invariants are

$$J_2 = -\frac{13}{3}$$

$$J_3 = -\frac{38}{27}$$

Solving the cubic equation we obtain $s_1 = -2.22$, $s_2 = 0.33$, $s_3 = 1.89$.

112. Given the stress tensor

$$
\begin{pmatrix}
\frac{1}{4} & -\frac{5}{4} & \left(\frac{1}{2}\right)^{\frac{3}{2}} \\[2mm]
-\frac{5}{4} & \frac{1}{4} & -\left(\frac{1}{2}\right)^{\frac{3}{2}} \\[2mm]
\left(\frac{1}{2}\right)^{\frac{3}{2}} & -\left(\frac{1}{2}\right)^{\frac{3}{2}} & \frac{3}{2}
\end{pmatrix}
$$

calculate:

(a) The principal stresses.

(b) The angles formed by the greatest of these stresses with the axes 1, 2, 3.

(a) As in the previous problem to find the principal stresses we calculate the eigenvalues of the stress matrix

$$
\begin{pmatrix}
\frac{1}{4} - \sigma & -\frac{5}{4} & \left(\frac{1}{2}\right)^{3/2} \\[2mm]
-\frac{5}{4} & \frac{1}{4} - \sigma & \left(\frac{1}{2}\right)^{3/2} \\[2mm]
\left(\frac{1}{2}\right)^{3/2} & -\left(\frac{1}{2}\right)^{3/2} & \frac{3}{2} - \sigma
\end{pmatrix} = 0 \Rightarrow \sigma^3 - 2\sigma^2 - \sigma + 2 = 0
$$

Solving the cubic equation, its three roots are

$$\sigma_1 = 2$$

$$\sigma_2 = 1$$

$$\sigma_3 = -1$$

The largest is σ_1. The associated eigenvector corresponds to the axis of greatest stress whose direction cosines are (v_1, v_2, v_3). They are found by solving the equation

$$
\begin{pmatrix}
-\frac{7}{4} & -\frac{5}{4} & \left(\frac{1}{2}\right)^{3/2} \\[2mm]
-\frac{5}{4} & -\frac{7}{4} & -\left(\frac{1}{2}\right)3/2 \\[2mm]
\left(-\frac{1}{2}\right)^{3/2} & -\left(\frac{1}{2}\right)^{3/2} & -\frac{1}{2}
\end{pmatrix}
\begin{pmatrix}
v_1 \\[1mm] v_2 \\[1mm] v_3
\end{pmatrix} = 0
$$

Solving this equation with the condition that $v_1^2 + v_2^2 + v_3^2 = 1$, we obtain,

$$v_1 = -v_2 = \frac{1}{2}$$

$$v_3 = \frac{1}{\sqrt{2}}$$

(b) From these values we obtain ϑ, the angle with the vertical axis (x_3) and φ, the angle which forms its projection on the horizontal plane with x_1:

$$\left.\begin{array}{l} v_1 = \sin\vartheta\cos\varphi = \dfrac{1}{2} \\[2mm] v_2 = \sin\vartheta\sin\varphi = -\dfrac{1}{2} \\[2mm] v_3 = \cos\vartheta = \dfrac{1}{\sqrt{2}} \end{array}\right\} \Rightarrow \varphi = 315°, \quad \vartheta = 45°$$

113. The stress tensor τ_{ij} in a continuous medium is

$$\begin{pmatrix} 3x_1x_2 & 5x_2^2 & 0 \\ 5x_2^2 & 0 & 2x_3 \\ 0 & 2x_3 & 0 \end{pmatrix}$$

Determine the stress vector T_i^ν acting at the point $(2, 1, \sqrt{3})$ through the plane tangential to the cylindrical surface $x_2^2 + x_3^2$ at that point.

First we calculate the value of the stress tensor at the given point:

$$\tau_{ij}(2, 1, \sqrt{3}) = \begin{pmatrix} 6 & 5 & 0 \\ 5 & 0 & 2\sqrt{3} \\ 0 & 2\sqrt{3} & 0 \end{pmatrix}$$

A unit vector normal to the surface $f = x_2^2 + x_3^2 - 4 = 0$ at the given point is

$$v_i = \frac{\operatorname{grad} f}{|\operatorname{grad} f|} = \left(\frac{\dfrac{\partial f}{\partial x_1}}{\left|\dfrac{\partial f}{\partial x_1}\right|}, \frac{\dfrac{\partial f}{\partial x_2}}{\left|\dfrac{\partial f}{\partial x_2}\right|}, \frac{\dfrac{\partial f}{\partial x_3}}{\left|\dfrac{\partial f}{\partial x_3}\right|} \right) = \left(0, \frac{1}{2}, \frac{\sqrt{3}}{2} \right)$$

Then, the stress vector acting at the point through that surface is given by

$$T_i^\nu = \tau_{ij}v_j = \begin{pmatrix} 6 & 5 & 0 \\ 5 & 0 & 2\sqrt{3} \\ 0 & 2\sqrt{3} & 0 \end{pmatrix} \begin{pmatrix} 0 \\ \dfrac{1}{2} \\ \dfrac{\sqrt{3}}{2} \end{pmatrix} = \left(\frac{5}{2}, 3, \sqrt{3} \right)$$

Wave propagation. Potentials and displacements

114. The amplitudes of P- and S-waves of frequency $4/2\pi$ Hz are

$$u^P = \left(\frac{4}{3\sqrt{2}}, \frac{4}{3\sqrt{2}}, \frac{4}{\sqrt{3}} \right)$$

$$u^S = \left(-\sqrt{3}, -\sqrt{3}, \sqrt{2} \right)$$

and their speeds of propagation are 6 km s^{-1} and 4 km s^{-1}, respectively. Find the scalar and vector potentials. Displacements are always given in μm.

The displacements of P-waves can be deduced from a scalar potential function φ such that $u_i^P = (\nabla\varphi)_i$. The general form of the potential function for P-waves for harmonic motion is

$$\varphi = A \exp ik_\alpha(v_i x_i - \alpha t) \tag{114.1}$$

where A is the amplitude, v_i the direction cosines of the ray or propagation direction, α the velocity of propagation, and k_α the wavenumber. If u_j is given in μm and k_α in km^{-1}, then A is given in 10^{-3} m^2. Taking the derivatives in (114.1) we obtain for the components of the displacement

$$u_j^P = \frac{\partial\varphi}{\partial x_j} = A i k_\alpha v_j \exp ik_\alpha(v_k x_k - \alpha t)$$

Their amplitude is

$$u_j^P = A k_\alpha v_j \tag{114.2}$$

and the wavenumber is

$$k_\alpha = \frac{\omega}{\alpha} = \frac{2\pi\dfrac{4}{2\pi}}{6} = \frac{2}{3} \text{ km}^{-1}$$

By substitution in (114.2), we obtain for the two horizontal components

$$u_1^P = A k_\alpha v_1 = \frac{4}{3\sqrt{2}} \text{ μm}$$

$$u_2^P = A k_\alpha v_2 = \frac{4}{3\sqrt{2}} \text{ μm}$$

Dividing these two expressions and writing the direction cosines of the ray in terms of the incident angle i and azimuth a_z,

$$\begin{aligned}
v_1 &= \sin i \cos a_z \\
v_2 &= \sin i \sin a_z \\
v_3 &= \cos i
\end{aligned} \tag{114.3}$$

we have

$$\frac{u_1^P}{u_2^P} = 1 = \frac{v_1}{v_2} = \frac{\sin i \cos a_z}{\sin i \sin a_z} \Rightarrow \alpha = 45°$$

Using the u_3^P and u_1^P components,

$$\frac{u_3^P}{u_1^P} = \frac{A k_\alpha v_3}{A k_\alpha v_1} = \frac{\cos i}{\sin i \cos a_z} = \sqrt{3} \Rightarrow i = 30°$$

From the values of the direction cosines and the amplitude of the displacement we calculate the amplitude of the potential A:

$$A = \frac{u_1^P}{k_\alpha v_1} \Rightarrow A = 4 \times 10^{-3} \text{ m}^2$$

Finally, the expression for the scalar potential of P-waves is

$$\varphi = 4 \exp \frac{2}{3} i \left(\frac{\sqrt{2}}{4} x_1 + \frac{\sqrt{2}}{4} x_2 + \frac{\sqrt{3}}{2} x_3 - 6t \right)$$

Displacements of the S-wave are obtained from a vector potential ψ_i of null divergence, whose general form is

$$\psi_i = B_i \exp i k_\beta \left(v_j x_j - \beta t \right)$$

where β is the velocity of propagation and k_β the wavenumber. The displacements are given by

$$u^S = \nabla \times \psi \tag{114.4}$$

The wavenumber is $k_\beta = \omega/\beta = 1 \ \text{km}^{-1}$. According to (114.4) the relation between the components of the displacement and of the amplitude of the potential is

$$u_1^S = B_3 \frac{\sqrt{2}}{4} - B_2 \frac{\sqrt{3}}{2} = -\sqrt{3} \ \mu\text{m}$$

$$u_2^S = B_1 \frac{\sqrt{3}}{2} - B_3 \frac{\sqrt{2}}{4} = -\sqrt{3} \ \mu\text{m}$$

$$u_3^S = B_2 \frac{\sqrt{2}}{4} - B_1 \frac{\sqrt{2}}{4} = \sqrt{2} \ \mu\text{m}$$

The potential must have null divergence,

$$\nabla \cdot \psi = \frac{\sqrt{2}}{4} B_1 + \frac{\sqrt{2}}{4} B_2 + \frac{\sqrt{3}}{2} B_3 = 0$$

From these equations we obtain, in units of $10^{-3} \ \text{m}^2$,

$$B_1 = -2$$
$$B_2 = 2$$
$$B_3 = 0$$

The S-wave vector potential is

$$\psi_j = (-2, 2, 0) \exp i \left(\frac{\sqrt{2}}{4} x_1 + \frac{\sqrt{2}}{4} x_2 + \frac{\sqrt{3}}{2} x_3 - 4t \right)$$

Note: These units will be used for all problems but not explicitly given.

115. The components of the S-wave with respect to the axes (x_1, x_2, x_3) are (6, 3.25, 3) where x_2 is the vertical axis, the azimuth is 60°, and the angle of incidence is 30°. Determine the amplitude and direction cosines of the SV and SH components.

From the azimuth and incident angles we calculate the direction cosines (x_2 is the vertical axis)

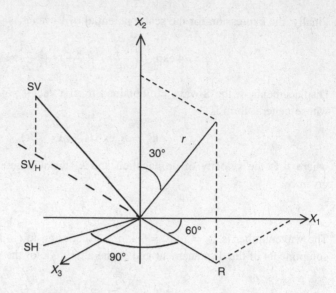

Fig. 115a

$$v_1 = \sin i \cos a_z = \frac{1}{4}$$

$$v_2 = \cos i = \frac{\sqrt{3}}{2}$$

$$v_3 = \sin i \sin a_z = \frac{\sqrt{3}}{4}$$

Since the SV and SH components are on a plane normal to the direction of the ray r (Fig. 115a) unit vectors in the direction of SV (a_1, a_2, a_3) and of SH $(b_1, 0, b_3)$ must satisfy the equations

$$\mathbf{SV} \cdot \mathbf{r} = 0 \Rightarrow \frac{a_1}{4} + \frac{a_2\sqrt{3}}{2} + \frac{a_3\sqrt{3}}{4} = 0$$

$$\mathbf{SH} \cdot \mathbf{r} = 0 \Rightarrow \frac{b_1}{4} + \frac{b_3\sqrt{3}}{4} = 0 \qquad\qquad (115.1)$$

$$\mathbf{SH} \cdot \mathbf{SV} = 0 \Rightarrow a_1 b_1 + a_3 b_3 = 0$$

The projections on the horizontal plane R of the ray r and SH are perpendicular (Fig. 115b). Then SH forms an angle of $180 - 30°$ with the x_1 axis. The direction cosines of SH are

$$b_1 = -\frac{\sqrt{3}}{2}$$

$$b_3 = \frac{1}{2}$$

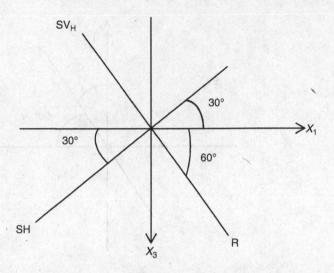

SV forms an angle of 60° with the vertical axis x_2 (Fig. 115a). Then $a_2 = -\sin i = -\frac{1}{2}$
From a_2 using Equations (115.1) and $a_1^2 + a_2^2 + a_3^2 = 1$, we calculate a_1 and a_2:

$$a_1 = \frac{\sqrt{3}}{4}$$

$$a_3 = \frac{3}{4}$$

116. Given the potential $\psi i = \left(\frac{\sqrt{7}}{\sqrt{5}}, \frac{5\sqrt{7}}{\sqrt{5}}, -6 \right) \exp\left[4i\left(\frac{1}{\sqrt{5}}x_1 + \frac{1}{\sqrt{3}}x_2 + \frac{\sqrt{7}}{\sqrt{15}}x_3 - 4t \right) \right]$,
calculate the polarization angle.

First we calculate the amplitudes of the components of the displacement of the S-wave
from the vector potential

$$u_i^S = \nabla \times \psi_i \Rightarrow \begin{cases} u_1^S = \psi_{3,2} - \psi_{2,3} = -4\dfrac{13}{\sqrt{3}} = -30.02 \\[2mm] u_2^S = \psi_{1,3} - \psi_{3,1} = 4\left(\dfrac{7}{5\sqrt{3}} + \dfrac{6}{\sqrt{5}} \right) = 13.97 \\[2mm] u_3^S = \psi_{2,1} - \psi_{1,2} = 4\left(\sqrt{7} - \dfrac{\sqrt{7}}{\sqrt{15}} \right) = 7.85 \end{cases}$$

The modulus is

$$u^S = \sqrt{u_1^2 + u_2^2 + u_3^2} = 34.03$$

From the vertical component u_3^S we calculate the SV component (Fig. 116) knowing that

$$v_3 = \cos i = \frac{\sqrt{7}}{\sqrt{15}} \quad \Rightarrow \sin i = \sqrt{1 - \frac{7}{15}} = \frac{\sqrt{8}}{\sqrt{15}}$$

Then, $u_3^S = u^{SV} \cos(90 - i) \quad \Rightarrow u^{SV} = 10.75$.

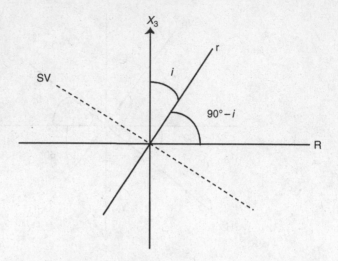

To find the SH component we use the relation

$$u^{\text{S}} = \sqrt{(u^{\text{SV}})^2 + (u^{\text{SH}})^2} \Rightarrow u^{\text{SH}} = 32.29$$

From the SV and SH components we obtain the polarization angle ε:

$$\tan \varepsilon = \frac{u^{\text{SH}}}{u^{\text{SV}}} \Rightarrow \varepsilon = 71.6°$$

117. Given the amplitudes of the P- and S-waves ($k_\alpha = 1$):

$$
\begin{aligned}
u_1^{\text{P}} &= 4 & u_1^{\text{S}} &= 8 \\
u_2^{\text{P}} &= 4 & u_2^{\text{S}} &= 2\sqrt{2} \\
u_3^{\text{P}} &= 8 & u_3^{\text{S}} &= -\left(4 + \sqrt{2}\right)
\end{aligned}
$$

determine the angle of incidence i, azimuth a_z, polarization angle ε of the S-wave, and apparent polarization angle γ.

Given that the displacements of the P-wave are on the incident plane, in the direction of the ray r, the angle of incidence i can be obtained from the modulus and the vertical component:

$$u^{\text{P}} = \sqrt{4^2 + 4^2 + 8^2} = 4\sqrt{6}$$

$$\cos i = \frac{u_3^{\text{P}}}{u^{\text{P}}} = \frac{8}{4\sqrt{6}} \Rightarrow i = 35.3°$$

The azimuth, the angle between the horizontal projection of the ray and the north (x_1), is obtained from the horizontal components, u_1^{P} and u_2^{P}:

$$\tan a_z = \frac{u_2^{\text{P}}}{u_1^{\text{P}}} \Rightarrow a_z = 45°$$

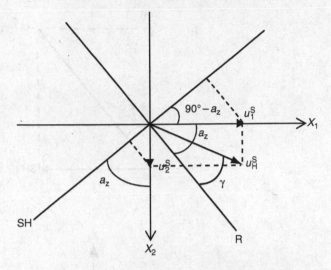

Fig. 117

We calculate the SV component from the vertical component u_3^S, as in the previous problem:

$$u^{SV} = \frac{u_3^S}{\cos(90 - i_0)} = \frac{u_3^S}{\sin i_0} = -9.37$$

The SH component is found from the horizontal components (Fig. 117)

$$u^{SH} = u_2^S \cos a_z - u_1^S \sin a_z = -3.66$$

From SV and SH we find the polarization angle ε (Fig. 117):

$$\tan \varepsilon = \frac{u^{SH}}{u^{SV}} = \frac{-3.66}{-9.37} \quad \Rightarrow \varepsilon = 21.3°$$

To calculate the apparent polarization angle γ, the angle between the horizontal component of S, (u_H^S), and the radial direction R (Fig. 117), we use the relation

$$\tan \varepsilon = \cos i_0 \tan \gamma \quad \Rightarrow \gamma = 25.5°$$

118. Given the values $a_z = 60°$, $\varepsilon = 30°$, $\gamma = 45°$, and $|u^\beta| = 5$, calculate the amplitudes of the components 1, 2, and 3 of the S-wave.

From the modulus of the displacement of S-waves and the polarization angle, we calculate the SV and SH components (Fig. 118a):

$$u_{SV} = u^S \cos \varepsilon = \frac{5\sqrt{3}}{2}$$

$$u_{SH} = u^S \sin \varepsilon = \frac{5}{2}$$

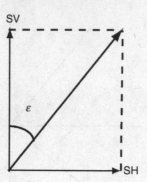

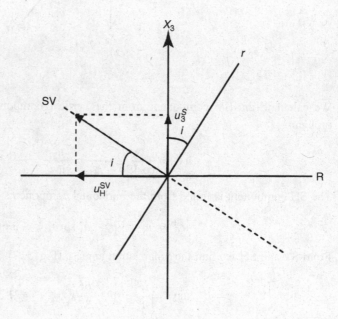

From the angles ε and γ we obtain the incidence angle i:

$$\tan \varepsilon = \tan \gamma \cos i \quad \Rightarrow \cos i = \frac{1}{\sqrt{3}} \quad \Rightarrow \sin i = \sqrt{1 - \frac{1}{3}} = \frac{\sqrt{2}}{\sqrt{3}}$$

The vertical component u_3 of the S-wave is obtained from the value of SV (Fig. 118b):

$$u_3^S = u^{SV} \cos(90 - i) = \frac{5\sqrt{3}}{2} \frac{\sqrt{2}}{\sqrt{3}} = 3.53$$

To calculate the horizontal components we have to take into account the horizontal component of SV (Fig. 118b):

$$u_H^{SV} = u^{SV} \cos i = \frac{5}{2}$$

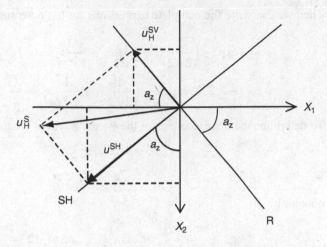

From SH and the horizontal component of SV we find the horizontal components of S (Fig. 118c):

$$u_1^S = -\left[u^{SH} \sin a_z + u_H^{SV} \cos a_z\right] = -\frac{5}{4}\left(1 + \sqrt{3}\right) = -3.42$$

$$u_2^S = u^{SH} \cos a_z - u_H^{SV} \sin a_z = \frac{5}{4}\left(1 - \sqrt{3}\right) = -0.92$$

119. For a scalar potential φ and a vector potential ψ_i, it is known that $A = 3$, $B_i = (-2, 2, 0)$, $k_a = 2/3$, $T = \pi/2$, and Poisson's ratio is $\sigma = 1/4$. Calculate the amplitudes on the free surface ($x_3 = 0$) of the components u_1, u_2, and u_3 of the P- and SV-waves. The direction cosines are $\left(\dfrac{1}{2\sqrt{2}}, \dfrac{1}{2\sqrt{2}}, \dfrac{\sqrt{3}}{2}\right)$.

The general expressions for the scalar and vector potentials are

$$\varphi = A \exp i k_\alpha (v_1 x_1 + v_2 x_2 + v_3 x_3 - \alpha t)$$
$$\psi_i = B_i \exp i k_\beta (v_1 x_1 + v_2 x_2 + v_3 x_3 - \beta t)$$

Since $k_\alpha = \omega/\alpha$ then $\alpha = \omega/k_\alpha = 6 \text{ km s}^{-1}$.

If Poisson's ratio is 0.25 then

$$\sigma = \frac{1}{4} = \frac{\lambda}{2(\lambda + \mu)} \quad \Rightarrow \lambda = \mu$$

Substituting this condition in the equation for the P-wave velocity α, we find for the velocity β of S-waves is given by

$$\alpha = \sqrt{\frac{\lambda + 2\mu}{\rho}} = \sqrt{\frac{3\mu}{\rho}} = \beta\sqrt{3} \Rightarrow \beta = \frac{\alpha}{\sqrt{3}} = 2\sqrt{3} \text{ km s}^{-1}$$

Then the wavenumber of S-waves is

$$k_\beta = \frac{\omega}{\beta} = \frac{2}{\sqrt{3}}$$

Then we can write the complete expressions for the potentials:

$$\varphi = 3 \exp i \frac{2}{3} \left(\frac{1}{2\sqrt{2}} x_1 + \frac{1}{2\sqrt{2}} x_2 + \frac{\sqrt{3}}{2} x_3 - 6t \right)$$

$$\psi_i = (-2, 2, 0) \exp i \frac{2}{\sqrt{3}} \left(\frac{1}{2\sqrt{2}} x_1 + \frac{1}{2\sqrt{2}} x_2 + \frac{\sqrt{3}}{2} x_3 - 2\sqrt{3}t \right)$$

(119.1)

To determine the displacements of the P- and S-waves we use the relations

$$u^P = \nabla \varphi$$
$$u^S = \nabla \times \psi$$

obtaining

$$u_1^P = \frac{1}{\sqrt{2}} \qquad u_1^S = -2$$

$$u_2^P = \frac{1}{\sqrt{2}} \qquad u_2^S = -2$$

$$u_3^P = \sqrt{3} \qquad u_3^S = \frac{2\sqrt{2}}{\sqrt{3}}$$

From the components of the displacement of the S-wave we can obtain the SV component. The values of the angles of incidence and azimuth are found from the direction cosines,

$$v_1 = \sin i \cos a_z = \frac{1}{2\sqrt{2}}$$

$$v_2 = \sin i \sin a_z = \frac{1}{2\sqrt{2}} \qquad \Rightarrow i = 30°, \qquad a_z = 45°$$

$$v_3 = \cos i = \frac{\sqrt{3}}{2}$$

and from the angle i we obtain the SV component from u_3^S:

$$u_3^S = u^{SV} \cos(90 - i) \Rightarrow u^{SV} = \frac{4\sqrt{2}}{\sqrt{3}} = 3.27$$

From the horizontal components of the S-waves and the azimuth we calculate the SH component:

$$u^{SH} = u_2^S \cos a_z - u_1^S \sin a_z = 0$$

120. In an elastic medium of density $\rho = 3$ g cm^{-3} and Poisson ratio 1/3 there propagate P- and S-waves of frequency 1 Hz in the direction $\left(\frac{1}{3}, \frac{1}{\sqrt{2}}, \frac{\sqrt{7}}{3\sqrt{2}} \right)$. Given that the pressure of the P-wave is 5000 dyn cm^{-2}, the magnitude of its displacement, 10 μm, is twice that of the S-wave, and the angle $\gamma = 45°$, find all the parameters involved in the expression of the potentials φ and ψ_i. (It is not necessary to solve the equations to obtain the coefficients B_i)

Given that Poisson's ratio is 1/3 the relation between the elastic coefficients λ and μ is,

$$\sigma = \frac{\lambda}{2(\lambda + \mu)} = \frac{1}{3} \Rightarrow \lambda = 2\mu$$

and between the velocities of P (α) and S (β)-waves is

$$\alpha = \sqrt{\frac{\lambda + 2\mu}{\rho}} = \sqrt{\frac{4\mu}{\rho}} \Rightarrow \alpha = 2\beta$$

The bulk modulus K is

$$K = \lambda + \frac{2}{3}\mu = \frac{8}{3}\mu$$

Expressing μ in terms of β and α, we get for K:

$$\beta = \sqrt{\frac{\mu}{\rho}} \Rightarrow \mu = \rho\beta^2 \Rightarrow K = \frac{8}{3}\rho\beta^2 = \frac{8}{3}\rho\frac{\alpha^2}{4} = \frac{2}{3}\rho\alpha^2$$

Taking into account that the bulk modulus K is defined as the applied pressure divided by the change in volume per unit volume θ,

$$P = K\theta = \frac{2}{3}\rho\alpha^2\theta$$

$$\theta = \nabla^2\varphi = Ak_\alpha^2 \tag{120.1}$$

$$\theta = \frac{3P}{2\rho\alpha^2} = Ak_\alpha^2 \Rightarrow \alpha^2 = \frac{3P}{2\rho Ak_\alpha^2}$$

The expression for the scalar potential is

$$\varphi = A \exp ik_\alpha(v_1x_1 + v_2x_2 + v_3x_3 - \alpha t) \tag{120.2}$$

where the wavenumber for P-waves is

$$k_\alpha = \frac{\omega}{\alpha} = \frac{2\pi f}{\alpha}$$

$$|u^P| = |\nabla\varphi| = k_\alpha A \Rightarrow A = \frac{|u^P|}{k_\alpha} = \frac{|u^P|\alpha}{2\pi f}$$

By substitution in (120.1) we obtain the values of α and β:

$$\alpha^2 = \frac{3P}{2\rho Ak_\alpha^2} = \frac{3P\alpha}{2\rho|u^P|2\pi f} \Rightarrow \alpha = \frac{3P}{2\rho|u^P|2\pi f}$$

where
$P = 5000$ dyn cm^{-2}
$\rho = 3$ g cm^{-3}
$f = 1$ Hz
$u^P = 10$ μm

We obtain $\alpha = 3.98$ km s^{-1} and $\beta = 1.99$ km s^{-1}.

Since we know A, α, and k_α we can write the complete expression for the scalar potential

$$\varphi = 10 \exp i 1.58 \left(\frac{1}{3} x_1 + \frac{1}{\sqrt{2}} x_2 + \frac{\sqrt{7}}{3\sqrt{2}} x_3 - 3.98t \right)$$

The vector potential of the S-waves is given by

$$\psi_i = B_i \exp i k_\beta (v_1 x_1 + v_2 x_2 + v_3 x_3 - \beta t) \qquad (120.3)$$

We calculate k_β:

$$k_\beta = \frac{\omega}{\beta} = 3.16 \, \text{km}^{-1}$$

The displacements are given by

$$u^S = \nabla \times \boldsymbol{\psi}$$

and

$$u_1^S = \frac{\partial \psi_3}{\partial x_2} - \frac{\partial \psi_2}{\partial x_3} = k_\beta (B_3 v_2 - B_2 v_3)$$

$$u_2^S = \frac{\partial \psi_1}{\partial x_3} - \frac{\partial \psi_3}{\partial x_1} = k_\beta (B_1 v_3 - B_3 v_1) \qquad (120.4)$$

$$u_3^S = \frac{\partial \psi_2}{\partial x_1} - \frac{\partial \psi_1}{\partial x_2} = k_\beta (B_2 v_1 - B_1 v_2)$$

The incidence angle i is found from v_3 and, using $\tan \varepsilon = \cos i \tan \gamma$, we find the polarization angle ε:

$$v_3 = \cos i = \frac{\sqrt{7}}{3\sqrt{2}} \Rightarrow i = 31.95° \Rightarrow \varepsilon = 31.95°$$

The azimuth is

$$a_z = \tan^{-1} \frac{v_2}{v_1} = 64.76°$$

Since the amplitude of the S-wave displacement is 5 μm, knowing the value of ε we can find the values of the SV and SH components:

$$u^{SV} = u^S \cos \varepsilon = 4.24 \, \mu\text{m}$$
$$u^{SH} = u^S \sin \varepsilon = 2.65 \, \mu\text{m}$$

From u^{SV} we calculate its vertical and horizontal components:

$$u_3^S = u^{SV} \cos(90 - i) = 2.25 \, \mu\text{m}$$
$$u_H^{SV} = u^{SV} \cos i = 3.61 \, \mu\text{m}$$

Using u_H^{SV} and u^{SH} we find the two horizontal components:

$$u_1^S = -\left[u^{SH} \sin a_z + u_H^{SV} \cos a_z\right] = -3.94 \,\mu m$$
$$u_2^S = u^{SH} \cos a_z - u_H^{SV} \sin a_z = -2.14 \,\mu m$$

Using the values found for the displacements and Equations (120.4) and $\nabla \cdot \boldsymbol{\psi} = 0$ $(v_1 B_1 + v_2 B_2 + v_3 B_3 = 0)$ we find the values of B_1, B_2, B_3. Substituting all the values in (120.3) we obtain for the vector potential

$$\psi_i = (1.17, 2.5, -3.44) \exp 3.16i \left(\frac{1}{3}x_1 + \frac{1}{\sqrt{2}}x_2 + \frac{\sqrt{7}}{3\sqrt{2}}x_3 - 1.99t\right)$$

121. At the origin in an infinite medium in which σ, Poisson's ratio, is 0.25, and the density is 3 g cm^{-3}, there is an emitter of elastic plane waves of frequency 0.5 cps. Calculate:

(a) The equation of the P- and S-waves in exponential form and with arbitrary amplitudes for the wave arriving at the point A(500, 300, 141) km.

(b) The arrival time.

(a) First we calculate the distance to point A and the direction cosines of the direction of the ray (r) (Fig. 121):

$$r = \sqrt{500^2 + 300^2 + 141^2} = 599.90 \approx 600 \,km$$
$$v_1 = \frac{500}{600} = \frac{5}{6}$$
$$v_2 = \frac{300}{600} = \frac{1}{2}$$
$$v_3 = \frac{141}{600} = \frac{\sqrt{2}}{6}$$

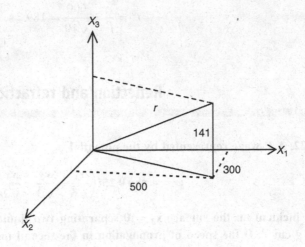

Fig. 121

The S-wave velocity is

$$\beta = \sqrt{\frac{\mu}{\rho}} = \sqrt{10}\,\text{km}\,\text{s}^{-1}$$

Given that Poisson's ratio is 0.25,

$$\sigma = 0.25 \Rightarrow \lambda = \mu \Rightarrow \alpha = \beta\sqrt{3} = \sqrt{30}\,\text{km}\,\text{s}^{-1}$$

The wavenumbers of the P- and S-waves are

$$k_\alpha = \frac{2\pi f}{\alpha} = \frac{2\pi \times 0.5}{5.5} = \frac{\pi}{\sqrt{30}}\,\text{km}^{-1}$$

$$k_\beta = \frac{2\pi f}{\beta} = \frac{2\pi \times 0.5}{\sqrt{10}} = \frac{\pi}{\sqrt{10}}\,\text{km}^{-1}$$

The general expressions for the scalar and vector potentials are

$$\varphi = A \exp ik_\alpha(v_1 x_1 + v_2 x_2 + v_3 x_3 - \alpha t)$$

$$\psi_i = B_i \exp ik_\beta(v_1 x_1 + v_2 x_2 + v_3 x_3 - \beta t)$$

Leaving the amplitudes A and B_i in arbitrary form and substituting the obtained values we have for the potentials,

$$\varphi = A \exp i\frac{\pi}{\sqrt{30}}\left(\frac{5}{6}x_1 + \frac{1}{2}x_2 + \frac{\sqrt{2}}{6}x_3 - \sqrt{30}t\right)$$

$$\psi_i = B_i \exp i\frac{\pi}{\sqrt{10}}\left(\frac{5}{6}x_1 + \frac{1}{2}x_2 + \frac{\sqrt{2}}{6}x_3 - \sqrt{10}t\right)$$

(b) The travel times for P- and S-waves from the origin to the given point are

$$t^\alpha = \frac{r}{\alpha} = \frac{600}{\sqrt{30}} = 109.5\,\text{s}$$

$$t^\beta = \frac{r}{\beta} = \frac{600}{\sqrt{10}} = 189.7\,\text{s}$$

Reflection and refraction

122. A P-wave represented by the potential

$$\varphi = 4 \exp 0.25i\left(\frac{x_1}{\sqrt{6}} + \frac{x_2}{\sqrt{3}} + \frac{x_3}{\sqrt{2}} - 4t\right)$$

is incident on the surface $x_3 = 0$ separating two liquids of densities 3 g cm^{-3} and 4 g cm^{-3}. If the speed of propagation in the second medium is 2 km s^{-1}, write the expressions for the potentials of the reflected and transmitted waves.

The potentials of the reflected and transmitted waves are given by

$$\varphi_{\text{refl}} = A \exp ik(-\tan e\, x_3 + x_1 - ct)$$
$$\varphi_{\text{trans}} = A' \exp ik(\tan e'\, x_3 + x_1 - ct)$$

$$(122.1)$$

where $e = 90 - i$ is the emergence angle and i the incidence angle, and $k = k_\alpha \cos e$ is the wavenumber corresponding to the apparent horizontal velocity, $c = \alpha /\cos e$. These expressions are written for rays contained on the incidence plane (x_1, x_3). Then, we have to rotate the given potential to refer it to the incidence plane. First, from the direction cosines we calculate the incidence angle i and the azimuth a_z:

$$\nu_1 = \sin i \cos a_z = \frac{1}{\sqrt{6}}$$

$$\nu_2 = \sin i \sin a_z = \frac{1}{\sqrt{3}}$$

$$\nu_3 = \cos i = \frac{1}{\sqrt{2}} \Rightarrow i = 45° = e$$

$$\cos a_z = \frac{1}{\sqrt{3}} \Rightarrow \sin a_z = \frac{\sqrt{2}}{\sqrt{3}}$$

Using the rotation matrix we obtain the direction cosines on the plane of incidence (x_1, x_3):

$$\begin{pmatrix} \nu'_1 \\ \nu'_2 \\ \nu'_3 \end{pmatrix} = \begin{pmatrix} \cos a_z & \sin a_z & 0 \\ -\sin a_z & \cos a_z & 0 \\ 0 & 0 & 1 \end{pmatrix} \begin{pmatrix} \nu_1 \\ \nu_2 \\ \nu_3 \end{pmatrix} \Rightarrow \begin{pmatrix} \frac{1}{\sqrt{2}} \\ 0 \\ \frac{1}{\sqrt{2}} \end{pmatrix}$$

The values of c and k are

$$c = \frac{\alpha}{\cos e} = \frac{\alpha'}{\cos e'} = \frac{8}{\sqrt{2}} = 4\sqrt{2}\,\text{km s}^{-1}$$

$$k = k_\alpha \cos e = k_{\alpha'} \cos e' = \frac{\sqrt{2}}{8}\,\text{km}^{-1}$$

Then the potential of the incident wave is now given by

$$\varphi_{\text{inc}} = 4 \exp i \frac{1}{4\sqrt{2}} \left(x_3 + x_1 - 4\sqrt{2}\,t \right)$$

The angle i' of the transmitted or refracted ray is found from Snell's law:

$$\frac{\sin i}{\alpha} = \frac{\sin i'}{\alpha'} \Rightarrow \sin i' = \frac{\sqrt{2}}{4} = \cos e'$$

$$\Rightarrow \cos i' = \sin e' = \frac{\sqrt{14}}{4} \Rightarrow \tan e' = \sqrt{7}$$

Using the expressions for the reflection and refraction coefficients, V and W, we can calculate the amplitude of the reflected and refracted potentials:

$$V = \frac{A}{A_0} = \frac{\rho' \tan e - \rho \tan e'}{\rho' \tan e + \rho \tan e'} = \frac{4 - 3\sqrt{7}}{4 + 3\sqrt{7}} \Rightarrow A = \frac{16 - 12\sqrt{7}}{4 + 3\sqrt{7}} = 1.07$$

$$W = \frac{A'}{A_0} = \frac{2\rho t g e'}{\rho' \tan e + \rho \tan e'} = \frac{6}{4 + 3\sqrt{7}} \Rightarrow A' = \frac{24}{4 + 3\sqrt{7}} = 2.01$$

By substitution in (122.1) we obtain

$$\varphi_{\text{refl}} = 1.07 \exp \frac{\sqrt{2}}{8} i \left(-x_3 + x_1 - \frac{8}{\sqrt{2}} t \right)$$

$$\varphi_{\text{trans}} = 2.01 \exp \frac{\sqrt{2}}{8} i \left(\sqrt{7} x_3 + x_1 - \frac{8}{\sqrt{2}} t \right)$$

123. A P-wave of amplitude $(5\sqrt{2}, 5\sqrt{6}, 10\sqrt{2})$ and frequency $\omega = 12$ rad s^{-1} in a semi-infinite medium of speed of propagation $\alpha = 6$ km s^{-1} and Poisson's ratio 0.25 is incident on the free surface. Calculate:

(a) The potential of the incident P-wave.
(b) The potential of the reflected S-wave.
(c) The components u_1, u_2, u_3 of the reflected S-wave.

(a) The displacements of the P-wave can be deduced from its scalar potential:

$$\varphi = A \exp i k_\alpha (v_1 x_1 + v_2 x_2 + v_3 x_3 - \alpha t)$$
$$\boldsymbol{u}^{\text{P}} = \nabla \varphi \tag{123.1}$$

where A is the amplitude, k_α is the wavenumber (P), v_i are the direction cosines, and α is the P-wave velocity. The wavenumber is found from the given angular frequency and velocity:

$$k_\alpha = \frac{\omega}{\alpha} = \frac{12}{6} = 2 \, \text{km}^{-1}$$

Since we know the amplitudes of the components of the displacements we can find the incidence angle i and the azimuth a_z:

$$u_1^{\text{P}} = \frac{\partial \varphi}{\partial x_1} = A k_\alpha v_1 = A2 \sin i \cos a_z = 5\sqrt{2}$$

$$u_2^{\text{P}} = \frac{\partial \varphi}{\partial x_2} = A k_\alpha v_2 = A2 \sin i \sin a_z = 5\sqrt{6}$$

$$u_3^{\text{P}} = \frac{\partial \varphi}{\partial x_3} = A k_\alpha v_3 = A2 \cos i = 10\sqrt{2}$$

Dividing the two first equations we obtain $\sqrt{3} = \tan a_z \Rightarrow a_z = 60°$, and dividing the last two,

$$\frac{5\sqrt{2}}{10\sqrt{2}} = \tan i \cos a_z = \frac{1}{2} \tan i \Rightarrow i = 45°$$

The amplitude A is given by

$$|u^{\text{P}}| = \sqrt{(u_1^{\text{P}})^2 + (u_2^{\text{P}})^2 + (u_3^{\text{P}})^2} = A k_\alpha \Rightarrow A = \frac{|u^{\text{P}}|}{k_\alpha} = 10^{-2} \, \text{m}^2$$

Finally the potential is given by

$$\varphi_{\text{inc}} = 10^{-2} \exp i2 \left(\frac{\sqrt{2}}{4} x_1 + \frac{\sqrt{6}}{4} x_2 + \frac{\sqrt{2}}{2} x_3 - 6t \right) \text{m}^2$$

If we express the potential referred to the plane (x_1, x_3) as the incidence plane, as we did in Problem 122, then

$$\varphi = A \exp ik(x_1 + \tan e \, x_3 - ct)$$

where

$$e = 90° - i = 45°, \qquad k = k_\alpha \cos e = 2\frac{1}{\sqrt{2}} = \sqrt{2}$$

$$c = \frac{\alpha}{\cos e} = \frac{6}{\frac{1}{\sqrt{2}}} = 6\sqrt{2} \, \text{km s}^{-1}$$

and the potential is

$$\varphi_{\text{inc}} = 10^{-2} \exp i\sqrt{2}\left(x_1 + x_3 - 6\sqrt{2}t\right) \text{m}^2 \qquad (123.2)$$

(b) Since Poisson's ratio is 1/4, $\lambda = \mu$, and

$$\beta = \frac{\alpha}{\sqrt{3}} = 2\sqrt{3} \, \text{km s}^{-1}$$

The angle f of the reflected S-wave is obtained by Snell's law:

$$\frac{\cos e}{\alpha} = \frac{\cos f}{\beta} \Rightarrow \cos f = \frac{\beta}{\alpha} \cos e = \frac{\sqrt{2}}{2\sqrt{3}} = \frac{1}{\sqrt{6}} \Rightarrow \sin f = \sqrt{\frac{5}{6}}$$

From the values of e and f we calculate the P-to-S reflection coefficient V_{PS}, using equation

$$V_{\text{PS}} = \frac{-4a(1 + 3a^2)}{4ab + (1 + 3a^2)^2}$$

where we substitute

$$a = \tan e = 1 \qquad \text{and} \qquad b = \tan f = \sqrt{5}$$

so

$$V_{\text{PS}} = \frac{-4(1 + 3)}{4\sqrt{5} + (1 + 3)^2} = -0.64$$

From this coefficient we calculate the proportion of the incident P-wave which is reflected as an S-wave (only with SV component; the negative sign indicates the opposite sense of the reflected ray):

$$B = AV_{\text{PS}} = -10 \times 0.64 = -6.4 \times 10^{-3} \, \text{m}^2$$

When the ray is contained in the (x_1, x_3) plane we use a scalar potential for the S-wave which in this case is given by

$$\psi = B \exp ik(x_1 - \tan f \, x_3 - ct)$$

$$= -6.4 \times 10^{-3} \exp i\sqrt{2}\left(x_1 - \sqrt{5}x_3 - 6\sqrt{2}t\right) \text{m}^2$$

(c) To calculate the amplitudes of the total displacements in terms of the two scalar potentials, we remember that for this orientation of the axes the displacements are given by

$$u_1 = \frac{\partial \varphi}{\partial x_1} - \frac{\partial \psi}{\partial x_3} = u_1^P + u_1^{SV}$$

$$u_3 = \frac{\partial \varphi}{\partial x_3} + \frac{\partial \psi}{\partial x_1} = u_3^P + u_3^{SV}$$

The displacements of the SV reflected wave in this case are

$$u_1^{SV} = -\frac{\partial \psi}{\partial x_3} = -6.4\sqrt{5}$$

$$u_3^{SV} = \frac{\partial \psi}{\partial x_1} = 6.4\sqrt{2}$$

If we want to determine the components 1 and 2, referred to the original system of axes, we project $u_1^{SV} = u_R^{SV}$ using the azimuth 60°:

$$u_1^{SV} = u_R^{SV} \cos a_z = -3.2\sqrt{5}$$

$$u_2^{SV} = u_2^{SV} \sin a_z = -3.2\sqrt{15}$$

124. An S-wave of vector potential

$$\psi i = \left(-10\sqrt{3}, 2, 4\right) \exp 5i\left(\frac{x_1}{4} + \frac{\sqrt{3}}{4}x_2 + \frac{\sqrt{3}}{2}x_3 - 4t\right)$$

is incident on the free surface $x_3 = 0$. Find the SV and SH components of the reflected S-wave referred to the same coordinate system as the incident wave, and the coefficient of reflection. Poisson's ratio is 3/8.

According to the value of Poisson's ratio the relation between λ and μ is

$$\sigma = \frac{\lambda}{2(\lambda + \mu)} = \frac{3}{8} \Rightarrow \lambda = 3\mu$$

and the relation between the velocities of the P-waves and S-waves is

$$\alpha = \sqrt{\frac{\lambda + 2\mu}{\rho}} = \sqrt{\frac{5\mu}{\rho}} = \sqrt{5}\beta$$

The incidence angle i and the azimuth a_z are obtained from the direction cosines:

$$v_3 = \cos i = \frac{\sqrt{3}}{2} \Rightarrow i = 30°$$

$$v_1 = \sin i \cos a_z = \frac{1}{4} = \sin 30 \cos a_z \Rightarrow a_z = 60°$$

Using Snell's law we find the value of the critical angle

$$\frac{\sin i_c}{\beta} = \frac{1}{\alpha} \Rightarrow \sin i_c = \frac{\beta}{\alpha} = \frac{1}{\sqrt{5}} \Rightarrow i_c = 26.5°$$

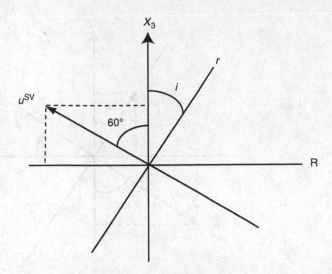

Fig. 124a

Since $i > i_c$, there is no reflected P-wave. The components of the incident S-wave are obtained from the potential

$$u_i^S = \nabla \times \psi \Rightarrow \begin{cases} u_1^S = 0 \\ u_2^S = -80 \\ u_3^S = 40 \end{cases}$$

The modulus of the displacement is $|u^S| = \sqrt{(-80)^2 + (40)^2} = 40\sqrt{5}$.

The SV component is given by (Fig. 124a)

$$u^{SV} = \frac{u_3}{\cos(90 - i_0)} = 80$$

and the SH component is

$$u^{SH} = \sqrt{(u^S)^2 - (u^{SV})^2} = -40$$

The minus sign corresponds to that of u_2^S

The amplitude of the reflected SH wave is equal to that of the incident SH wave. We find the 1 and 2 components using the azimuth 60° (Fig. 124b):

$$u_i^{SH} = -40\left(-\frac{\sqrt{3}}{2}, \frac{1}{2}, 0\right) \exp 5i\left(\frac{1}{4}x_1 + \frac{\sqrt{3}}{4}x_2 - \frac{\sqrt{3}}{2}x_3 - 4t\right)$$

For total reflection the amplitude of the reflected SV is equal to that of the incident one, but with a phase shift δ. The components are given by (Figs 124a and 124b)

$$u_1^{SV} = u^{SV} \cos i \cos a_z$$
$$u_2^{SV} = u^{SV} \cos i \sin a_z$$
$$u_3^{SV} = u^{SV} \sin i$$

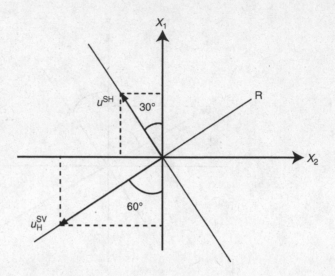

The components of the reflected SV wave are

$$u_i^{\mathrm{SV}} = 80\left(-\frac{\sqrt{3}}{4}, -\frac{3}{4}, \frac{1}{2}\right) \exp\left[5i\left(\frac{1}{4}x_1 + \frac{\sqrt{3}}{4}x_2 - \frac{\sqrt{3}}{2}x_3 - 4t\right) + i\delta\right]$$

To determine the phase shift δ we have to determine the reflection coefficients for a free surface. Textbooks usually give those for Poisson's ratio $\sigma = 1/4$, but since in this problem $\sigma = 3/8$ we have to calculate them. On a free surface the boundary conditions are that stresses are null, which in terms of the scalar potentials φ and ψ, are given by

$$\tau_{31} = 0 = \mu(u_{3,1} + u_{1,3}) = 2\varphi_{,31} - \psi_{,11} + \psi_{,33}$$
$$\tau_{33} = 0 = -\lambda(u_{1,1} + u_{3,3}) + 2\mu u_{3,3} = 3\varphi_{,11} - 5\varphi_{,33} - 2\psi_{,13} \tag{124.1}$$

where, using (x_1, x_3) as the plane of incidence, the scalar potential φ of the reflected P-waves is

$$\varphi = A \exp ik(ax_3 + x_1 - ct)$$

and the scalar potential of the incident and reflected S-wave is

$$\psi = B_0 \exp ik(bx_3 + x_1 - ct) + B \exp ik(-bx_3 + x_1 - ct)$$

Substituting in (124.1) we obtain for the coefficient of the reflected S-waves,

$$V_{\mathrm{SS}} = \frac{B}{B_0} = \frac{i4\hat{a}b - (1 - b^2)(3 + 5\hat{a}^2)}{i4ab + (1 - b^2)(3 + 5a^2)}$$

The phase shift is given by

$$\delta = \tan^{-1}\left(\frac{4\hat{a}b}{(1 - b^2)(3 - 5\hat{a}^2)}\right)$$

By substitution of the values of the problem,

$$b = \tan f = \sqrt{\frac{c^2}{\beta^2} - 1} = \frac{1}{\sqrt{3}}$$

$$\hat{a} = \sqrt{1 - \frac{c^2}{\alpha^2}} = \sqrt{\frac{22}{30}}$$

and finally we obtain $\delta = 77.33°$.

125. An S-wave represented by the potential

$$\psi = 10 \exp i3 \left(\frac{1}{2}x_1 + \frac{\sqrt{3}}{2}x_3 - 4\sqrt{2}t \right)$$

is incident from an elastic medium with $\lambda = 0$ onto a liquid with velocity $\alpha' = 4$ km s^{-1} (the two media have the same density). Derive the equations relating the amplitudes of the potentials of the incident, reflected, and transmitted waves.

Given that the wave is incident from an elastic medium onto a liquid medium, there are reflected S- and P-waves in the elastic medium and transmitted P-waves in the liquid (Fig. 125).
If $\lambda = 0$, the P-wave velocity in the elastic medium is

$$\alpha = \sqrt{\frac{\lambda + 2\mu}{\rho}} = \sqrt{\frac{2\mu}{\rho}} = \beta\sqrt{2} = 4\sqrt{2}\sqrt{2} = 8 \text{ km s}^{-1}$$

Assuming (x_1, x_3) is the incidence plane, we use the scalar S-wave potential which, for the incident and reflected waves in the solid medium, is given by

$$\psi = B_0 \exp ik_\beta(x_1 \cos f + x_3 \sin f - \beta t)$$
$$+ B \exp ik_\beta(x_1 \cos f - x_3 \sin f - \beta t)$$

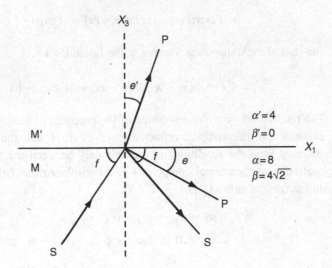

Fig. 125

From the potential of the incident wave given in the problem,

$$\cos f = \frac{1}{2} \Rightarrow f = 60°; \qquad B_0 = 10; \qquad k_\beta = 3$$

The potential can also be written in the form (Problem 122)

$$\psi = B_0 \exp ik(x_1 + \tan f x_3 - ct) + B \exp ik(x_1 - \tan f x_3 - ct)$$

where

$$k = k_\beta \cos f = \frac{3}{2} \text{ km}^{-1}$$

$$c = \frac{\beta}{\cos f} = 8\sqrt{2} \text{ km s}^{-1}$$

and the potential is given by

$$\psi = 10 \exp i\frac{3}{2}\left(x_1 + x_3\sqrt{3} - 8\sqrt{2}t\right) + B \exp i\frac{3}{2}\left(x_1 - x_3\sqrt{3} - 8\sqrt{2}t\right)$$

Applying Snell's law we determine the angle e of the reflected P-wave in the solid medium and the angle e' of the transmitted P-wave onto the liquid (Fig. 125):

$$\frac{\cos f}{\beta} = \frac{\cos e}{\alpha} \Rightarrow \frac{\cos 60}{4\sqrt{2}} = \frac{\cos e}{8} \Rightarrow e = 45°$$

$$\frac{\cos e'}{\alpha'} = \frac{\cos f}{\beta} \Rightarrow \cos e' = \frac{1}{2\sqrt{2}} \Rightarrow e' = 69°$$

$$\sin e' = \sqrt{1 - \frac{1}{8}} = \frac{\sqrt{7}}{2\sqrt{2}} \Rightarrow \tan e' = \sqrt{7}$$

The potential of the reflected P-wave in the solid medium (M) is

$$\varphi = A \exp ik(x_1 - x_3 \tan e - ct) = A \exp i\frac{3}{2}\left(x_1 - x_3 - 8\sqrt{2}t\right)$$

and that of the transmitted P-wave in the liquid (M′) is

$$\varphi' = A' \exp ik(x_1 + x_3 \tan e' - ct) = A' \exp i\frac{3}{2}\left(x_1 + x_3\sqrt{7} - 8\sqrt{2}t\right)$$

The relation between the amplitude of the potential of the incident S-wave ($B_0 = 10$) and those of the reflected and refracted P-waves A, A' and the reflected S-wave B can be obtained from the conditions at the boundary between the two media ($x_3 = 0$), that is, continuity of the normal component of the displacements (u_3) and of the stress (τ_{33}) and null tangential stress (τ_{31}):

$$u_3 = u_3' \Rightarrow \varphi_{,3} + \psi_{,1} = \varphi_{,3}'$$

$$\tau_{31} = 0 \Rightarrow 2\varphi_{,13} - \psi_{,33} + \psi_{,11} = 0$$

$$\tau_{33} = \tau_{33}' \Rightarrow \lambda'\left(\varphi_{,33}' + \varphi_{,11}'\right) = 2\mu\left(\varphi_{,33} + \psi_{,13}\right)$$

By substitution of the potentials we obtain the equations (in units of 10^{-3} m^2)

$$-A + 10 + B = A'\sqrt{7}$$
$$A + B + 10 = 0$$
$$2A' = A + 10\sqrt{3} - \sqrt{3}B$$

Solving the system of equations we obtain

$$A' = \frac{40\sqrt{3}}{4 + \sqrt{7} + \sqrt{21}} = 6.2$$

$$A = \frac{-20\sqrt{21}}{4 + \sqrt{7} + \sqrt{21}} = -8.2$$

$$B = \frac{10(-4 - \sqrt{7} + \sqrt{21})}{4 + \sqrt{7} + \sqrt{21}} = -1.8$$

126. An S-wave incident on the free surface of a semi-infinite medium with $\sigma = 0.25$ is given by (in units of 10^{-3} m^2)

$$\psi_i = \left(-10\sqrt{3}, 2, 4\right) \exp\left[5i\left(\frac{1}{4}x_1 + \frac{\sqrt{3}}{4}x_2 + \frac{\sqrt{3}}{2}x_3 - 4t\right)\right]$$

Calculate:

(a) The amplitude of the components of the reflected P-wave.
(b) The components of the reflected S-wave.

(a) From the direction cosines we calculate the incidence i and emergence f angles and azimuth a_z of the incident S-wave:

$$v_1 = \frac{1}{4} = \sin i \cos a_z$$

$$v_2 = \frac{\sqrt{3}}{4} = \sin i \sin a_z$$

$$v_3 = \frac{\sqrt{3}}{2} = \cos i \Rightarrow i = 30° \Rightarrow f = 60°$$

$$\frac{1}{4} = \frac{1}{2}\cos a_z \Rightarrow a_z = 60°$$

Since Poisson's ratio is 0.25 then $\lambda = \mu \Rightarrow \alpha = \sqrt{3}\beta = 4\sqrt{3}$, and

$$\frac{\sin i_c}{\alpha} = \frac{1}{\beta} \Rightarrow \sin i_c = \frac{1}{\sqrt{3}} \Rightarrow i_c = 35°$$

Since $i < i_c$ we have a reflected P-wave. The reflection coefficient at a free surface for a reflected P-wave from an incident S-wave is given by

$$V_{SP} = \frac{4b(1 + 3a^2)}{4ab + (1 + 3a^2)^2} \tag{126.1}$$

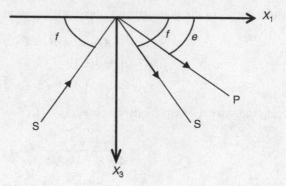

Fig. 126

where $a = \tan e$ and $b = \tan f$, and f is the emergence angles of the incident S-wave and e is that of the reflected P-wave (Fig. 126). The relation between f and e according to Snell's law is

$$\frac{\cos f}{\beta} = \frac{\cos e}{\alpha} \Rightarrow \cos e = \frac{\alpha}{\beta}\cos f = \sqrt{3}\frac{1}{2} \Rightarrow e = 30°$$

Substituting in Equation (126.1):

$$V_{SP} = \frac{4\sqrt{3}(1+1)}{4+(1+1)^2} = \sqrt{3}$$

We can write the potential of the incident S-wave referred to the incidence plane (x_1, x_3) by means of the rotation matrix

$$\begin{pmatrix} \cos a_z & \sin a_z & 0 \\ -\sin a_z & \cos a_z & 0 \\ 0 & 0 & 1 \end{pmatrix}$$

and substituting

$$\begin{pmatrix} \dfrac{1}{2} & \dfrac{\sqrt{3}}{2} & 0 \\ -\dfrac{\sqrt{3}}{2} & \dfrac{1}{2} & 0 \\ 0 & 0 & 1 \end{pmatrix} \begin{pmatrix} -10\sqrt{3} \\ 2 \\ 4 \end{pmatrix} = \begin{pmatrix} B_1 \\ B_2 \\ B_3 \end{pmatrix} \Rightarrow \begin{cases} B_1 = -4\sqrt{3} \\ B_2 = 16 = B_0 \\ B_3 = 4 \end{cases}$$

the potential is

$$\psi_i' = (B_1, B_2, B_3)\exp ik_\beta(\cos f x_1 + \sin f x_3 - 4t)$$

$$= \left(-4\sqrt{3}, 16, 4\right)\exp 5i\left(\frac{1}{2}x_1 + \frac{\sqrt{3}}{2}x_3 - 4t\right) \qquad (126.2)$$

The scalar potential of the SV component is

$$\psi_{SV} = B_0\exp ik_\beta(\cos f x_1 + \sin f x_3 - 4t)$$

$$= 16\exp 5i\left(\frac{1}{2}x_1 + \frac{\sqrt{3}}{2}x_3 - 4t\right)$$

The amplitude of the potential of the reflected P-wave is

$$A = B_0 V_{\text{SP}} = 16\sqrt{3}$$

From

$$k_\beta = \frac{\omega}{\beta} = 5 = \frac{\omega}{4} \qquad \text{we have} \qquad \omega = 20\,\text{s}^{-1}$$

and

$$k_\alpha = \frac{\omega}{\alpha} = \frac{20}{4\sqrt{3}}\,\text{km}^{-1}$$

The scalar potential of the reflected P-wave referred to the original system of axes is given by

$$\varphi = A \exp ik_\alpha\left(v_j x_j - \alpha t\right)$$

where the direction cosines are now

$$v_1 = \sin i \cos a_z = \cos e \cos a_z = \frac{\sqrt{3}}{2}\frac{1}{2} = \frac{\sqrt{3}}{4}$$

$$v_2 = \sin i \sin a_z = \cos e \sin a_z = \frac{\sqrt{3}}{2}\frac{\sqrt{3}}{2} = \frac{3}{4}$$

$$v_3 = \cos i = \sin e = \frac{1}{2}$$

Substituting these values we obtain

$$\varphi = 16\sqrt{3} \exp i\frac{5}{\sqrt{3}}\left(\frac{\sqrt{3}}{4}x_1 + \frac{3}{4}x_2 - \frac{1}{2}x_3 - 4\sqrt{3}t\right)$$

The components of the displacement in μm are

$$\mathbf{u}^P = \nabla\varphi \Rightarrow \begin{cases} u_1^P = 20\sqrt{3} \\ u_2^P = 60 \\ u_3^P = -40 \end{cases}$$

(b) For the reflected S-wave we have to separate the SV and SH components. The SV component can be deduced from the scalar potential

$$\psi_{\text{SV}}^r = B \exp ik_\beta(\cos f\, x_1 - \sin f\, x_3 - 4t) = B \exp 5i\left(\frac{1}{2}x_1 - \frac{\sqrt{3}}{2}x_3 - 4t\right)$$

We obtain B by means of the reflection V_{ss}:

$$V_{\text{SS}} = \frac{4ab - (1 + 3a^2)^2}{4ab + (1 + 3a^2)^2} = 0 = \frac{B}{B_0} \Rightarrow B = 0$$

The reflected S-wave doesn't have an SV component.

For the reflected SH component we use the displacement of the u_2 instead of the potential. The displacements of the SH component of the incident wave are obtained from (126.2):

$$u^i_{SH} = \psi'_{1,3} - \psi'_{3,1} = -40$$

Referred to the reference of the plane of incidence, the displacement is given by

$$u^i_{SH} = 40 \exp 5i\left(\frac{1}{2}x_1 + \frac{\sqrt{3}}{2}x_3 - 4t\right)$$

The amplitude of the reflected SH wave is equal to that of the incident SH wave. Referred to the incidence plane system of reference,

$$u^r_{SH} = 40 \exp 5i\left(\frac{1}{2}x_1 - \frac{\sqrt{3}}{2}x_3 - 4t\right)$$

The displacement of the reflected S-wave referred to the original system of axes is

$$u^r_i = (B^r_1, B^r_2, B^r_3) \exp i5\left(\frac{1}{4}x_1 + \frac{\sqrt{3}}{4}x_2 - \frac{\sqrt{3}}{2}x_3 - 4t\right)$$

Since the SV component is zero, $B^r_3 = 0$, B^r_1 and B^r_2 are found using the equations

$$|u^r| = u^r_{SH} \Rightarrow (B^r_1)^2 + (B^r_2)^2 = 1600$$

$$u^r_i v_i = 0 \Rightarrow \frac{1}{4}B^r_1 + \frac{\sqrt{3}}{4}B^r_2 = 0$$

resulting in

$$B^r_1 = 20\sqrt{3}$$
$$B^r_2 = -20$$

127. A wave represented by the potential

$$\varphi = 4 \exp 0.25i\left(\frac{x_1}{\sqrt{6}} + \frac{x_2}{\sqrt{3}} + \frac{x_3}{\sqrt{2}} - 4t\right)$$

is incident on the surface $x_3 = 0$ of separation between two liquids. If the speed of propagation in the second medium is 2 km s^{-1}, the pressure exerted by the incident wave on the surface of separation is 5×10^9 Pa, and the transmitted energy is four times greater than the reflected energy, calculate:

(a) The energy transmitted to the second medium.

(b) The potentials of the transmitted and reflected waves referred to the same coordinate system as the incident potential.

(a) The intensity or energy per unit surface area of the wavefront of an incident P-wave is given in units of J m^{-2} by

$$I_{inc} = A^2_0 \omega^2 k^2_\alpha \alpha \rho \tag{127.1}$$

From the given potential we have

$$A_0 = 4 \times 10^{-3} \, \text{m}^2$$

$$k_\alpha = 0.25 \, \text{km}^{-1}$$

$$\alpha = 4 \, \text{km s}^{-1}$$

$$k_\alpha = \frac{\omega}{\alpha} \Rightarrow \omega = 1 \, \text{s}^{-1}$$

We need to know the value of the density ρ. Since the medium is liquid $\lambda = K$ (bulk modulus) and then $\lambda = P/\theta$, where P is the pressure and θ the cubic dilatation (change of volume per unit volume). For liquids the shear modulus μ is zero and from the velocity of P-waves we obtain,

$$\alpha = \sqrt{\frac{\lambda + 2\mu}{\rho}} = \sqrt{\frac{\lambda}{\rho}} \Rightarrow \lambda = \alpha^2 \rho$$

The cubic dilatation is obtained from the potential φ:

$$\theta = \nabla^2 \varphi = k_\alpha^2 A = \frac{1}{16} 4 = \frac{1}{4}$$

Then, we obtain

$$\lambda = \alpha^2 \rho = \frac{P}{\theta} = \frac{5 \times 10^9}{\dfrac{1}{4} \times 16 \times 10^6} \frac{\text{Pa}}{\text{m}^2 \, \text{s}^{-2}} \Rightarrow \rho = \frac{5}{4} \, \text{g cm}^{-3}$$

By substitution in (127.1) the incident energy is

$$I_{\text{inc}} = 16 \times 1 \times \frac{1}{16} \times 4 \times \frac{5}{4} = 5 \, \text{J m}^{-2}$$

The energy transmitted into the second medium is

$$I_{\text{tras}} = A'^2 \omega^2 k_\alpha^2 \alpha' \rho' = W^2 A_0^2 k_\alpha^2 \alpha' \rho' \tag{127.2}$$

and the energy reflected is

$$I_{\text{refl}} = A \omega^2 k_\alpha^2 \alpha' \rho' = V^2 A_0^2 k_\alpha^2 \alpha' \rho' \tag{127.3}$$

where W and V are the transmission and reflection coefficients, respectively:

$$\begin{aligned} W &= \frac{A'}{A_0} = \frac{2\rho \tan e}{\rho' \tan e + \rho \tan e'} \\ V &= \frac{A}{A_0} = \frac{\rho' \tan e - \rho \tan e'}{\rho' \tan e + \rho \tan e'} \end{aligned} \tag{127.4}$$

The emergence angle e of the incident wave is

$$v_3 = \frac{1}{\sqrt{2}} = \cos i = \sin e \Rightarrow i = e = 45°$$

and from Snell's law the emergence angle of the transmitted wave e' is

$$\frac{\cos e}{\alpha} = \frac{\cos e'}{\alpha'} \Rightarrow \cos e' = \frac{\alpha'}{\alpha}\cos e = \frac{2}{4}\frac{1}{\sqrt{2}} = \frac{1}{2\sqrt{2}} \Rightarrow \sin e' = \sqrt{1 - \frac{1}{8}} = \frac{\sqrt{7}}{2\sqrt{2}}$$

Given that the transmitted energy is four times the reflected energy,

$$\frac{I_{tras}}{I_{ref}} = 4 = \frac{\dfrac{W^2 A_0^2 \rho' \omega^4}{\alpha'}}{\dfrac{V^2 A_0^2 \rho \omega^4}{\alpha}} \Rightarrow \frac{V^2}{W^2} = \frac{2\rho'\alpha}{5\rho\alpha'}$$

If we substitute in (127.4)

$$V = \frac{\rho' - \dfrac{5}{4}\sqrt{7}}{\rho' + \dfrac{5}{4}\sqrt{7}}$$

$$W = \frac{2\dfrac{5}{4}}{\rho' + \dfrac{5}{4}\sqrt{7}}$$

We have three equations for ρ', V, and W. The solution for positive values of the variables is

$$\rho' = 7.7 \Rightarrow \begin{cases} W = 0.23 \\ V = 0.40 \end{cases}$$

(b) The potential of the reflected P-wave is

$$\varphi_{ref} = VA_0 \exp ik_\alpha(v_1 x_1 + v_2 x_2 - v_3 x_3 - \alpha t)$$

$$= 1.6 \exp i\frac{1}{4}\left(\frac{1}{\sqrt{6}}x_1 + \frac{1}{\sqrt{3}}x_2 - \frac{1}{\sqrt{2}}x_3 - 4t\right)$$

To determine the potential of the transmitted wave we have to calculate the direction cosines of the transmitted ray. The azimuth is the same as that of the incident wave which can be deduced from the direction cosines and the value of i:

$$v_1 = \sin i \cos a_z = \frac{1}{\sqrt{6}} \Rightarrow \cos a_z = \frac{1}{\sqrt{3}}$$

$$v_2 = \sin i \sin a_z = \frac{1}{\sqrt{3}} \Rightarrow \sin a_z = \frac{\sqrt{2}}{\sqrt{3}}$$

The direction cosines of the transmitted ray in the medium M' are

$$v_1' = \sin i' \cos a_z = \frac{1}{2\sqrt{2}}\frac{1}{\sqrt{3}} = \frac{1}{2\sqrt{6}}$$

$$v_2' = \sin i' \sin a_z = \frac{1}{2\sqrt{3}}$$

$$v_3' = \cos i' = \frac{\sqrt{7}}{2\sqrt{2}}$$

and the potential of the transmitted wave is

$$\varphi_{\text{tras}} = 0.92 \exp i\frac{1}{2}\left(\frac{1}{2\sqrt{6}}x_1 + \frac{1}{2\sqrt{3}}x_2 + \frac{\sqrt{7}}{2\sqrt{2}}x_3 - 2t\right)$$

128. Two liquid media are separated at $x_3 = 0$, the first of volumetric coefficient $K = \frac{1}{2} \times 10^9$ Pa and density 1 g cm^{-3}. The amplitudes of the components of an incident wave of frequency 3 Hz are $u_i = 18\pi(1, 1, \sqrt{6})$ μm and those of the wave transmitted to medium 2 are $\frac{63\sqrt{2}\pi}{7}\left(\sqrt{2}, \sqrt{2}, \sqrt{3}\right)$ μm. Given that the amplitude of the transmitted potential is twice that of the reflected potential, find expressions for the incident, reflected, and transmitted potentials.

In liquids only P-waves are propagated and their displacements can be deduced from the scalar potential

$$\varphi = A_0 \exp ik_\alpha(v_1x_1 + v_2x_2 + v_3x_3 - \alpha t) \Rightarrow \mathbf{u}^P = \nabla\varphi$$

Then in our case the components of the displacement in μm are

$$u_1^P = \frac{\partial\varphi}{\partial x_1} = A_0k_\alpha v_1 = A_0k_\alpha \sin i \cos a_z = 18\pi$$

$$u_2^P = \frac{\partial\varphi}{\partial x_2} = A_0k_\alpha v_2 = A_0k_\alpha \sin i \sin a_z = 18\pi$$

$$u_3^P = \frac{\partial\varphi}{\partial x_3} = A_0k_\alpha v_3 = A_0k_\alpha \cos i = 18\pi\sqrt{6}$$

and we find $a_z = 45°$; $i = 30°$; $e = 60°$, $A_0 = 6 \times 10^{-3}$ m^2.

The emergence angle of the refracted wave, e', can be found from its displacements,

$$u_{\text{tras}}^P = \frac{63\pi\sqrt{2}}{\sqrt{7}}\left(\frac{\sqrt{2}}{\sqrt{7}}, \frac{\sqrt{2}}{\sqrt{7}}, \frac{\sqrt{3}}{\sqrt{7}}\right) \Rightarrow v_3' = \sin e' = \frac{\sqrt{3}}{\sqrt{7}} \Rightarrow \cos e' = \frac{2}{\sqrt{7}}$$

The P-wave velocity in the medium of the incident wave is

$$\alpha = \sqrt{\frac{\lambda + 2\mu}{\rho}} = \sqrt{\frac{K}{\rho}} = \frac{1}{\sqrt{2}} \text{ km s}^{-1}$$

Using Snell's law we find the velocity of the medium of the refracted wave,

$$\frac{\cos e}{\cos e'} = \frac{\alpha}{\alpha'} \Rightarrow \alpha' = \frac{2\sqrt{2}}{\sqrt{7}}$$

From the values of the velocities in the two media we calculate their densities:

$$\left.\begin{array}{l} \dfrac{\alpha^2}{\alpha'^2} = \dfrac{7}{16} = \dfrac{K\rho'}{K'\rho} = \dfrac{\frac{1}{2}\rho'}{K'} \\[4mm] \alpha' = \sqrt{\dfrac{K'}{\rho'}} = \dfrac{2\sqrt{2}}{\sqrt{7}} \Rightarrow K' = \dfrac{8}{7}\rho' \end{array}\right\} \Rightarrow \rho' = \frac{3}{2}\rho = \frac{3}{2} \text{ g cm}^{-3}$$

The reflection V and transmission W coefficients are found using their expressions and from them we get the relation between the amplitude A_0 of the incident wave potential and those of the reflected A and refracted A' waves, and substituting the value for $A_0 = 6$, we obtain

$$A' = 6 \Rightarrow A = 3 \times 10^{-3}\,\mathrm{m}^2$$

From these values we can write the potentials of the incident, reflected, and transmitted waves:

$$\varphi_{\mathrm{inc}} = 6 \exp i6\pi\sqrt{2}\left(\frac{1}{2\sqrt{2}}x_1 + \frac{1}{2\sqrt{2}}x_2 + \frac{\sqrt{3}}{2}x_3 - \frac{1}{\sqrt{2}}t\right)$$

$$\varphi_{\mathrm{ref}} = 3 \exp i6\pi\sqrt{2}\left(\frac{1}{2\sqrt{2}}x_1 + \frac{1}{2\sqrt{2}}x_2 - \frac{\sqrt{3}}{2}x_3 - \frac{1}{\sqrt{2}}t\right)$$

$$\varphi_{\mathrm{tras}} = 6 \exp i3\pi\frac{\sqrt{7}}{\sqrt{2}}\left(\frac{\sqrt{2}}{\sqrt{7}}x_1 + \frac{\sqrt{2}}{\sqrt{7}}x_2 + \frac{\sqrt{3}}{\sqrt{7}}x_3 - 2\frac{\sqrt{2}}{\sqrt{7}}t\right)$$

129. Two liquids in contact have speeds of propagation of 4 and 6 km s^{-1}. The density of the first is 2 g cm^{-3} and is less than that of the second. For waves of normal incidence, the reflected and transmitted energies are equal. A wave of $\omega = 1$ s^{-1} and with a potential of amplitude $A_0 = 2\times10^3$ cm^2 is incident from the first onto the second at an angle of 30°. Calculate:

(a) The transmitted and reflected energies.
(b) An expression for the transmitted potential.

(a) For normal incidence, the reflection and transmission coefficients in terms of the refractive index $m = \alpha/\alpha'$ and the density contrast $m = \rho'/\rho$ are given by

$$V_{\mathrm{n}} = \frac{m-n}{m+n}$$

$$W_{\mathrm{n}} = \frac{2}{m+n}$$

If the reflected energy is equal to the transmitted energy, then

$$V_{\mathrm{n}}^2 = mnW_{\mathrm{n}}^2 \Rightarrow \frac{(m-n)^2}{(m+n)^2} = \frac{4mn}{(m+n)^2} \tag{129.1}$$

Substituting $n = \alpha/\alpha' = 2/3$, from (129.1) we obtain the value of m:

$$m^2 - 4m + \frac{4}{9} = 0 \Rightarrow \begin{cases} m = 3.9 \approx 4 \\ m = 0.1 \end{cases}$$

Trying both values we obtain for ρ'

$$m = \frac{\rho'}{\rho} = 0.1 \Rightarrow \rho' = 0.2\,\mathrm{g\,cm}^{-3}$$

$$m = 4 \Rightarrow \rho' = 8\,\mathrm{g\,cm}^{-3}$$

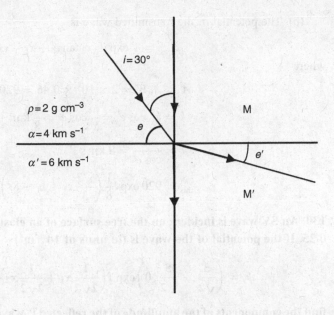

$i = 30°$

$\rho = 2 \text{ g cm}^{-3}$

$\alpha = 4 \text{ km s}^{-1}$

$\alpha' = 6 \text{ km s}^{-1}$

e

M

e'

M'

Fig. 129

But the problem states that $\rho = 4 < \rho'$, so $\rho' = 8 \text{ g cm}^{-3}$.

For a wave with incidence angle 30° ($e = 60°$) the emergence angle of the transmitted wave e' is, according to Snell's law (Fig. 129), given by

$$\frac{\cos e}{\alpha} = \frac{\cos e'}{\alpha'} \Rightarrow \cos e' = \frac{3}{4} \Rightarrow \sin e' = \sqrt{1 - \frac{9}{16}} = \frac{\sqrt{7}}{4}$$

The partition of energy between the reflected and refracted waves is given by

$$mn \frac{\sin e'}{\sin e} W^2 + V^2 = 1$$

The reflection V and transmission W coefficients are

$$V = \frac{m \sin e - n \sin e'}{m \sin e + n \sin e'} = \frac{4 \dfrac{\sqrt{3}}{2} - \dfrac{2}{3} \dfrac{\sqrt{7}}{4}}{4 \dfrac{\sqrt{3}}{2} + \dfrac{2}{3} \dfrac{\sqrt{7}}{4}} = 0.8$$

$$W = \frac{2 \sin e}{m \sin e + n \sin e'} = \frac{\sqrt{3}}{4 \dfrac{\sqrt{3}}{2} + \dfrac{2}{3} \dfrac{\sqrt{7}}{4}} = 0.46$$

The incident, reflected, and transmitted energies per unit time and surface area are (Problem 127)

$$E_{\text{inc}} = \frac{\rho \omega^4}{\alpha} A_0^2 \sin e = 17.3 \text{ erg cm}^{-2} \text{ s}$$

$$E_{\text{ref}} = \frac{\rho \omega^4}{\alpha} A^2 \sin e = \frac{\rho \omega^4}{\alpha} A_0^2 V^2 \sin e = 11.1 \text{ erg cm}^{-2} \text{ s}$$

$$E_{\text{trans}} = \frac{\rho' \omega^4}{\alpha'} A'^2 \sin e' = \frac{\rho' \omega^4}{\alpha'} A_0^2 W^2 \sin e' = 7.5 \text{ erg cm}^{-2} \text{ s}$$

(b) The potential of the transmitted wave is

$$\varphi_{tras} = A' \exp ik(x_3 \tan e' + x_1 - ct)$$

where

$$A' = A_0 W = 2 \times 10^3 \times 0.46 = 920 \text{ cm}^2$$

$$k = k'_{\alpha'} \cos e' = \frac{\omega}{\alpha'} \cos e' = \frac{1}{8} \text{ km}^{-1}$$

$$c = \frac{\alpha'}{\cos e'} = 8 \text{ km s}^{-1}$$

$$\varphi_{tras} = 920 \exp i\frac{1}{8}\left(\frac{\sqrt{7}}{3}x_3 + x_1 - 8t\right)$$

130. An SV wave is incident on the free surface of an elastic medium of Poisson ratio 0.25. If the potential of the wave is (in units of 10^{-3}m^2)

$$\psi_i = \left(\frac{5}{\sqrt{2}}, -\frac{5}{\sqrt{2}}, 0\right) \exp i\left(\frac{1}{2\sqrt{2}}x_1 + \frac{1}{2\sqrt{2}}x_2 + \frac{\sqrt{3}}{2}x_3 - 4t\right)$$

find the components of the amplitude of the reflected P-wave referred to this set of axes.

From the direction cosines we find the incidence angle i, the emergence angle f, and the azimuth a_z of the incident SV wave (Fig. 130):

$$v_3 = \cos i = \sin f = \frac{\sqrt{3}}{2} \Rightarrow i = 30° \text{ and } f = 60°$$

$$v_1 = \sin i \cos a_z = \frac{1}{2}\cos a_z = \frac{1}{2\sqrt{2}} \Rightarrow a_z = 45°$$

Bearing in mind that Poisson's ratio is 0.25, from Snell's law we find the emergence angle e of the reflected P-wave:

$$\sigma = 0.25 \Rightarrow \lambda = \mu \Rightarrow \alpha = \sqrt{3}\beta = 4\sqrt{3} \text{ km s}^{-1}$$

$$\frac{\cos f}{\beta} = \frac{\cos e}{\alpha} \Rightarrow \cos e = \frac{\sqrt{3}}{2} \Rightarrow e = 30°$$

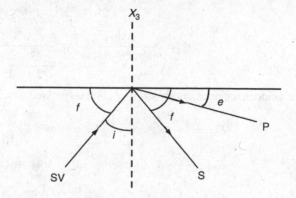

Fig. 130

The reflection coefficient for the reflected P-wave gives us the relation between the amplitude of the potential, B_0, of the incident SV wave and A, that of the reflected P wave:

$$V_{SP} = \frac{A}{B_0} = \frac{-4\tan f\,(1 + 3\tan^2 e)}{4\tan e\,\tan f + (1 + 3\tan^2 e)^2} = -\sqrt{3}$$

To find B_0 we write the potential of the incident SV wave referred to the (x_1, x_3) plane of incidence using the rotation matrix

$$\begin{pmatrix} \cos a_z & \sin a_z & 0 \\ -\sin a_z & \cos a_z & 0 \\ 0 & 0 & 1 \end{pmatrix} \begin{pmatrix} \dfrac{5}{\sqrt{2}} \\ -\dfrac{5}{\sqrt{2}} \\ 0 \end{pmatrix} = \begin{pmatrix} 0 \\ -5 \\ 0 \end{pmatrix}$$

$$B_0 = -5 \times 10^{-3}\,\mathrm{m}^2 \Rightarrow A = 5\sqrt{3} \times 10^{-3}\,\mathrm{m}^2$$

Referred to this system of axes the potential of the reflected P-wave is given by

$$\varphi = A \exp ik(x_3 \tan e + x_1 - \alpha t)$$

$$k = \frac{k_\beta}{\cos f} = \frac{1}{\frac{1}{2}} = 2\,\mathrm{km}^{-1}$$

$$\varphi = 5\sqrt{3} \exp i2\left(x_3\,\frac{1}{\sqrt{3}} + x_1 - 4\sqrt{3}t\right)$$

The amplitudes of the displacements of the reflected P-wave referred to this set of axes are

$$u_1 = \frac{\partial\varphi}{\partial x_1} = 10\sqrt{3}\,\mu\mathrm{m}$$

$$u_3 = \frac{\partial\varphi}{\partial x_3} = 10\,\mu\mathrm{m}$$

Referred to the original set of axes the horizontal components are

$$u_{1'} = u_1 \cos a_z = \frac{10\sqrt{3}}{\sqrt{2}}\,\mu\mathrm{m}$$

$$u_{2'} = u_1 \sin a_z = \frac{10\sqrt{3}}{\sqrt{2}}\,\mu\mathrm{m}$$

Ray theory. Constant and variable velocity

131. Assume that the Earth's crust consists of a single layer of thickness H and a constant speed of propagation of seismic waves of v_1 on top of a mantle of velocity of propagation 20% greater than the crust. Given that a focus on the surface produces a reflected wave that takes 17.2 s to reach a distance of 99 km, and that this is the

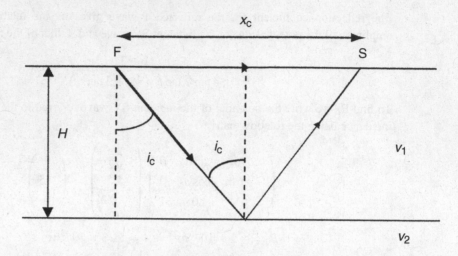

critical distance, calculate the values of H, v_1, and v_2. Plot the travel-time curve (t, x) for this specific case with numerical values.

The critical distance x_c is the distance at which a ray that is reflected with the critical angle at the top of the mantle arrives at the surface and is given by the equation (Fig. 131a)

$$x_c = 2H \tan i_c = 99 \, \text{km} \tag{131.1}$$

where H is the thickness of the crust. Since we know the relation between the velocities in the crust and the mantle, we can calculate the critical angle

$$v_2 = 1.2 v_1 \Rightarrow \frac{\sin i_c}{v_1} = \frac{1}{v_2} \Rightarrow \sin i_c = \frac{1}{1.2} \Rightarrow i_c = 56.44°$$

If we substitute in Equation (131.1) we obtain the thickness of the crust, H:

$$99 = 2H \tan 56.44° \Rightarrow H = 32.8 \, \text{km}$$

The travel time of the critically reflected ray is

$$t = 2 \frac{H}{v_1 \cos i_c} = 2 \frac{H}{v_1 \cos 56.44} = 17.2$$

$$\Rightarrow v_1 = 2 \frac{32.84}{17.2 \cos 56.44} = 6.9 \, \text{km s}^{-1} \Rightarrow v_2 = 8.3 \, \text{km s}^{-1}$$

To draw the travel-time curve for different distances of the direct, reflected, and critically refracted waves we use the equations

$$t_1 = \frac{x}{v_1}$$

$$t_2 = \frac{2}{v_1} \sqrt{\frac{x^2}{4} + H^2}$$

$$t_3 = \frac{x}{v_2} + \frac{2H \sqrt{v_2^2 - v_1^2}}{v_1 v_2}$$

We obtain the following values

x(km)	t_1 (s)	t_2 (s)	t_3 (s)
0	0	9.5	–
30	4.3	10.4	–
60	8.7	12.9	–
90	13.0	16.1	–
99	14.3	17.2	17.2
120	17.4	19.8	19.7
150	21.7	23.7	23.4

The travel-time curves are drawn in Fig. 131b.

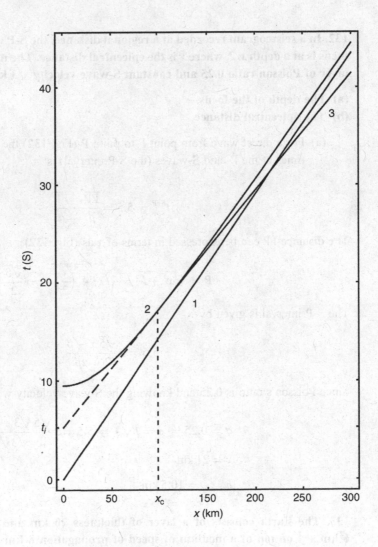

Fig. 131b

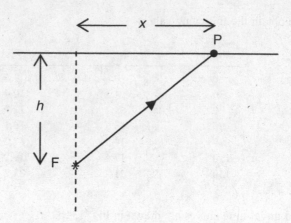

Fig. 132

132. In a seismogram recorded at a regional distance, the S-P time lag is 5.5 s, and the focus is at a depth $x/2$, where x is the epicentral distance. The model Earth has a single layer of Poisson ratio 0.25 and constant S-wave velocity $\sqrt{3}$ km s^{-1}. Calculate:

(a) The depth of the focus.
(b) The epicentral distance.

(a) For a direct wave from point F to point P (Fig. 132) the difference of the arrival times of the P- and S-waves (the S-P interval) is

$$t^{\text{S-P}} = 5.5 = \frac{\overline{\text{FP}}}{\beta} - \frac{\overline{\text{FP}}}{\alpha}$$

The distance FP can be expressed in terms of x as (Fig. 132)

$$\overline{\text{FP}} = \sqrt{x^2 + h^2} = \sqrt{x^2 + \left(\frac{x}{2}\right)^2} = x\frac{\sqrt{5}}{2}$$

The S-P interval is given by

$$5.5 = x\frac{\sqrt{5}}{2}\frac{\alpha - \beta}{\alpha\beta}$$

Since Poisson's ratio is 0.25 and knowing the S-wave velocity we obtain

$$\sigma = 0.25 \Rightarrow \alpha = \beta\sqrt{3} \Rightarrow 5.5 = x\frac{\sqrt{5}}{2}\frac{\sqrt{3} - 1}{\sqrt{3}\sqrt{3}}$$

$$x = 21 \,\text{km}$$
$$h = \frac{x}{2} = 10.5 \,\text{km}$$

133. The Earth consists of a layer of thickness 20 km and seismic wave velocity 6 km s^{-1} on top of a medium of speed of propagation 8 km s^{-1}. A seismic focus is

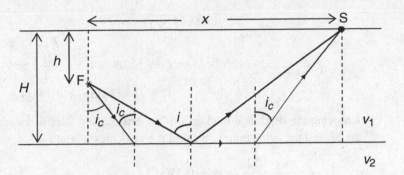

Fig. 133

located at a depth of 10 km. **Calculate the difference in travel times between the reflected and the critical refracted waves observed on the surface at a distance of 150 km from the epicentre.**

This problem is similar to Problem 131, but now the focus is at depth $h = 10$ km.

The critical distance in this case is given by (Fig. 133)

$$x_c = (2H - h)\tan i_c$$

$$\sin i_c = \frac{v_1}{v_2} \Rightarrow i_c = \sin^{-1}\left(\frac{6}{8}\right) = 48.6°$$

$$\Rightarrow x_c = (2 \times 20 - 10)\tan(48.6) = 34.0\,\text{km}$$

Since the distance 150 km is greater than the critical distance there arrive critically refracted rays. The travel times of the reflected (t_2), and critically refracted (t_3) rays at that distance are

$$t_2 = \frac{\sqrt{x^2 + (2H - h)^2}}{v_1} = \frac{\sqrt{150^2 + (2 \times 20 - 10)^2}}{6} = 25.5\,\text{s}$$

$$t_3 = \frac{x}{v_2} + \frac{(2H - h)\sqrt{v_2^2 - v_1^2}}{v_1 v_2} = \frac{150}{8} + \frac{(2 \times 20 - 10)\sqrt{8^2 - 6^2}}{8 \times 6} = 22.1\,\text{s}$$

The time difference between the travel times of the two rays is

$$t_3 - t_2 = 22.06 - 24.49 = -3.4\,\text{s}$$

134. Consider a crust of thickness H and constant speed of propagation v_1 on a mantle of constant speed of propagation v_2. A seismic focus is located at depth $H/2$, the critical distance is 51.09 km, the delay time is 4.96 s, and the critical angle is 48.59°. Calculate the values of H, v_1 and v_2, and the depth of the focus.

For a focus at depth $h = H/2$, the travel times of the critically refracted (t_3) rays and the critical distance are given by the expressions (Fig. 133).

$$t_3 = \frac{x}{v_2} + \frac{(2H - h)\sqrt{v_2^2 - v_1^2}}{v_1 v_2}$$

$$x_c = (2H - h)\tan i_c \Rightarrow 51.09 = \left(2H - \frac{H}{2}\right)\tan(48.59)$$

$$\Rightarrow H = 30\,\text{km}, \ h = 15\,\text{km}$$

Knowing the depth and thickness of the crust, using Snell's law, the value of the critical angle, and the delay time t_i, we find the velocities v_1 and v_2:

$$\frac{\sin i_c}{v_1} = \frac{1}{v_2} \Rightarrow v_1 = 0.75 v_2$$

$$t_i = \frac{(2H - h)\sqrt{v_2^2 - v_1^2}}{v_1 v_2} \Rightarrow 4.96 = \frac{(2 \times 30 - 15)\sqrt{v_2^2 - (0.75 v_2)^2}}{0.75 v_2^2}$$

$$\Rightarrow v_2 = 8\,\text{km s}^{-1}$$

$$v_1 = 6\,\text{km s}^{-1}$$

135. In a seismogram, the S-P time difference is equal to 5.31 s, and corresponds to a regional earthquake that occurred at a depth $h = 2H$, where H is the thickness of the crust. Given that the crust is formed by a layer of constant P-wave velocity of 3 km s^{-1}, that below it there is a semi-infinite mantle of double that speed of propagation, and that Poisson's ratio is 0.25, determine:

(a) **An expression for the travel-time of the P- and S-waves.**

(b) **The epicentral distance for an emerging P-wave with a take-off angle of 30° at the focus.**

(a) The travel time corresponding to the ray given in Fig. 135 is given by

$$t = \frac{\overline{FA}}{2v} + \frac{\overline{AS}}{v}$$

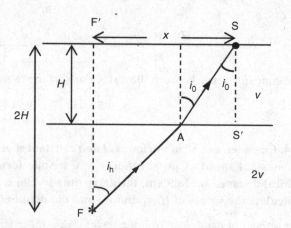

Fig. 135

where $v = \alpha$ for P-waves and $v = \beta$ for S-waves. Using Snell's law we find the relation between the incidence angle at the focus i_h and at the station i_0:

$$\frac{\sin i_h}{2v} = \frac{\sin i_0}{v} \Rightarrow \sin i_0 = \frac{1}{2}\sin i_h$$

From Fig. 135 we obtain

$$\cos i_h = \frac{H}{\overline{FA}} \Rightarrow \overline{FA} = \frac{H}{\cos i_h}$$

$$\cos i_0 = \frac{H}{\overline{AS}} \Rightarrow \overline{AS} = \frac{H}{\cos i_0}$$

From these equations we deduce the expression for the travel time:

$$t = \frac{H}{2v\cos i_h} + \frac{H}{v\cos i_0} \tag{135.1}$$

For the epicentral distance we obtain

$$x = \overline{F'A} + \overline{AS'} = H\tan i_h + H\tan i_0 \tag{135.2}$$

Using Equations (135.1) and (135.2) and putting $v = \alpha$ we obtain the travel time and epicentral distance for P-waves and putting $v = \beta$ for S-waves.

(b) For a P-wave with take-off angle at the focus (i_h) of 30°, we first find the value of the angle at the station i_0,

$$\frac{\sin i_h}{2\alpha} = \frac{\sin i_0}{\alpha} \Rightarrow i_0 = 14.47°$$

Substituting in (135.1) we obtain

$$t^P = \frac{H}{2\alpha\dfrac{\sqrt{3}}{2}} + \frac{H}{\alpha 0.97} = 1.61\frac{H}{\alpha}$$

The travel time of the S-wave with take-off angle $j_h = 30°$ can also be calculated using (135.1). Since Poisson's ratio is 0.25, the velocity of the S-wave is

$$\sigma = 0.25 \Rightarrow \alpha = \sqrt{3}\beta \Rightarrow \beta = \frac{\alpha}{\sqrt{3}} = \sqrt{3}\,\text{km s}^{-1}$$

The incidence angle at the station, j_0, using Snell's law, is given by

$$\frac{\sin j_h}{2\beta} = \frac{\sin j_0}{\beta} \Rightarrow j_0 = 14.47°$$

and the travel time is

$$t^S = \frac{H}{2\beta\cos j_h} + \frac{H}{\beta\cos j_0} = 2.79\frac{H}{\alpha}$$

Since we know the S-P time interval we can obtain the value of h:

$$t^{S-P} = 5.31 = 2.79\frac{H}{3} - 1.61\frac{H}{3} \Rightarrow h = 13.61\,\text{km}$$

The epicentral distance is found using (135.2):

$$x = 13.61 \tan 30° + 13.61 \tan 14.47° = 11.41 \text{ km}$$

136. Consider a crust composed of two layers of thickness 12 and 18 km, and constant P-wave speeds of propagation of 7 and 6 km s^{-1}, respectively, on top of a semi-infinite mantle of constant speed of propagation 8 km s^{-1}. There is a seismic focus at a depth of 6 km below the surface. For a station located at 100 km epicentral distance, calculate the travel time of the direct, reflected, and critical refracted waves (neglecting waves with more than a single reflection or critical refraction).

The travel times of the direct ray t_1 and the ray reflected on the bottom of the first layer t_2 are given by (Fig.136)

$$t_1 = \frac{\sqrt{h^2 + x^2}}{v_1} = \frac{\sqrt{6^2 + 100^2}}{7} = 14.3 \text{ s}$$

$$t_2 = \frac{\sqrt{(2H_1 - h)^2 + x^2}}{v_1} = \frac{\sqrt{(2 \times 12 - 6)^2 + 100^2}}{7} = 14.5 \text{ s}$$

As the velocity of the second layer is less than that of the first layer there is no critical refraction at that boundary. There is critical refraction at the boundary between the second layer and the mantle where the velocity is greater. Using Snell's law, we can calculate the

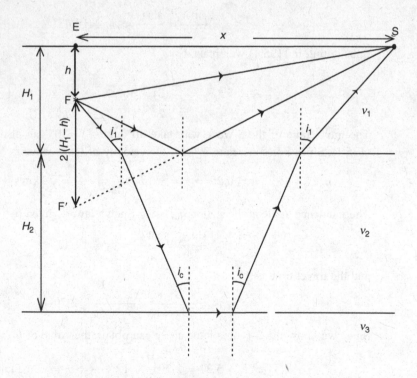

Fig. 136

critical angle i_c and from this value the incidence angle at the focus i_1 for the critically refracted ray:

$$\frac{\sin i_1}{v_1} = \frac{\sin i_c}{v_2} = \frac{1}{v_3} \Rightarrow \sin i_c = \frac{6}{8} \Rightarrow i_c = 48.6°$$

$$\sin i_1 = \frac{7}{8} \Rightarrow i_1 = 61.0°$$

The travel time t_3 of the critically refracted ray at the bottom of the second layer is given by

$$t_3 = \frac{\overline{FA}}{v_1} + \frac{\overline{AB}}{v_2} + \frac{\overline{BC}}{v_3} + \frac{\overline{CD}}{v_2} + \frac{\overline{DS}}{v_1} \tag{136.1}$$

If the epicentral distance x is 100 km, the different segments of (136.1) are

$$\cos i_1 = \frac{H_1 - h}{\overline{FA}} = \frac{H_1}{\overline{DS}} \Rightarrow \overline{FA} = 12.4\,\text{km}; \quad \overline{DS} = 24.8\,\text{km}$$

$$\cos i_c = \frac{H_2}{\overline{AB}} \Rightarrow \overline{AB} = \overline{CD} = 27.2\,\text{km}$$

$$\overline{BC} = x - 2H_2 \tan i_c - (H_1 - h)\tan i_1 - H_1 \tan i_1 = 26.8\,\text{km}$$

Finally, by substitution in (136.1) we obtain,

$$t_3 = 17.7\,\text{s}$$

137. Consider a two-layered structure of thickness H and speed of propagation v and $3v$ on top of a half-space medium of speed of propagation $2v$. At a depth $3H$ below the surface there is a seismic focus. Write the expressions (as functions of H, v, and i_h) for the travel times of waves that reach the surface without being reflected. Give the range of values of i_h.

In this problem the focus is located at the half-space medium at depth $h=3H$ under its boundary. Applying Snell's law we can find the relation between the velocities, the incidence angles at the focus and at the bottom of each layer, and the critical angle at the boundary between the second layer and the half-space (Fig. 137):

$$\frac{\sin i_h}{2v} = \frac{\sin i_2}{3v} = \frac{\sin i_1}{v}$$

$$\frac{\sin i_c}{2v} = \frac{1}{3v} \Rightarrow i_c = 41.8° \tag{137.1}$$

The rays which leave the focus and arrive at the surface at a distance x are only those with angles less than the critical angle (Fig. 137). The travel time for these rays is

$$t = \frac{\overline{FA}}{2v} + \frac{\overline{AB}}{3v} + \frac{\overline{BS}}{v} = \frac{H}{2v\cos i_h} + \frac{H}{3v\cos i_2} + \frac{H}{v\cos i_1} \tag{137.2}$$

According to Equation (137.1) we have the relation between the incidence angles:

$$\sin i_2 = \frac{3}{2}\sin i_h \Rightarrow \cos i_2 = \sqrt{1 - \sin^2 i_h} = \frac{1}{2}\sqrt{4 - 9\sin^2 i_h}$$

$$\sin i_1 = \frac{1}{2}\sin i_h \Rightarrow \cos i_1 = \sqrt{1 - \sin^2 i_h} = \frac{1}{2}\sqrt{4 - \sin^2 i_h}$$

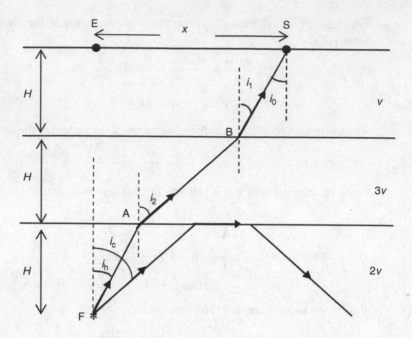

Fig. 137

Substituting in (137.2) we write the travel time as function of the take-off angle i_h:

$$t = \frac{H}{v}\left(\frac{1}{2\cos i_h} + \frac{2}{3\sqrt{4 - 9\sin^2 i_h}} + \frac{2}{\sqrt{4 - \sin^2 i_h}}\right)$$

We find a similar expression for the epicentral distance x:

$$x = H\tan i_h + H\tan i_2 + H\tan i_1$$

$$x = H\left(\tan i_h + \frac{\sin i_h}{\sqrt{4 - \sin^2 i_h}} + \frac{3\sin i_h}{\sqrt{4 - 9\sin^2 i_h}}\right)$$

The range of values of the take-off angle for rays which arrive at the surface is $0° < i_h < 42°$.

138. A semi-infinite medium consists of two media of velocities v and $3v$ separated by a vertical surface. In the first medium there is a focus of seismic waves at a depth a below the free surface and at the same distance a from the surface separating the two media. Write the expressions for the direct, reflected, and transmitted waves arriving at the free surface, and plot the travel time curve (t, x) in units of a/v and a (neglecting waves with more than a single reflection).

In this situation we have the following rays arriving at the surface: direct in the first medium, reflected at the boundary, and critically refracted and refracted to the second medium. We consider two cases for rays arriving at distances $0 < x < a$ and distances $x > a$.

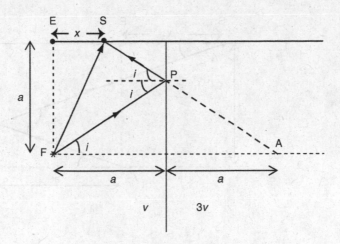

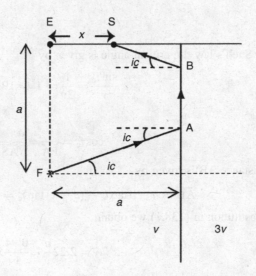

(a) For $0 < x < a$, the travel time t_1 of the direct wave is (Fig. 138a)

$$t_1 = \frac{\overline{FS}}{v} = \frac{\sqrt{x^2 + a^2}}{v}$$

The travel time t_2 of the reflected ray is

$$t_2 = \frac{\overline{FP} + \overline{PS}}{v} = \frac{\overline{AP} + \overline{PS}}{v} = \frac{\overline{AS}}{v} = \frac{\sqrt{(2a - x)^2 + a^2}}{v}$$

The reflected rays exist also for negative distances, but we will not consider them.

The travel time t_3 of the critically refracted ray (Fig. 138b) is

$$t_3 = \frac{\overline{FA}}{v} + \frac{\overline{AB}}{3v} + \frac{\overline{AS}}{v} \qquad (138.1)$$

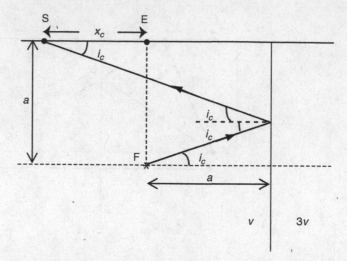

Using Snell's law the critical angle is given by

$$\frac{\sin i_c}{v} = \frac{1}{3v} \Rightarrow i_c = 19.47°$$

and

$$\cos i_c = \frac{a}{\overline{FA}} = \frac{a - x}{\overline{AS}}$$

The distance AB is given by

$$\overline{AB} = a - a \tan i_c - (a - x)\tan i_c = a - \tan i_c(x - 2a)$$

By substitution in (138.1) we obtain

$$t_3 = 2.22\frac{a}{v} - \frac{0.94}{v}x$$

The critical distance (distance of the ray reflected with the critical angle) is given by (Fig. 138c)

$$x_c = a - \overline{SB} \tag{138.2}$$

From Fig. 138c we obtain

$$\tan i_c = \frac{\overline{BP}}{\overline{SB}} = \frac{a - a\tan i_c}{\overline{SB}} \Rightarrow \overline{SB} = 1.83a$$

and substituting in (138.2)

$$x_c = -0.83a$$

Critically refracted rays exists for distances $-0.83a < x < a$. Here we consider only those for $0 < x < a$.

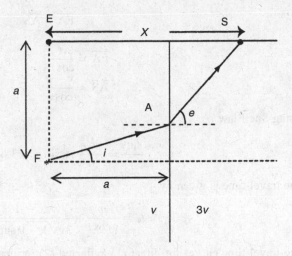

Fig. 138d

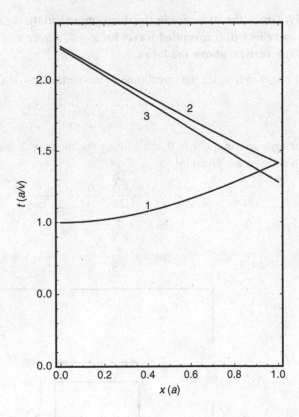

Fig. 138e

(b) For distances $x > a$, we have the rays refracted at the boundary between the two media when the incidence angle is less than the critical angle (Fig. 138d), that is, $i < 19.47°$:

$$t_4 = \frac{\overline{FA}}{v} + \frac{\overline{AS}}{3v}$$

$$\overline{FA} = \frac{a}{\cos i}$$

$$\overline{AS} = \frac{x - a}{\cos e}$$

Using Snell'law

$$\frac{\sin i}{v} = \frac{\sin e}{3v} \Rightarrow \sin e = 3 \sin i$$

The travel-time is given by

$$t_4 = \frac{a}{v \cos i} + \frac{x - a}{3v\sqrt{1 - 9 \sin^2 i}}$$

The travel-time curves for direct (1), reflected (2), critically refracted (3) and transmitted (4) waves are given in Fig. 138e.

139. Given the structure in the diagram, calculate the arrival times of the direct and (non-reflected) transmitted waves for $x \geq 0$, where $x = 0$ is a point on the free surface in the vertical above the focus.

At $x = 0 \Rightarrow i_h = 0°$, the travel-time of the vertical ray is (Fig. 139a)

$$t = \frac{a}{2v} + \frac{a}{v} = \frac{3a}{2v}$$

For rays arriving at $x > 0$ and leaving the focus with take-off angles $0 < i_h < 45°$, the travel-times are given by

$$t = \frac{\overline{FA}}{2v} + \frac{\overline{AS}}{v} = \frac{a}{2v \cos i_h} + \frac{a}{v \cos r}$$

$$\cos i_h = \frac{a}{\overline{FA}}$$

$$\cos r = \frac{a}{\overline{AS}}$$

(139.1)

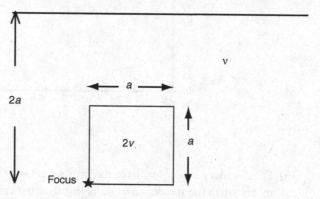

Fig. 139

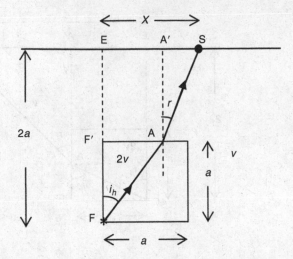

Applying Snell's law,

$$\frac{\sin i_h}{2v} = \frac{\sin r}{v} \Rightarrow \sin r = \frac{1}{2}\sin i_h \qquad (139.2)$$

Substituting in (139.1):

$$t = \frac{a}{v}\left(\frac{1}{2\cos i_h} + \frac{2}{\sqrt{4 - \sin^2 i_h}}\right) \qquad (139.3)$$

The relation between the epicentral distance x and the incidence angle i_h is

$$x = \overline{F'A} + \overline{A'S}$$

$$\overline{F'A} = a\tan i_h$$

$$\overline{A'S} = a\tan r$$

$$x = a(\tan i + \tan r) = a\left(\tan i + \frac{\sin i_h}{\sqrt{4 - \sin^2 i_h}}\right)$$

From Fig. 139a we deduce:

$$\tan i_h = \frac{\overline{F'A}}{a}$$

$$\tan r = \frac{x - \overline{F'A}}{a} = \frac{x}{a} - \tan i_h$$

Using Equation (139.2),

$$\frac{\sin r}{\cos r} = \frac{\frac{\sin i_h}{2}}{\cos r} = \frac{x}{a} - \tan i_h \Rightarrow \frac{1}{\cos r} = \frac{2\frac{x}{a}}{\sin i_h} - \frac{2}{\cos i_h}$$

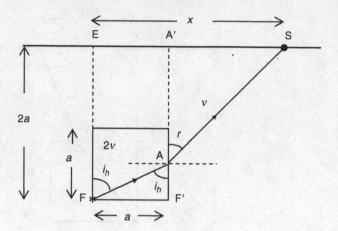

Fig. 139b

By substitution in (139.3),

$$t = \frac{2x}{v \sin i_h} - \frac{3a}{2v \cos i_h}$$

For $i_h = 45°$ the epicentral distance is

$$x = a\left(1 + \frac{1}{\sqrt{7}}\right)$$

This is the limit of the epicentral distance at which these rays arrive. The corresponding time limit is

$$t = \frac{a}{v}\left(\frac{1}{\sqrt{2}} + \frac{4}{\sqrt{14}}\right)$$

For angles $i_h > 45°$ (Fig. 139b), the travel time and epicentral distance, as a function of the take-off angle i_h, are

$$t = \frac{\overline{FA}}{2v} + \frac{\overline{AS}}{v} = \frac{a}{2v \sin i_h} + \frac{x - a}{v \sin r} = \frac{a}{2v \sin i_h} + \frac{x - a}{v \dfrac{\sin i_h}{2}}$$

$$x = a + \overline{A'S}$$

$$\tan r = \frac{\overline{A'S}}{2a - F'A}$$

$$\overline{F'A} = \frac{a}{\tan i_h}$$

$$x = a + \left(2a - \frac{a}{\tan i_h}\right)\frac{\sin i_h}{\sqrt{4 - \sin^2 i_h}}$$

Using

$$\frac{\cos i_h}{2v} = \frac{\cos r}{v} \Rightarrow \cos r = \frac{1}{2}\cos i_h$$

$$\sin r = \frac{1}{2}\sqrt{4 - \cos^2 i_h}$$

$$\tan r = \frac{\sqrt{4 - \cos^2 i_h}}{\cos i_h}$$

we obtain for x and t,

$$x = a + \left(2a - \frac{a}{\tan i_h}\right)\frac{\sqrt{4 - \cos^2 i_h}}{\cos i_h}$$

$$t = \frac{a}{2v \sin i_h} + \frac{2(x-a)}{v\sqrt{4 - \cos^2 i_h}}$$

For $i = 90°$, as expected the ray doesn't arrive at the free surface.

140. For the structure in Fig. 140a, write the equations of the travel times of the direct, reflected, and transmitted waves (neglecting waves with more than a single reflection) as a function of the epicentral distance. Determine the times of intersection, and the minimum and maximum distances in each case in terms of a/v and v. Plot the travel-time curves.

The travel-time of the direct wave for distance $0 < x < \infty$ is (Fig. 140a)

$$t = \frac{\sqrt{4x^2 + a^2}}{2v}$$

For the ray reflected on the horizontal surface at depth a the travel time is

$$t = \frac{\sqrt{x^2 + \left(\frac{3a}{2}\right)^2}}{v} = \frac{\sqrt{4x^2 + 9a^2}}{2v}$$

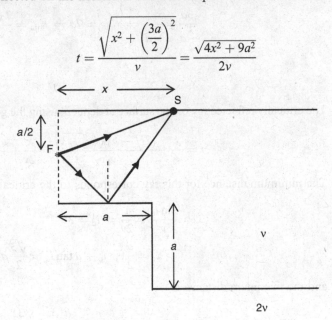

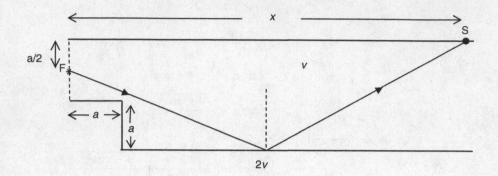

The range of distances for this ray is

$$x_{\min} = 0$$

$$x_{\max} \Rightarrow \frac{a}{\frac{a}{2}} = \frac{x_{\max}}{\frac{3a}{2}} \Rightarrow x_{\max} = 3a$$

For the reflected ray on the surface at depth $2a$ the travel time is (Fig. 140b)

$$t = \frac{\sqrt{x^2 + \left(\frac{7a}{2}\right)^2}}{v} = \frac{\sqrt{4x^2 + 49a^2}}{2v}$$

The range of distances is

$$\frac{3a}{\frac{3}{2}a} = \frac{x_{\min}}{\frac{7}{2}a} \Rightarrow x_{\min} = 7a \Rightarrow t_{\min} = \frac{7\sqrt{5}}{2}\frac{a}{v}$$

$$x_{\max} = \infty$$

The critically refracted ray on the surface at depth a, using the general expression, is given by

$$t = \frac{x}{v_2} + \frac{(2H - h)\sqrt{v_2^2 - v_1^2}}{v_1 v_2} = \frac{x}{2v} + \frac{a}{v}\frac{3\sqrt{3}}{4}$$

The minimum distance for this ray corresponds to the critical distance:

$$\frac{\sin i_c}{v} = \frac{1}{2v} \Rightarrow i_c = 30°$$

$$x_c = \frac{a}{2}\tan i_c + a\tan i_c = \frac{\sqrt{3}}{2}a$$

and the maximum distance is

$$x_{\max} = a + \frac{a}{\sqrt{3}} = \frac{a(3 + \sqrt{3})}{3}$$

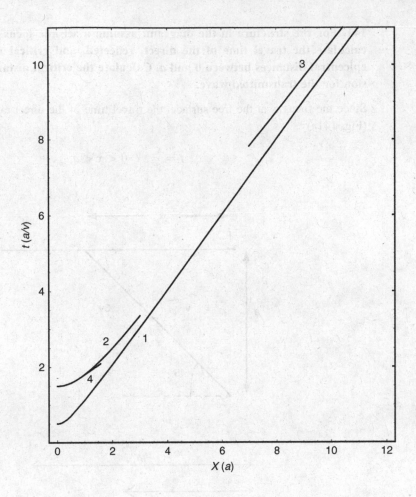

On the surface at depth $2a$ there is no critically refracted ray, since the minimum take-off angle i_h at that surface is

$$\tan i_h = \frac{a}{\dfrac{a}{2}} = 2 \Rightarrow i_h = 63.4°$$

greater than the critical angle $30°$.

The travel-time curves are drawn after rewriting the equations in units of a/v and a, and are represented in Fig. 140c

1. Direct ray: $\left(\dfrac{tv}{a}\right)^2 - \left(\dfrac{x}{a}\right)^2 = \frac{1}{4}; \quad 0 \leq \dfrac{x}{a} < \infty$ (a hyperbola)

2. Reflected ray on the surface at depth a: $\left(\dfrac{tv}{a}\right)^2 - \left(\dfrac{x}{a}\right)^2 = \dfrac{9}{4}; \quad 0 \leq \dfrac{x}{a} \leq 3$ (a hyperbola)

3. Reflected ray on the surface at depth $2a$: $\left(\dfrac{tv}{a}\right)^2 - \left(\dfrac{x}{a}\right)^2 = \dfrac{49}{4}; \quad 7 \leq \dfrac{x}{a} \leq \infty$ (a hyperbola)

4. Critically refracted ray on surface at depth a: $\dfrac{tv}{a} = \dfrac{1}{2}\dfrac{x}{a} + \dfrac{3\sqrt{3}}{4}; \quad 0.87 \leq \dfrac{x}{a} \leq 1.58$ (due to the short range of distances this is not noticeable in the figure)

141. For the structure in the diagram, assume a seismic focus at the surface, and calculate the travel time of the direct, reflected, and critical refracted waves for epicentral distances between 0 and *a*. Calculate the critical distance, and the expression for the transmitted wave.

Since the focus is at the free surface, the travel time of the direct ray is simply given by (Fig. 141a)

$$t = \frac{x}{v} \qquad 0 < x < a$$

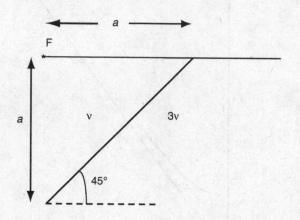

Fig. 141

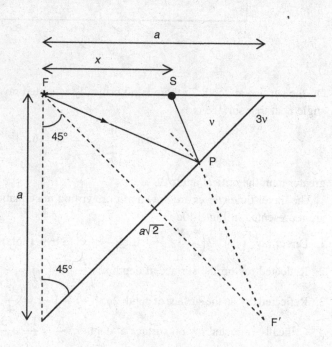

Fig. 141a

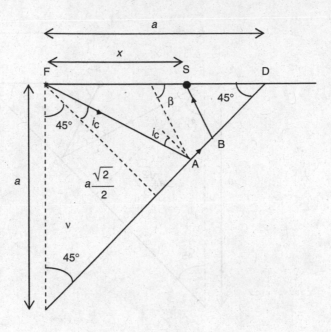

Fig. 141b

The travel time of the reflected wave is (Fig. 141a)

$$t = \frac{\overline{FP}}{v} + \frac{\overline{PS}}{v} = \frac{\overline{F'P}}{v} + \frac{\overline{PS}}{v} = \frac{1}{v}\sqrt{x^2 + 2a^2 - 2ax}$$

The critical angle is given by

$$\frac{\sin i_c}{v} = \frac{1}{3v} \Rightarrow \sin i_c = \frac{1}{3} \Rightarrow \cos i_c = \frac{2\sqrt{2}}{3} \Rightarrow i_c = 19.47°$$

The travel-time of the critically refracted ray is (Fig. 141b)

$$t = \frac{\overline{FA}}{v} + \frac{\overline{AB}}{3v} + \frac{\overline{BS}}{v}$$

$$\overline{FA} = \frac{a\frac{}{\sqrt{2}}}{\cos i_c} = \frac{3}{4}a$$

The distance BS can be found using the sine law in triangle SBD:

$$\frac{\sin(90 - i_c)}{a - x} = \frac{\sin 45}{\overline{BS}} \quad \Rightarrow \quad \overline{BS} = (a - x)\frac{3}{4}$$

and the distance AB (calling $d = BD$) is

$$\overline{AB} = a\sqrt{2} - a\cos° 45 - a\frac{\sqrt{2}}{2}\tan i_c - d = a\frac{\sqrt{2}}{2} - a\frac{\sqrt{2}}{2}\tan i_c - d$$

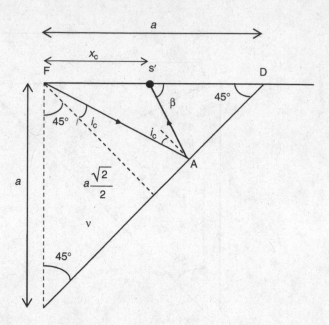

Fig. 141c

The distance d, using the sine law, is given by

$$\frac{\sin(45 + i_c)}{d} = \frac{\sin(90 - i_c)}{a - x} \Rightarrow d = (a - x)\frac{\sin(45 + i_c)}{\sin(90 - i_c)}$$

$$AB = a\frac{\sqrt{2}}{2}(1 - \tan i_c) - (a - x)\frac{\sin(45 + i_c)}{\sin(90 - i_c)} = 0.96x - 0.50a$$

The travel time is given by

$$t = -0.43\frac{x}{v} + 1.33\frac{a}{v}$$

Finally we determine the critical distance x_c from the triangle AS'D (Fig. 141c):

$$\beta + (90° - i_c) + 45° = 180° \Rightarrow \beta = 64.47°$$

$$\frac{\sin(90° - i_c)}{a - x_c} = \frac{\sin \beta}{\dfrac{a}{\sqrt{2}} - \dfrac{a}{\sqrt{2}}\tan i_c} \Rightarrow x_c = 0.52a$$

142. A medium consists of a flat crust of thickness H and constant speed of propagation v_1 on a semi-infinite mantle of constant speed of propagation v_2. For a focus at the surface, at a distance x the direct wave arrives at a time $t_1 = x/a$, the critical distance is $x_c = \dfrac{2a}{\sqrt{3}}$, and the direct and critical refracted waves intersect at the distance $x = 2a\sqrt{3}$.

(a) Calculate the crust's thickness, its speed of propagation, the mantle's speed of propagation, and the critical angle.

(b) Assume now that this is a layer which dips downwards at 45° with the parameters of the model being those determined in the previous part. Calculate the travel times of the reflected and critical refracted waves at $x = a$, $3a$, and $5a$.

(a) We determine the velocity of the crust from the travel time of the direct ray:

$$t_1 = \frac{x}{v_1} = \frac{x}{a} \Rightarrow v_1 = a$$

The critical distance is given by

$$x_c = 2H \tan i_c = 2H \frac{v_1}{\sqrt{v_2^2 - v_1^2}} = 2H \frac{a}{\sqrt{v_2^2 - a^2}} = 2 \frac{a}{\sqrt{3}} \qquad (142.1)$$

Equating the travel times of the direct and critically refracted ray for the value of the distance $x = 2a\sqrt{3}$ we obtain

$$t_1 = t_3 \Rightarrow \frac{x}{v_1} = \frac{x}{v_2} + \frac{2H\sqrt{v_2^2 - v_1^2}}{v_1 v_2} \qquad (142.2)$$

From Equations (142.1) and (142.2) we obtain the values of H and v_2:

$$H = a$$
$$v_2 = 2a$$

The critical angle may be estimated from

$$x'_c = 2H \tan i_c \Rightarrow \frac{2a}{\sqrt{3}} = 2a \tan i_c \Rightarrow i_c = 30°$$

(b) We now consider the case of a dipping layer with dip angle $\theta = 45°$. The critical distance is now given by the equation (Fig. 142)

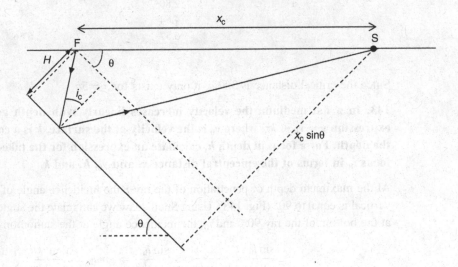

Fig. 142

$$x_c \cos \theta = H \tan i_c + (H + x_c \sin \theta) \tan i_c \qquad (142.3)$$

so

$$x_c = \frac{2H \tan i_c}{\cos \theta - \sin \theta \tan i_c}$$

The critical distance along the horizontal free surface x_c is found by substituting H, θ, and i_c in (142.3)

$$x_c = 3.86a$$

The travel times of the reflected and critically refracted rays for a dipping layer are given by the equations

$$t_2 = \frac{\sqrt{4H^2 + x^2 + 4Hx \sin \theta}}{v_1}$$

$$t_3 = \frac{x \cos \theta}{v_2} + \frac{x \sin \theta + 2H}{v_1 v_2} \sqrt{v_2^2 - v_1^2}$$

By substitution of the values of H, v_1, v_2, and θ, we obtain

$$t_2 = \frac{\sqrt{4a^2 + x^2 + 2\sqrt{2}ax}}{a}$$

$$t_3 = \frac{x\sqrt{2}}{4a} + \frac{x\frac{\sqrt{2}}{2} + 2a}{2a} \sqrt{3} = \frac{x\sqrt{2}}{4a} \left(1 + \sqrt{3}\right) + \sqrt{3}$$

For the required values of x, we obtain the following values of the travel time in units of a/v:

x	t_2 (a/v)	t_3 (a/v)
a	2.80	–
3a	4.64	–
5a	6.57	6.56

Since the critical distance is $3.86a$, it only exists for $x = 5a$.

143. In a flat medium, the velocity increases linearly with depth according to the expression $v = v_0 + kz$, where v_0 is the velocity at the surface, k is a constant, and z is the depth. For a focus at depth h, calculate an expression for the take-off angle at the focus i_h in terms of the epicentral distance x, and v_0, h, and k.

At the maximum depth of penetration of the ray r the incidence angle of the ray with the vertical is equal to $90°$ (Fig. 143). Using Snell's law we can relate the angles at the focus i_h, at the bottom of the ray $90°$, and i_0, the incidence angle at the station on the surface:

$$\frac{\sin i_0}{v_0} = \frac{1}{v_0 + kr} = \frac{\sin i_h}{v_0 + kh} \Rightarrow \sin i_h = \frac{v_0 + kh}{v_0 + kr} \qquad (143.1)$$

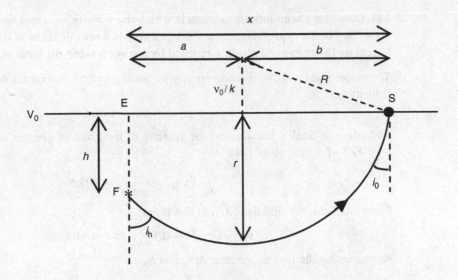

Fig. 143

The problem is solved if we can express r as a function of x, v_0, h, and k. The epicentral distance x is the sum of a and b (Fig. 143):

$$b = \sqrt{R^2 - \left(\frac{v_0}{k}\right)^2}$$

But we know that for a distribution of velocities which increases linearly with depth the rays are circular and their radius R is

$$R = \frac{v_0}{k} + r \Rightarrow b = \sqrt{\left(\frac{v_0}{k} + r\right)^2 - \left(\frac{v_0}{k}\right)^2}$$

and

$$a = \sqrt{\left(\frac{v_0}{k} + r\right)^2 - \left(\frac{v_0}{k} + h\right)^2}$$

Therefore,

$$b = x - a \Rightarrow \sqrt{\left(\frac{v_0}{k} + r\right)^2 - \left(\frac{v_0}{k}\right)^2} = x - \sqrt{\left(\frac{v_0}{k} + r\right)^2 - \left(\frac{v_0}{k} + h\right)^2}$$

Solving for r we obtain

$$r = \sqrt{\left(\frac{x^2 - h^2 - 2\frac{v_0}{k}h}{2x}\right)^2 + \left(\frac{v_0}{k} + h\right)^2} - \frac{v_0}{k}$$

By substitution of this expression for r in (143.1) we obtain the required expression for i_h.

144. Consider a semi-infinite medium in which the velocity increases linearly with depth according to the expression $v = 4 + 0.1\,z$. There is a seismic focus at a depth of 10 km. Calculate the epicentral distance reached by a wave leaving the focus at an angle of 30°.

The velocity at the focus is found directly by putting in the equation for the distribution of velocity, $z = h$:

$$v_h = 4 + 0.1 \times 10 = 5\,\text{km s}^{-1}$$

According to Snell's law we find the velocity at the point of greatest depth penetration ($i = 90°$) of the ray (Fig. 144):

$$\frac{\sin i_h}{v_h} = \frac{1}{v_m} \Rightarrow v_m = \frac{5}{\sin 30} = 10\,\text{km s}^{-1}$$

From this value we find the depth to that point:

$$v_m = 10 = 4 + 0.1 \times r \Rightarrow r = 60\,\text{km}$$

Knowing that the rays are circular of radius R,

$$R = r + \frac{v_0}{k} = 60 + 40 = 100\,\text{km}$$

As in the previous problem, the epicentral distance (from point E to S) is (Fig. 144):

$$x = a + b$$

where

$$a = \sqrt{R^2 - \left(\frac{v_0}{k} + h\right)^2} = 86.60\,\text{km}$$

$$b = \sqrt{R^2 - \left(\frac{v_0}{k}\right)^2} = 91.65\,\text{km}$$

so

$$x = 178.25\,\text{km}$$

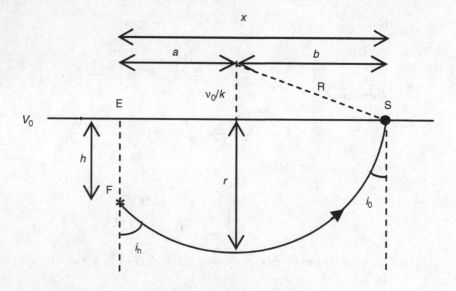

Fig. 144

145. A flat medium consists of a layer of thickness H and constant speed of propagation v on top of a medium of variable speed of propagation $v = v_0 + k(z - H)$ where z is the depth and k is a constant. If there is a focus at the surface:

(a) Write expressions for the epicentral distance x and the travel time t as functions of the angle of incidence i_0 at the surface.

(b) If $H = 10$ km, $k = 0.1$ s^{-1}, and $v_0 = 6$ km s^{-1}, calculate the angle of incidence of a wave that reaches an epicentral distance of 140 km.

(a) As we saw in Problem 143, for a distribution with a linear increase of velocity with depth, now in the medium under the layer, $v = v_0 + k(z - H)$, the rays are circular with radius $R = \dfrac{v_0}{k} + r$ where r is the maximum depth of penetration of the ray (Fig. 145). The travel-time of the ray that crosses the layer and penetrates the medium is given by

$$t = 2\frac{\overline{FP}}{v_0} + \frac{2}{k}\sinh^{-1}\frac{kx'}{2v_0} \qquad (145.1)$$

In the layer of constant velocity the path is a straight line and in the medium it is circular. The epicentral distance x (from F to S) is (Fig. 145)

$$x = x' + 2H \tan i_0$$

The length of the straight ray in the layer is

$$\overline{FP} = \frac{H}{\cos i_0}$$

Substituting in (145.1):

$$t = \frac{2H}{v_0 \cos i_0} + \frac{2}{k}\sinh^{-1}\left(\frac{kx'}{2v_0}\right) \qquad (145.2)$$

Since the layer has constant velocity the angle i_0 is the same at the focus as at the bottom of the layer at the boundary with the medium. According to Snell's law

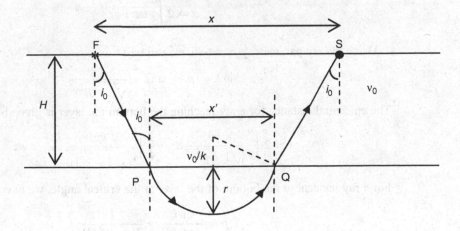

Fig. 145

$$\frac{\sin i_0}{v_0} = \frac{1}{v_0 + kr} \Rightarrow v_m = v_0 + kr = \frac{v_0}{\sin i_0}$$

where r is the maximum depth reached by the ray in the medium and v_m the velocity at that depth. According to Fig. 145,

$$\left(\frac{x'}{2}\right)^2 + \left(\frac{v_0}{k}\right)^2 = \left(\frac{v_0}{k} + r\right)^2 = \frac{v_m^2}{k^2} = \frac{v_0^2}{k^2 \sin^2 i_0} \Rightarrow x' = 2\frac{v_0}{k} \cot i_0 \qquad (145.3)$$

The epicentral distance x is given by

$$x = x' + 2H \tan i_0 = \frac{2v_0}{k} \cot i_0 + 2H \tan i_0 \qquad (145.4)$$

Substituting in (145.2) the expression for x' in terms of i_0 (143.3) we obtain

$$t = \frac{2H}{v_0 \cos i_0} + \frac{2}{k} \sinh^{-1}(\cot i_0)$$

(b) By substituting the given values in (145.4), we obtain

$$140 = \frac{2 \times 6}{0.01} \cot i_0 + 2 \times 10 \tan i_0 \Rightarrow \begin{cases} i_0 = 45° \\ i_0 = 80.5 \end{cases}$$

146. Beneath a layer of thickness H of velocity distribution $v = v_0 + kz$ there is a semi-infinite medium of speed of propagation $v_1 = 2(v_0 + kH)$.

(a) **Determine expressions (as functions of the above parameters) for the critical distance, the time of intersection of the reflected wave, and the maximum distance of the direct wave.**

(b) **For $H = 10$ km, $v_0 = 1$ km s^{-1}, and $k = 0.1$ s^{-1}, calculate these parameters and plot the travel-time curves.**

(a) In a layer of thickness H with variable velocity the epicentral distance x for a reflected ray is given by

$$x = 2 \int_0^H \tan i \, dz$$

Using the ray parameter $p = \sin i/v$ we can write

$$\sin i = vp \Rightarrow \tan i = \frac{pv}{\sqrt{1 - p^2 v^2}}$$

The epicentral distance for a ray reaching the bottom the layer is given by

$$x = 2 \int_0^H \frac{vp}{\sqrt{1 - p^2 v^2}} \, dz = \frac{2}{k} \int_0^H \frac{kp(v_0 + kz)}{\sqrt{1 - p^2(v_0 + kz)^2}} \, dz \qquad (146.1)$$

For a ray incident at the bottom of the layer at the critical angle, we have

$$p = \frac{\sin i_c}{v_0 + kH} = \frac{1}{2(v_0 + kH)}$$

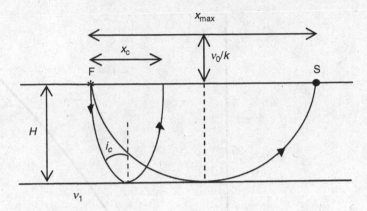

Fig. 146a

Substituting this expression in (146.1) and evaluating the integral, making the change of variable $u = v_0 + kz$, we obtain the critical distance

$$x_c = -\frac{4(v_0 + kH)}{k}\left[\frac{\sqrt{3}}{2} - \sqrt{1 - \frac{v_0^2}{4(v_0 + kH)^2}}\right] \tag{146.2}$$

The intercept time for $x = 0$, corresponding to the time of the reflected vertical ray ($p = 0$), is given by

$$t_i = 2\int_0^H \frac{dz}{v_0 + kz} = \frac{2}{k}\ln\frac{v_0 + kH}{v_0} \tag{146.3}$$

The maximum distance x_{max} corresponds to the last ray propagated inside the layer without penetrating into the medium and has a circular path of radius $R = H + \dfrac{v_0}{k}$ (Fig. 146a):

$$\left(\frac{x_{max}}{2}\right)^2 + \left(\frac{v_0}{k}\right)^2 = \left(H + \frac{v_0}{k}\right)^2 \Rightarrow x_{max} = 2H\sqrt{1 + \frac{2v_0}{kH}} \tag{146.4}$$

(b) For the particular case with the values, $H = 10$ km, $v_0 = 1$ km s^{-1}, and $k = 0.1$ s^{-1}, the velocity at the bottom of the layer H is

$$v_H = v_0 + kH = 1 + 10 \times 0.1 = 2\,\text{km s}^{-1}$$

The velocity of the medium is

$$v_1 = 2(v_0 + kH) = 2(1 + 10 \times 0.1) = 4\,\text{km s}^{-1}$$

The critical distance, using Equation (146.2), is

$$x_c = -\frac{4 \times 2}{0.1}\left[\frac{\sqrt{3}}{2} - \sqrt{1 - \frac{1}{4(2)^2}}\right] = -20\sqrt{3}\left(2 - \sqrt{5}\right) = 8.2\,\text{km}$$

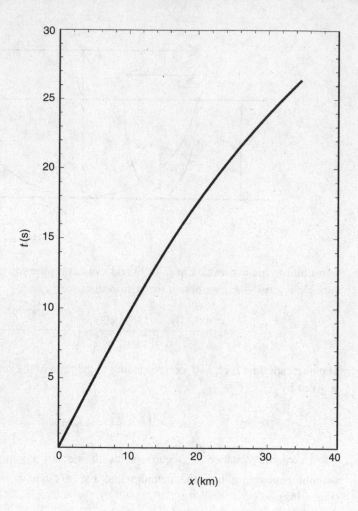

The intercept time (146.3) of the reflected ray is

$$t_i = \frac{2}{k}\ln\frac{v_0 + kH}{v_0} = \frac{2}{0.1}\ln\frac{1 + 0.1 \times 10}{1} = 13.9\,\text{s}$$

and the maximum distance for the ray in the layer (146.4) is

$$x_{max} = 2 \times 10\sqrt{1 + \frac{2 \times 1}{0.1 \times 10}} = 20\sqrt{3} = 34.6\,\text{km}$$

The travel-time curve for rays inside the layer is calculated using the expression

$$t = \frac{2}{k}\sinh^{-1}\left(\frac{kx}{2v_0}\right) = 20\sinh^{-1}(0.05x)$$

and is represented in Fig. 146b.

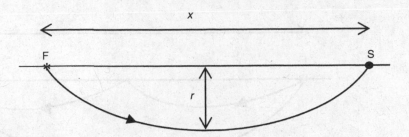

Fig. 147

147. A medium has a distribution of velocity with depth of the form $v = v_0\,e^{\alpha z}$, with $0 < \alpha < 1$. Write as functions of the epicentral distance x the expressions for the ray parameter, travel-time, and maximum depth reached.

If r is the maximum depth reached for a ray with ray parameter p (Fig. 147), the epicentral distance x is given by

$$x = \int_0^r \frac{pv\,dz}{\sqrt{1 - p^2 v^2}} = \int_0^r \frac{pv_0 e^{\alpha z}\,dz}{\sqrt{1 - p^2 v_0^2 e^{2\alpha z}}}$$

$$= \left[\frac{2}{\alpha}\sin^{-1}(pv_0 e^{\alpha z})\right]_0^r = \frac{2}{\alpha}\left[\sin^{-1}(pv_0 e^{\alpha r}) - \sin^{-1}(pv_0)\right]$$

and, as $p = \dfrac{1}{v_0 e^{\alpha r}}$, we have

$$x = \frac{2}{\alpha}\left[\frac{\pi}{2} - \sin^{-1}(pv_0)\right]$$

From this expression we obtain

$$p = \frac{1}{v_0}\cos\frac{\alpha x}{2}$$

The travel-time is given by

$$t = \int_0^x p\,dx = \frac{1}{v_0}\int_0^x \cos\frac{\alpha x}{2}\,dx = \frac{2}{v_0\alpha}\sin\frac{\alpha x}{2}$$

To find the maximum depth of penetration r of a ray arriving at distance x, we write

$$p = \frac{1}{v_0 e^{\alpha r}} = \frac{1}{v_0}\cos\frac{\alpha x}{2} \Rightarrow r = -\frac{1}{\alpha}\ln\left(\cos\frac{\alpha x}{2}\right)$$

148. In a semi-infinite medium of speed of propagation $v = 6\exp\left(\dfrac{z}{2}\right)$, the P-wave emerges with an angle of incidence of $30°$. Calculate the difference in arrival times at a given station of the P-wave and the PP-wave (the wave reflected once at the free surface). At what angle of incidence does the PP-wave emerge?

For a velocity distribution increasing with depth of the type $v = v_0 e^{\alpha\,z}$ (in our case with $v_0 = 6$, $\alpha = 1/2$) rays follow a curved path. For a focus on the free surface the ray parameter p and the travel times t are given by (Problem 147)

$$p = \frac{1}{v_0}\cos\left(\frac{\alpha x}{2}\right)$$

$$t = \frac{2}{\alpha v_0}\sin\left(\frac{\alpha x}{2}\right)$$

$$(148.1)$$

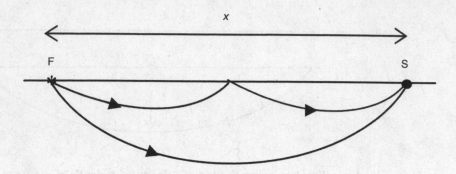

For a ray with incidence angle at the surface $i_0 = 30°$, the ray parameter of the direct P-wave is given by

$$p = \frac{\sin i_0}{v_0} = \frac{1}{2 \times 6} = \frac{1}{12}$$

Substituting this value in (148.1) we obtain, for the epicentral distance x,

$$\frac{1}{12} = \frac{1}{6}\cos\left(\frac{x}{2 \times 2}\right) \Rightarrow x = \frac{4\pi}{3} \text{ km}$$

The corresponding travel time is

$$t^P = \frac{2}{\frac{1}{2}6}\sin\left(\frac{1}{2}\frac{4\pi}{3}\frac{1}{2}\right) = 0.58 \text{ s}$$

The travel time of the reflected PP-wave (Fig. 148) is double that of the direct P-wave arriving at the distance $x/2$:

$$t^{PP} = 2\frac{2}{\frac{1}{2}6}\sin\left(\frac{1}{2}\frac{4\pi}{6}\frac{1}{2}\right) = 0.67 \text{ s}.$$

The difference between the two times is

$$t^{PP} - t^P = 0.67 - 0.58 = 0.09 \text{ s}.$$

To calculate the incidence angle of the PP-wave, we determine first the ray parameter p corresponding to the distance $x/2$:

$$\frac{x}{2} = \frac{4\pi}{3 \times 2} = \frac{4\pi}{6}$$

so

$$p = \frac{1}{6}\cos\left(\frac{1}{2}\frac{4\pi}{6}\frac{1}{2}\right) = \frac{\sqrt{3}}{12}$$

From the value of p, using Snell's law, we obtain i_0:

$$p = \frac{\sin i_0}{v_0} \Rightarrow \sin i_0 = \frac{\sqrt{3}}{12} 6 \Rightarrow i_0 = 60°$$

149. A layer of thickness H has a velocity distribution $v = v_0 \exp(\alpha z)$ where $\alpha < 1$. Beneath it there is a semi-infinite medium of speed of propagation $v_1 = 2v_0 \exp(\alpha H)$. Determine in terms of v_0, v_1, α, and H:

(a) The distance and critical angle.
(b) The time of intersection of the reflected wave.
(c) The maximum distance of the direct wave.
(d) Calculate the values of these parameters if $v = 1\,\text{km}\,\text{s}^{-1}$, $H = 10\,\text{km}$, and $\alpha = 0.1\,\text{km}^{-1}$.

(a) The distance for a ray reaching a depth H is given by (Fig. 149)

$$x = 2 \int_0^H \tan i\, dz \qquad (149.1)$$

and using the definition of the ray parameter p,

$$p = \frac{\sin i}{v} \Rightarrow \sin i = vp$$

$$\cos i = \sqrt{1 - v^2 p^2}$$

$$\tan i = \frac{vp}{\sqrt{1 - v^2 p^2}}$$

Substituting in (149.1) we obtain

$$x = 2 \int_0^H \frac{vp}{\sqrt{1 - v^2 p^2}}\, dz = 2 \int_0^H \frac{p v_0 e^{\alpha z}}{\sqrt{1 - p^2 v_0^2 e^{2\alpha z}}}\, dz \qquad (149.2)$$

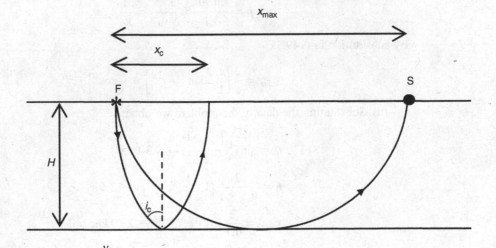

Fig. 149

Introducing the change of variable

$$u = v_0 e^{\alpha z}$$
$$du = v_0 e^{\alpha z} \alpha \, dz$$

we find

$$x = \frac{2}{\alpha} \left[\sin^{-1} p v_0 e^{\alpha H} - \sin^{-1} p v_0 \right]$$

For the critical angle i_c at the bottom of the layer we have

$$p = \frac{\sin i_0}{v_0} = \frac{\sin i_c}{v_H} = \frac{1}{v_1} \Rightarrow i_c = \sin^{-1} \frac{v_H}{v_1} = \sin^{-1} \frac{v_0 e^{\alpha H}}{v_1}$$

By substitution in (149.2) we find, for the critical distance x_c,

$$x_c = \frac{2}{\alpha} \left[\sin^{-1} \frac{v_0 e^{\alpha H}}{v_1} - \sin^{-1} \frac{v_0}{v_1} \right] = \frac{2}{\alpha} \left[\sin^{-1} \left(\frac{v_H}{v_1} \right) - \sin^{-1} \left(\frac{v_0}{v_1} \right) \right] = \frac{2}{\alpha} (i_c - i_0)$$

(b) The intercept time of the reflected ray corresponding to the vertical ray ($x = 0$ and $p = 0$) is

$$t_i = 2 \int_0^H \frac{dz}{v} = 2 \int_0^H \frac{1}{v_0} e^{-\alpha z} dz = \frac{2}{\alpha v_0} \left[1 - e^{-\alpha H} \right]$$

(c) The maximum distance of a ray contained in the layer is given by (149.2)

$$x_{max} = 2 \int_0^H \frac{vp}{\sqrt{1 - p^2 v^2}} dz = \frac{2}{\alpha} \left[\sin^{-1} p v_0 e^{\alpha H} - \sin^{-1} p v_0 \right] \qquad (149.3)$$

At the point of greatest depth penetration the incidence angle is 90° and, according to Snell's law,

$$p = \frac{\sin 90°}{v_H} = \frac{e^{-\alpha H}}{v_0} = \frac{1}{v_H} = \eta_p$$

By substitution in (149.3),

$$x_{max} = \frac{2}{\alpha} \left[\frac{\pi}{2} - \sin^{-1} e^{-\alpha H} \right] = \frac{2}{\alpha} \cos^{-1} e^{-\alpha H}$$

(d) Substituting the data of the problem we obtain

$$i_c = \sin^{-1} \frac{1 \times e^{0.1 \times 10}}{5.62} = 30°$$

$$t_i = \frac{2}{0.1 \times 1} \left[1 - e^{-0.1 \times 10} \right] = 12.6 \, s$$

$$x_{max} = \frac{2}{0.1} \cos^{-1} e^{-0.1 \times 10} = 23.9 \, km$$

$$x_c = \frac{2}{0.1} \left[\sin^{-1} \frac{1}{2} - \sin^{-1} \frac{1}{5.62} \right] = 6.8 \, km$$

Ray theory. Spherical media

150. Assume that the Earth consists of two concentric regions of constant velocity: the core of radius $R/2$ and the mantle. The speed of propagation in the core is twice that of the mantle. Calculate:

(a) The maximum angular distance of the direct ray in the mantle.
(b) The critical angular distance of the refracted ray in the core.
(c) Plot the paths of the waves that propagate through the Earth's interior, and the travel-time curves of these waves in units of R/v, where v is the speed of propagation in the mantle.

   (a) The travel time of the direct ray in the mantle in terms of the angular distance is given by

$$t_1 = 2\frac{R}{v}\sin\frac{\Delta}{2}$$

where $0 \le \Delta \le \Delta_{max}$ and Δ_{max} is the maximum distance for a ray contained in the mantle. According to Fig. 150a the last ray which propagates in the mantle without entering the core corresponds to angular distance Δ_{max} which in our case is

$$\cos\frac{\Delta_{max}}{2} = \frac{\frac{R}{2}}{R} \Rightarrow \Delta_{max} = 120°$$

   (b) The critical angle for a ray incident at the core is

$$\frac{\sin i_c}{v} = \frac{1}{2v} \Rightarrow i_c = 30°$$

To calculate the critical distance Δ_c we consider the relation (Fig. 150b)

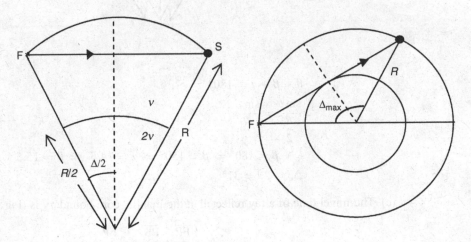

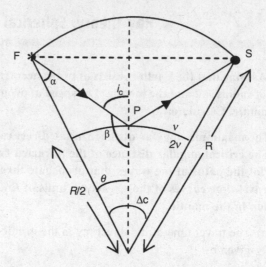

Fig. 150b

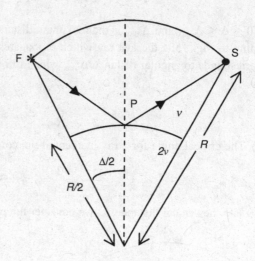

Fig. 150c

$$\theta + \beta + \alpha = 180°$$

$$\frac{\sin \alpha}{\dfrac{R}{2}} = \frac{\sin \beta}{R}$$

$$i_c + \beta = 180° \Rightarrow \beta = 150° \Rightarrow \alpha = 14.5° \Rightarrow \theta = 15.5°$$

$$\Delta_c = 2\theta = 31°$$

(c) The travel time of a ray reflected at the mantle–core boundary is (Fig. 150c)

$$t_2 = \frac{\overline{FP}}{v} + \frac{\overline{PS}}{v} = 2\frac{\overline{FP}}{v} \tag{150.1}$$

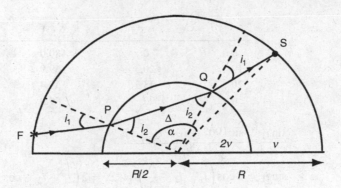

According to the cosine law,

$$\overline{FP}^2 = \left(\frac{R}{2}\right)^2 + R^2 - 2\frac{R}{2}R\cos\frac{\Delta}{2}$$

Substituting in (150.1):

$$t_2 = \frac{R}{v}\sqrt{5 - 4\cos\frac{\Delta}{2}}$$

The travel times for the minimum and maximum angular distances are

$$\Delta_{\min} = 0 \Rightarrow \quad t = \frac{R}{v}$$

$$\Delta_{\max} = 120° \Rightarrow t = \sqrt{3}\frac{R}{v}$$

The travel time for a ray which enters the core, that is, with $i_1 < i_c$, can be determined according to Fig. 150d using Snell's law:

$$\frac{\sin i_1}{v} = \frac{\sin i_2}{2v}$$

$$\sin i_2 = 2\sin i_1$$

(150.2)

Adding the times of the paths through the mantle and the core:

$$t_3 = \frac{\overline{FP}}{v} + \frac{\overline{PQ}}{2v} + \frac{\overline{QS}}{v} = \frac{2\overline{FP}}{v} + \frac{\overline{PQ}}{2v}$$

$$\overline{FP}^2 = R^2 + \frac{R^2}{4} - 2R\frac{R}{2}\cos\left(\frac{\Delta - \alpha}{2}\right)$$

$$\overline{PQ} = 2\frac{R}{2}\sin\frac{\alpha}{2}$$

Because $\alpha = 180° - 2i_2$, we obtain

$$t_3 = \frac{2R}{v}\sqrt{\frac{5}{4} - \sin\left(\frac{\Delta}{2} + i_2\right)} + \frac{R}{2v}\cos i_2$$

(150.3)

for values of the incidence angle $0 < i < i_c$, corresponding to distances $\Delta_c < \Delta < 180°$.

The relation between the incidence angle i_1 and angular distance Δ is given by

i_1 (°)	i_2 (°)	Δ (°)	t_3 (R/v)
0	0	180.0	1.50
10	20.3	169.9	1.53
20	43.2	153.4	1.60
30	90	89.0	1.46

$$\frac{\sin i_1}{R} = \frac{\sin\left(90° + i_1 - i_2 - \frac{\Delta}{2}\right)}{R/2}$$

$$\sin i_1 = 2\cos\left(i_1 - i_2 - \frac{\Delta}{2}\right) \Rightarrow \Delta = 2\left(i_1 - i_2 - \cos^{-1}\left(\frac{\sin i_1}{2}\right)\right) \tag{150.4}$$

Using Equations (150.2), (150.3), and (150.4) we can calculate the travel times of the direct ray in the mantle, the reflected ray, and the transmitted ray through the core. Some values for the transmitted rays in the core are given in the table.

The travel-time curves for direct rays (1), reflected rays (2), and rays refracted in the core (3) are shown in Fig. 150e.

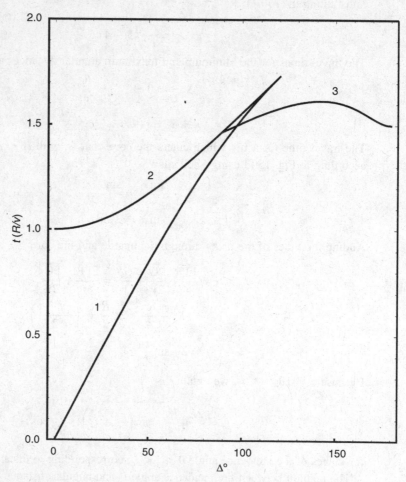

Fig. 150e

151. Assume that the Earth consists of two concentric regions of constant velocity: the core of radius $R/2$ and the mantle. The speed of propagation in the core is half that of the mantle. Plot the travel-time curves of the waves that propagate in the interior of the Earth in units of R/v where v is the speed of propagation in the mantle.

This problem is similar to Problem 150, but now the velocity of the core is less than that of the mantle. In the mantle we have direct and reflected rays. As in Problem 150 the maximum angular distance for the direct wave is 120°. The travel times for the direct (t_1) and reflected (t_2) rays are

$$t_1 = 2\frac{R}{v}\sin\frac{\Delta}{2} \tag{151.1}$$

$$t_2 = \frac{R}{v}\sqrt{5 - 4\cos\frac{\Delta}{2}} \tag{151.2}$$

Since the velocity of the core is less than that of the mantle there is no critical angle. All rays incident at the core are refracted into it. According to Snell's law the refracted angle i_2 is less than the incident angle i_1 (Fig. 151a):

$$\frac{\sin i_1}{v} = \frac{\sin i_2}{\frac{v}{2}}$$

$$\sin i_2 = \frac{1}{2}\sin i_1$$

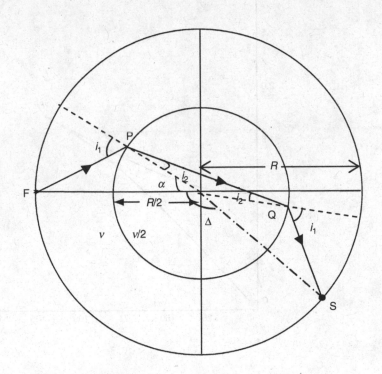

The travel-time for a ray crossing the mantle and the core is (Fig. 151a)

$$t_3 = \frac{\overline{FP}}{v} + \frac{\overline{PQ}}{\frac{v}{2}} + \frac{\overline{QS}}{v} = \frac{2\overline{FP}}{v} + \frac{2\overline{PQ}}{v}$$

where

$$\overline{FP}^2 = R^2 + \frac{R^2}{4} - 2R\frac{R}{2}\cos\alpha = R^2 + \frac{R^2}{4} - R^2\cos\left(90 + i_2 - \frac{\Delta}{2}\right)$$

$$\overline{FP}^2 = \frac{5R^2}{4} + R^2\sin\left(i_2 - \frac{\Delta}{2}\right)$$

$$\overline{PQ} = 2\frac{R}{2}\cos i_2 = R\cos i_2$$

so

$$t_3 = \frac{2R}{v}\sqrt{\frac{5}{4} + \sin\left(i_2 - \frac{\Delta}{2}\right)} + \frac{2R}{v}\cos i_2 \tag{151.3}$$

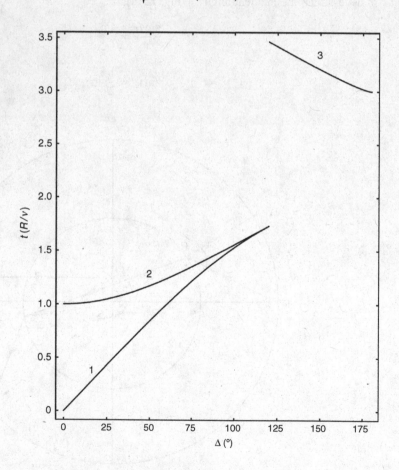

The relation between the incidence angle at the mantle–core boundary, i_1, and the angular distance, Δ, of a ray which crosses the core is

$$\frac{\sin(180° - i_1)}{R} = \frac{\sin\left(i_1 - i_2 + \dfrac{\Delta}{2} - 90°\right)}{R/2}$$

$$\sin i_1 = -2\cos\left(i_1 - i_2 + \frac{\Delta}{2}\right) \Rightarrow \Delta = 2\left(\cos^{-1}\left(\frac{-\sin i_1}{2}\right) - i_1 + \sin^{-1}\left(\frac{1}{2}\sin i_1\right)\right) \tag{151.4}$$

The range of distance for this ray is $120° < \Delta < 180°$.

From Equations (151.1), (151.2), and (151.3) we can calculate the values for the travel-times of the direct, reflected, and refracted rays. Some values for t_3 are given in the table.

i_1 (°)	i_2 (°)	Δ (°)	t_3 (R/v)
0	0	180.0	1.50
10	20.3	149.4	1.48
20	43.2	114.0	1.40
30	90	31.0	1.07

The travel-time curves for rays that are direct (1), reflected (2), and refracted in the core (3) are shown in Fig. 151b.

152. Consider a spherical Earth of radius R formed by two hemispherical media of constant velocities of propagation v and $2v$. For a focus on the surface of the hemisphere of velocity v at the point of intersection of the diameter perpendicular to the plane that separates the two media, calculate the travel times and travel-time curves of the direct, reflected, and critical refracted waves at the surface of separation of the two media, in units of R/v. Calculate the expression for the travel time of waves that propagate through the medium of speed of propagation $2v$.

The travel time for angular distances $\Delta \leq 90°$ are given by (Fig. 152a)

$$t_1 = \frac{\overline{FS}}{v} = 2\frac{R}{v}\sin\frac{\Delta}{2} \tag{152.1}$$

The travel time of the ray reflected at the plane boundary between the two hemispheres is (Fig. 152a)

$$t_2 = \frac{\overline{FP}}{v} + \frac{\overline{PS}}{v} = \frac{\overline{F'S}}{v}$$

According to Fig. 152a (triangle OSF') the relation between the angles α and Δ is

$$(180° - \Delta) + 2\alpha = 180° \Rightarrow \alpha = \frac{\Delta}{2}$$

$$\sin\alpha = \frac{\overline{SS'}}{\overline{F'S}} \Rightarrow \overline{F'S} = \frac{\overline{SS'}}{\sin\alpha}$$

$$\overline{SS'} = R\sin\Delta \Rightarrow \overline{F'S} = \frac{R\sin\Delta}{\sin\alpha} = 2R\cos\frac{\Delta}{2}$$

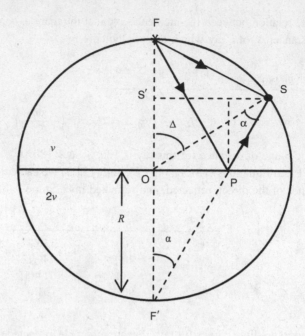

Fig. 152a

Then for $\Delta < 90°$

$$t_2 = \frac{2R}{v} \cos \frac{\Delta}{2} \tag{152.2}$$

The critical angle, according to Snell's law, is given by

$$\frac{\sin i_c}{v} = \frac{1}{2v} \Rightarrow i_c = 30°$$

The travel-time of the critically refracted ray is (Fig. 152b)

$$t_3 = \frac{\overline{FA}}{v} + \frac{\overline{AB}}{2v} + \frac{\overline{BS}}{v} \tag{152.3}$$

According to Fig. 152b

$$\overline{FA} = \frac{R}{\cos i_c}$$

$$\overline{BS} = \frac{SS'}{\cos i_c} = \frac{R \cos \Delta}{\cos i_c}$$

$$\overline{AB} = \overline{OS'} - \overline{OA} - \overline{BS}'$$

$$\overline{OS'} = R \sin \Delta$$

$$\overline{OA} = R \tan i_c$$

$$\overline{BS}' = \overline{BS} \sin i_c$$

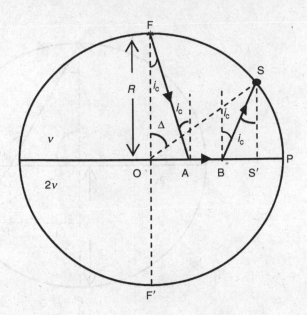

Substituting $i_c = 30°$:

$$\overline{FA} = \frac{2R}{\sqrt{3}}; \quad \overline{BS} = \frac{2R\cos\Delta}{\sqrt{3}}; \quad \overline{AB} = R\sin\Delta - \frac{R}{\sqrt{3}} - \frac{R\cos\Delta}{\sqrt{3}}$$

and substituting in (152.3) we obtain, for $\Delta_c \le \Delta \le 90°$,

$$t_3 = \frac{R}{2v}\left[\sin\Delta + \sqrt{3}(1 + \cos\Delta)\right] \tag{152.4}$$

The critical distance can be calculated from $i_c = 30°$ using Fig. 152a (triangle OSP) and $\alpha = \Delta/2$

$$90° - \Delta_c + \alpha + i_c + 90° = 180° \Rightarrow \Delta_c = i_c + \alpha \Rightarrow \Delta_c = 60°$$

The travel time of the rays that cross the boundary and penetrate in to the medium of velocity $2v$ is given by (Fig. 152c)

$$t_4 = \frac{\overline{FP}}{v} + \frac{\overline{PS}}{2v} \tag{152.5}$$

According to Snell's law,

$$\frac{\sin i}{v} = \frac{\sin i'}{2v} \Rightarrow \sin i' = 2\sin i$$

and we have that

$$\overline{FP} = \frac{R}{\cos i}$$

$$\overline{PS} = \frac{SP'}{\sin i'}$$

$$\overline{SP}' = R\sin(180° - \Delta) - R\tan i$$

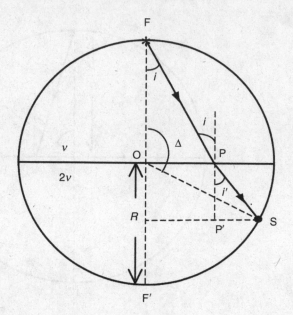

Fig. 152c

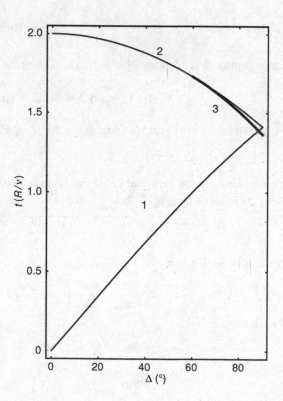

Fig. 152d

Substituting in Equation (152.5) we obtain, for $\Delta > 90°$,

$$t_4 = \frac{3R}{4v\cos i} + \frac{R\sin\Delta}{4v\sin i} \qquad (152.6)$$

Travel-time curves for t_1, t_2, and t_3 are shown in Fig. 152d.

153. Assume an Earth formed by a mantle of thickness R and a core of radius $R/2$, with velocities v and $2v$. There is a seismic focus at depth $R/4$ below the surface. Calculate the travel-time curves of the direct and reflected waves.

The travel time of the direct ray is (Fig. 153a)

$$t_1 = \frac{\overline{FP}}{v}$$

The distance $\overline{FP}$ can be expressed in terms of R and Δ using the cosine law in triangle FOP:

$$\overline{FP} = \sqrt{\left(\frac{3R}{4}\right)^2 + R^2 - 2R\frac{3R}{4}\cos\Delta}$$

Then, we obtain

$$t_1 = \frac{R}{v}\sqrt{\frac{25}{16} - \frac{3}{2}\cos\Delta} \qquad (153.1)$$

The maximum distance for the direct ray is

$$\Delta_{\max} = \Delta_1 + \Delta_2$$

$$\cos\Delta_1 = \frac{\dfrac{R}{2}}{\dfrac{3R}{4}} \Rightarrow \Delta_1 = 48.2°$$

$$\cos\Delta_2 = \frac{\dfrac{R}{2}}{R} \Rightarrow \Delta_2 = 60.0°$$

$$\Delta_{\max} = 48.2 + 60.0 = 108.2°$$

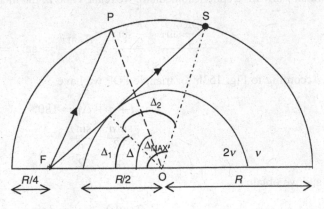

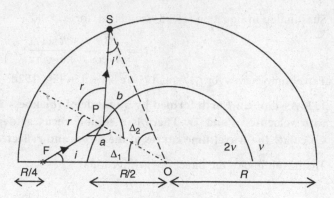

The travel time for the reflected ray (Fig. 153b) is

$$t_2 = \frac{\overline{FP}}{v} + \frac{\overline{PS}}{v}$$

The distances $\overline{FP}$ and $\overline{PS}$ are expressed in terms of R, Δ_1, and Δ_2 using the cosine law in triangles FOP and SOP (Fig. 153b):

$$\overline{FP} = \sqrt{\left(\frac{3}{4}R\right)^2 + \left(\frac{R}{2}\right)^2 - 2\frac{R}{2}\frac{3R}{4}\cos\Delta_1}$$

$$\overline{PS} = \sqrt{R^2 + \left(\frac{R}{2}\right)^2 - 2\frac{R}{2}\cos\Delta_2}$$

Then, we obtain

$$t_2 = \frac{R\sqrt{\frac{13}{16} - \frac{3}{4}\cos\Delta_1}}{v} + \frac{R\sqrt{\frac{5}{4} - \cos\Delta_2}}{v} \tag{153.2}$$

Now we need to express Δ_1 and Δ_2 in terms of the take-off angle i at the focus (F). Using Snell's law for a spherical medium, we relate i and i', the incidence angle at the station (S):

$$\frac{\frac{3R}{4}\sin i}{v} = \frac{\frac{R}{2}\sin r}{v} = \frac{R\sin i'}{v} \Rightarrow \sin i' = \frac{3}{4}\sin i \tag{153.3}$$

According to Fig. 153b for triangle FOP we have

$$i + a + \Delta_1 = 180°$$

$$\frac{\sin a}{\frac{3R}{4}} = \frac{\sin i}{\frac{R}{2}}$$

and we obtain

$$2\sin(\Delta_1 + i) = 3\sin i \tag{153.4}$$

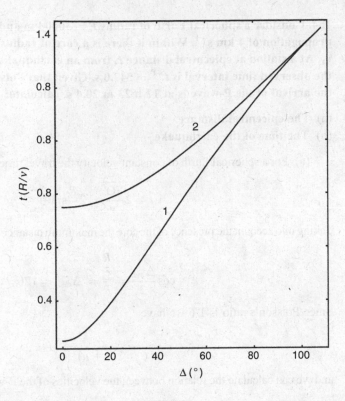

Fig. 153c

and for triangle POS

$$b = 180° - \Delta_2 - i'$$

$$\frac{\sin i'}{\dfrac{R}{2}} = \frac{\sin b}{R}$$

and

$$2 \sin i' = \sin(\Delta_2 + i') \tag{153.5}$$

Equations (153.3), (153.4), and (153.5) allow us to calculate Δ_1 and Δ_2 from the take-off angle i at the focus. The travel-times are given in the following table.

i (°)	i' (°)	Δ_1 (°)	Δ_2 (°)	Δ (°)	t_2 (R/v)
0	0	0	0	0	0.75
10	7.5	5.1	7.6	12.7	0.76
30	22.0	18.6	26.6	45.2	0.92
40	28.8	34.6	45.8	80.4	1.19
41.8	30.0	47.1	58.9	106.0	1.41

For angular distance Δ greater than 108.2° there are no reflected rays. The travel-time curves corresponding to the direct (1) and reflected (2) rays are shown in Fig. 153c.

154. Consider a spherical Earth of radius $R = 3000$ km and constant P-wave speed of propagation of 4 km s^{-1}. Within it there is a core of radius $R/2$ and constant velocity v_1. At a station at epicentral distance Δ from an earthquake with focus at the surface, the observed time interval is $t^{S-P} = 547.0$ s. Given that Poisson's ratio is 1/6, and that the arrival of the P-wave is at 12 h 23 m 20.4 s, calculate:

(a) The epicentral distance.
(b) The time of the earthquake.

(a) For a spherical Earth of constant velocity the travel time of the direct ray is given by

$$t = 2\frac{\overline{FO'}}{v} = 2\frac{R}{v}\sin\frac{\Delta}{2} \tag{154.1}$$

Taking into account the presence of the core the maximum distance for the direct ray is (Fig. 154)

$$\cos\frac{\Delta_{max}}{2} = \frac{\frac{R}{2}}{R} \Rightarrow \Delta_{max} = 120°$$

Since Poisson's ratio is 1/6 we have

$$\sigma = \frac{1}{6} = \frac{\lambda}{2(\lambda+\mu)} \Rightarrow \mu = 2\lambda$$

and we can calculate the relation between the velocities of the P-wave (α) and the S-wave (β),

$$\left.\begin{array}{c} \alpha = \sqrt{\dfrac{\lambda+2\mu}{\rho}} \\[2mm] \beta = \sqrt{\dfrac{\mu}{\rho}} \end{array}\right\} \Rightarrow \alpha = \sqrt{\frac{5}{2}}\beta \Rightarrow \beta = 2.53\,\text{km s}^{-1}$$

Using (154.1) and assuming the same path for P- and S-waves, from the time interval S-P we obtain the distance:

$$t^{S-P} = 2R\sin\frac{\Delta}{2}\left(\frac{1}{\beta} - \frac{1}{\alpha}\right) \Rightarrow \Delta = 77.7°$$

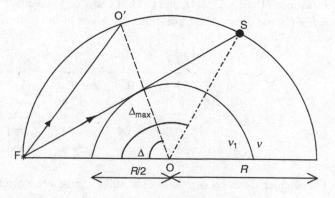

Fig. 154

Notice that $\Delta < \Delta_{max}$.

(b) To calculate the time of origin we subtract from the arrival time of the P-wave the value of the travel time for that distance:

$$t^P = \frac{2R}{\alpha} \sin \frac{\Delta}{2} = 2\frac{3000}{4} \sin \frac{77.7}{2} = 940.9 \, s$$

The time of origin is then given by

$$t_0 = 12\,h\,23\,m\,20.4\,s - 940.9\,s = 12\,h\,07\,m\,39.5\,s$$

155. Consider the Earth formed by a sphere of radius $R = 4000$ km, Poisson's ratio 1/8, and constant S-wave speed of propagation 3 km s^{-1}. Within it, there is a liquid core of radius $R/2$. There occurs an earthquake with a focus in the interior of the Earth. At a station of epicentral distance Δ the observed time interval is $t^{S-P} = 600$ s. This focus may be at a depth of either $R/10$ or $2R/5$. Calculate the correct depth of the focus, and the epicentral distance.

First we determine the maximum distance for direct rays which don't penetrate into the core, which according to Fig. 155 corresponds to $\Delta_{max} = \Delta_1 + \Delta_2$:

$$\cos \Delta_1 = \frac{\frac{R}{2}}{R - h}$$

$$\cos \Delta_2 = \frac{\frac{R}{2}}{R} \Rightarrow \Delta_2 = 60°$$

If the depth of the focus is $R/10$ then

$$h = \frac{R}{10} \Rightarrow \cos \Delta_1 = \frac{5}{9} \Rightarrow \Delta_1 = 56° \Rightarrow \Delta_{max} = 116°$$

and if it is $2R/5$, then

$$h = \frac{2}{5}R \Rightarrow \cos \Delta_1 = \frac{5}{6} \Rightarrow \Delta_1 = 33.56° \Rightarrow \Delta_{max} = 94°$$

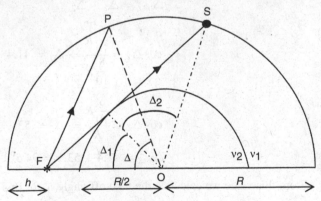

Fig. 155

For a point on the surface at distance Δ, the S-P time interval implies, assuming the same path for P- and S-waves,

$$_t\text{S-P} = \frac{\overline{FP}}{\beta} - \frac{\overline{FP}}{\alpha} = \frac{\overline{FP}}{\alpha\beta}(\alpha - \beta) \tag{155.1}$$

From the value of Poisson's ratio the relation between the P and S velocities is given by

$$\sigma = \frac{1}{8} = \frac{\lambda}{2(\lambda + \mu)} \Rightarrow \lambda = \frac{\mu}{3} \Rightarrow \alpha = \sqrt{\frac{\lambda + 2\mu}{\rho}} = \sqrt{\frac{7\mu}{3\rho}} = \sqrt{\frac{7}{3}}\beta$$

Substituting in (155.1) the S-P interval equal to 600 s we obtain the length of the ray:

$$\overline{FP} = 600\frac{\alpha\beta}{(\alpha - \beta)} = 5212\,\text{km}$$

Using the cosine law for triangle FOP

$$\overline{FP}^2 = (R - h)^2 + R^2 - 2R(R - h)\cos\Delta$$

$$\cos\Delta = \frac{\overline{FP}^2 - (R - h)^2 - R^2}{2R(R - h)}$$

If $h = \frac{2}{5}R$ then $\Delta = 106°$, but this result is not possible because the maximum distance of the direct ray for that depth is 94°. If $h = \frac{R}{10}$ then $\Delta = 86°$, this result is possible because this distance is less than the maximum distance. The depth is, then, 400 km.

156. Consider the Earth of radius R and constant velocity v with a core of radius $6R/10$ and constant speed of propagation $2v$. An earthquake occurs with focus at $8R/10$ from the centre of the Earth. A wave emerges from that focus with a take-off angle of 15°.

(a) Will it pass through the core?
(b) What epicentral distance will it reach?
(c) What will be the travel time of the wave (in units of R/v)?

(a) First we calculate the maximum epicentral distance for a ray which doesn't penetrate the core. According to Fig. 156a the maximum distance is

$$\Delta_{max} = \Delta_1 + \Delta_2$$

$$\cos\Delta_1 = \frac{\frac{6}{10}R}{\frac{8}{10}R} \Rightarrow \Delta_1 = 41.4°$$

$$\cos\Delta_2 = \frac{\frac{6}{10}R}{R} \Rightarrow \Delta_2 = 53.1°$$

$$\Delta_{max} = 41.4 + 53.1 = 94.5°$$

From this value we calculate the take-off angle i_h for this ray:

$$\Delta_1 + i_h = 90° \Rightarrow i_h = 48.6°$$

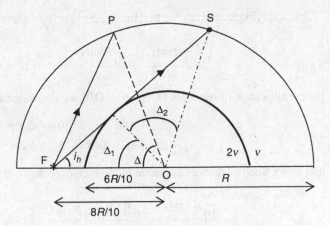

Fig. 156a

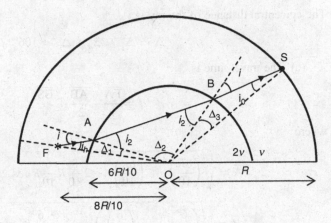

Fig. 156b

For take-off angles less than 48.6° the rays travel through the core.

Since the velocity in the core is greater than in the mantle, to find out which rays penetrate into the core, we also need to know the critical angle. Rays with incidence angle at the core–mantle boundary with $i > i_c$ are totally reflected and don't penetrate into the core. According to Snell's law the critical angle is given by

$$\frac{\sin i_c}{v} = \frac{1}{2v} \Rightarrow i_c = 30.0°$$

We calculate, using Snell's law, the angle of incidence i corresponding to the take-off angle of 15° (Fig. 156b):

$$\frac{\frac{8}{10}R \sin i_h}{v} = \frac{\frac{6}{10}R \sin i}{v} \quad \Rightarrow \quad i = 20.2°$$

Since the incidence angle i (20.2°) is less than the critical angle (30°) and less than the angle corresponding to the maximum distance (48.6°), the ray with take-off angle of 15° penetrates into the core.

(b) Applying Snell's law we find the angle of the transmitted ray in the core i_2 (Fig. 156b):

$$\frac{\frac{8}{10}R \sin i_h}{v} = \frac{\frac{6}{10}R \sin i_2}{2v} \quad \Rightarrow \quad i_2 = 43.7°$$

By consideration of triangles FOA and AOB, we determine Δ_1 and Δ_2 (Fig. 156b):

$$i_h + \Delta_1 + 180 - i = 180 \Rightarrow \Delta_1 = 5.2°$$
$$i_2 + \Delta_2 + i_2 = 180 \Rightarrow \Delta_2 = 92.6°$$

and using Snell's law we determine i_0 the incidence angle at the surface and Δ_3:

$$\frac{\frac{6}{10}R \sin i_2}{2v} = \frac{\frac{6}{10}R \sin i}{v} = \frac{R \sin i_0}{v} \quad \Rightarrow i_0 = 11.9°$$
$$\Delta_3 = i - i_0 = 8.3°$$

The epicentral distance of the ray is

$$\Delta = \Delta_1 + \Delta_2 + \Delta_3 = 106°$$

(c) The travel time is

$$t = \frac{\overline{FA}}{v} + \frac{\overline{AB}}{2v} + \frac{\overline{BS}}{v}$$

where

$$\overline{FA} = \sqrt{\left(\frac{8R}{10}\right)^2 + \left(\frac{6R}{10}\right)^2 - 2\frac{8}{10}R\frac{6}{10}R \cos \Delta_1} = 0.21 R$$

$$\overline{AB} = \sqrt{\left(\frac{6}{10}R\right)^2 + \left(\frac{6}{10}R\right)^2 - 2\frac{6}{10}\frac{6}{10}R^2 \cos \Delta_2} = 0.87 R$$

$$\overline{BS} = \sqrt{\left(\frac{6}{10}R\right)^2 + R^2 - 2\frac{6}{10}RR \cos \Delta_3} = 0.42 R$$

so

$$t = 1.07\frac{R}{v}$$

157. Assume a spherical Earth of radius $R = 6000$ km and constant S-wave speed of propagation 4.17 km s^{-1}. Poisson's ratio is 1/4. At a station at epicentral distance 60° an earthquake is recorded with a time interval $t^{S-P} = 554$ s. Calculate the depth of the earthquake.

Given that Poisson's ratio is 0.25, the P-wave velocity is

$$\sigma = \frac{1}{4} = \frac{\lambda}{2(\lambda + \mu)} \Rightarrow \lambda = \mu \Rightarrow \alpha = \sqrt{3}\beta = 7.22 \text{ km s}^{-1}$$

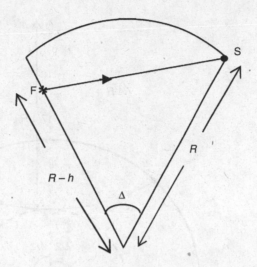

Fig. 157

From the time interval S-P we can calculate the length of the ray $\overline{FS}$ (Fig. 157):

$$t^{S\text{-}P} = \frac{\overline{FS}}{\beta} - \frac{\overline{FS}}{\alpha} = \overline{FS}\,\frac{\alpha - \beta}{\alpha\beta}$$

$$\overline{FS} = \frac{t^{S\text{-}P}\alpha\beta}{\alpha - \beta} = 5469 \text{ km} \tag{157.1}$$

The distance along the ray in terms of the angular epicentral distance Δ, using the cosine law, is

$$\overline{FS} = \sqrt{R^2 + (R - h)^2 - 2R(R - h)\cos\Delta}$$

Substituting $\overline{FS}$ from (157.1):

$$5370 = \sqrt{R^2 + (R - h)^2 - 2R(R - h)\cos\Delta} \tag{157.2}$$

We substitute in (157.2) the values $R = 6000$ km and $\Delta = 60°$ and solve for h, finding two possible solutions:

$$h_1 = 4706 \text{ km}$$
$$h_2 = 1294 \text{ km}$$

158. Assume a spherical Earth of radius R and P-wave velocity which can be expressed by the equation $v(r) = a - br^2$. The speed of propagation at the surface of the Earth is v_0 and at the centre of the Earth it is $2v_0$. What angular distance Δ does a wave reach which penetrates to a depth equal to half the Earth's radius?

If the velocity distribution inside the Earth is $v(r) = a - br^2$, the ray paths are circular with radius given by (Fig. 158)

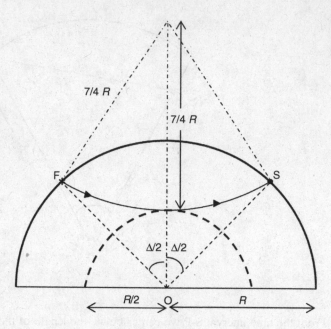

Fig. 158

$$\rho = \frac{r}{p\dfrac{dv}{dr}} \qquad\qquad (158.1)$$

From the conditions of the problem

$$r = R \Rightarrow v = v_0 = a - bR^2$$
$$r = 0 \Rightarrow v = 2v_0 = a$$

and

$$\left.\begin{array}{l} b = \dfrac{v_0}{R^2} \\[2mm] a = 2v_0 \end{array}\right\} \Rightarrow v = v_0\left(2 - \frac{r^2}{R^2}\right)$$

The radius of curvature of the ray which penetrates to $r = R/2$ is that corresponding to the ray parameter

$$p' = \frac{r'}{v'} = \frac{R}{2v'} \qquad\qquad (158.2)$$

The velocity at depth $R/2$ is

$$v' = v_0\left(2 - \left(\frac{\dfrac{R}{2}}{R}\right)^2\right) = v_0\frac{7}{4}$$

Substituting the ray parameter in (158.2):

$$p' = \frac{R}{v_0\dfrac{7}{2}} = \frac{2R}{7v_0}$$

The derivative of the velocity is

$$\frac{dv}{dr} = \frac{d}{dr}\left(a - br^2\right) = -2br = -2\frac{v_0}{R^2}r$$

Substituting in (158.1) we obtain, for the radius of curvature (Fig. 158),

$$\rho = \frac{r}{p(-2br)} = -\frac{7R}{4}$$

The epicentral (angular) distance corresponding to this ray is found by applying the cosine law to the triangle POS:

$$\left(\frac{7}{4}R\right)^2 = \left(\frac{9}{4}R\right)^2 + R^2 - 2\frac{9R}{4}R\cos\frac{\Delta}{2} \Rightarrow \Delta = 96.4°$$

159. The Earth consists of a mantle of radius R and speed of propagation $v = a/\sqrt{r}$, and a core of radius $R/2$ and speed of propagation $4v_0$, where v_0 is the speed of propagation at the Earth's surface. Calculate:

(a) The maximum epicentral distance corresponding to a wave that travels only through the mantle.

(b) The critical angle of the wave reflected at the core, and the angle at which it leaves the surface.

(c) The epicentral distance Δ_c corresponding to the critical angle.

(d) Plot the travel-time curve, specifying Δ_c and Δ_m.

(a) The value of a in the velocity distribution is found from the value of velocity at the surface:

$$r = R \rightarrow v = v_0 = aR^{-\frac{1}{2}} \Rightarrow a = v_0 R^{\frac{1}{2}}$$

The velocity distribution is

$$v = v_0 \left(\frac{R}{r}\right)^\alpha = v_0 \left(\frac{R}{r}\right)^{\frac{1}{2}}$$

For this general type of distribution of velocity with depth $\alpha < 1$, the angular distance for a surface focus (Fig. 159a) is given by

$$\Delta = \frac{2}{1+\alpha}\cos^{-1}\left(\frac{p}{\eta_0}\right) \tag{159.1}$$

where

$$\eta = \frac{r}{v} \Rightarrow \eta_0 = \frac{R}{v_0}$$

The maximum distance for a ray which travels only through the mantle, that is that reaches depth $R/2$, can be calculated from the velocity at that depth, v_m:

$$v_m = v_0 \left(\frac{R}{\frac{R}{2}}\right)^{\frac{1}{2}} = v_0\sqrt{2} \Rightarrow \eta_p = \frac{\frac{R}{2}}{v_m} = \frac{\frac{R}{2}}{v_0\sqrt{2}} = \frac{1}{2\sqrt{2}}\frac{R}{v_0} = p$$

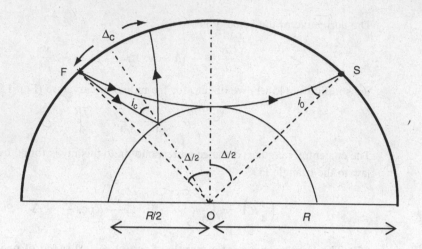

By substitution in (159.1) we obtain, for the maximum distance,

$$\Delta_m = \frac{2}{1+\frac{1}{2}} \cos^{-1}\left(\frac{\frac{1}{2\sqrt{2}}\frac{R}{v_0}}{\frac{R}{v_0}}\right) = 92.4°$$

(b) The critical angle of a reflected ray at the mantle-core boundary applying Snell's law is

$$\frac{\sin i_c}{v_0\sqrt{2}} = \frac{1}{4v_0} \Rightarrow i_c = 20.7°$$

The take-off angle at the surface i_0 for this ray is found by again applying Snell's law:

$$\frac{R\sin i_0}{v_0} = \frac{\frac{R}{2}\sin(20.7°)}{v_0\sqrt{2}} \Rightarrow i_0 = 7.2°$$

(c) To find the critical distance we use the expression

$$\Delta_c = 2\int_{r_p}^{r_0} \frac{p}{r}\frac{dr}{\sqrt{\eta^2 - p^2}} \tag{159.2}$$

where

$$r_p = \frac{R}{2}$$

$$r_0 = R$$

$$p = \frac{r_0\sin i_0}{v_0} = \frac{R}{8v_0}$$

$$\eta = \frac{r}{v} = \frac{r}{v_0\left(\frac{R}{r}\right)^{\frac{1}{2}}} = \frac{r^{3/2}}{v_0\sqrt{R}}$$

Substituting the values of the problem in (159.2):

$$\Delta_c = \frac{1}{4}\int_{R/2}^{R} \frac{dr}{\frac{r}{R^{3/2}}\sqrt{r^3 - \frac{R^3}{64}}}$$

This integral is of the type

$$\int \frac{dx}{x\sqrt{x^n - a^n}} = \frac{2}{n\sqrt{a^n}} \cos^{-1} \sqrt{\frac{a^n}{x^n}}$$

so we can write the solution

$$\Delta_c = \frac{8}{6} \left[\cos^{-1}\left(\frac{1}{8}\right) - \cos^{-1}\left(\frac{1}{\sqrt{8}}\right) \right] = 18°$$

(d) The travel time of the rays in the mantle is given by

$$t = \frac{2\eta_0}{1 + \alpha} \sin\left[(1 + \alpha)\frac{\Delta}{2} \right] = \frac{R}{v_0}\frac{4}{3} \sin\frac{3\Delta}{4} \qquad \text{for } 0° \leq \Delta \leq 92.4°$$

By substitution of values of Δ we find the travel time curve given in Fig. 159b.

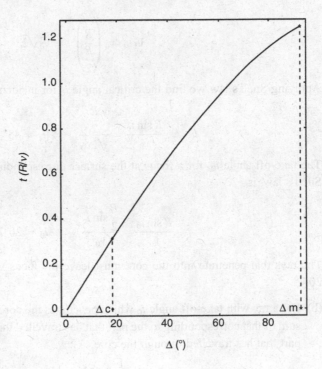

Fig. 159b

160. A spherical medium of radius R has a constant speed of propagation v_0 from the surface down to $R/2$, and from $R/2$ to the centre a core of variable speed of propagation $v = v_0 \left(\dfrac{R}{r}\right)^{1/2}$.

(a) What value should i_0 have for the waves to penetrate into the core?

(b) Calculate the epicentral distance reached by a wave leaving a focus at the surface at angle i_0.

(a) The velocity at the top of the core ($r = R/2$) is

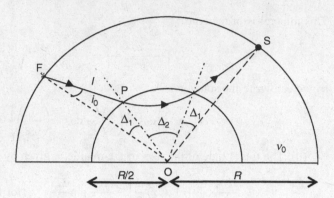

Fig. 160

$$v_1 = v_0 \left(\frac{R}{\dfrac{R}{2}}\right)^{\frac{1}{2}} = v_0\sqrt{2}$$

Applying Snell's law we find the critical angle i_c for incident rays at the core (Fig. 160):

$$\frac{R \sin i_c}{v_0} = \frac{\dfrac{R}{2}}{\sqrt{2}v_0} \quad \Rightarrow i_c = 45°$$

The take-off angle i_0, for a focus at the surface corresponding to the critical angle, using Snell's law, is

$$\frac{R \sin i_0}{v_0} = \frac{\dfrac{R}{2}\sin i_c}{v_0} \quad \Rightarrow i_0 = 20.7°$$

The rays that penetrate into the core must leave the focus with take-off angles less than 20.7°.

(b) For a ray with take-off angle i_0 which penetrates the core the epicentral distance is the sum of that corresponding to the part that has travelled through the mantle, Δ_1, plus the part that has travelled through the core, Δ_2:

$$\Delta = 2\Delta_1 + \Delta_2$$

Since in the core the velocity varies with depth with the law given in the problem, the epicentral distance is given by

$$\Delta_2 = \frac{2}{1+\alpha}\cos^{-1}\left(\frac{p}{\eta_1}\right)$$

where α is the exponent of the velocity distribution

$$v = v_0 r^{-\alpha} \Rightarrow \alpha = \frac{1}{2}$$

and p is the ray parameter, which can be obtained using Snell's law:

$$p = \frac{R \sin i_0}{v_0} = \frac{\frac{R}{2} \sin i_1}{\sqrt{2} v_0}$$

and

$$\eta = \frac{r}{v} \Rightarrow \eta_1 = \frac{\frac{R}{2}}{\sqrt{2} v_0}$$

Then, we find

$$\Delta_2 = \frac{2}{1 + \frac{1}{2}} \cos^{-1}\left(\frac{\frac{R \sin i_0}{v_0}}{\frac{R}{2\sqrt{2} v_0}}\right) = \frac{4}{3} \cos^{-1}\left(2\sqrt{2} \sin i_0\right)$$

The distance Δ_1 can be determined using the sine and cosine laws for the triangle FOP (Fig. 160):

$$\frac{\frac{R}{2}}{\sin i_0} = \frac{l}{\sin \Delta_1} \Rightarrow \Delta_1 = \sin^{-1}\left(\frac{2l \sin i_0}{R}\right)$$

$$l^2 + R^2 - 2Rl \cos i_0 = \frac{R^2}{4} \Rightarrow l = R \cos i_0 \pm \frac{1}{2} \sqrt{4R^2 \cos^2 i_0 - 3R^2}$$

and we find the expression in terms of i_0:

$$\Delta = 2\sin^{-1}\left[2\left(\cos i_0 \pm \frac{1}{2}\sqrt{4\cos^2 i_0 - 3}\right)\sin i_0\right] + \frac{4}{3}\cos^{-1}\left(2\sqrt{2}\sin i_0\right)$$

161. Consider a spherical Earth of radius 6000 km and surface velocity of 6 km s^{-1}, with a velocity distribution of the type $v(r) = a/\sqrt{r}$. At the distance reached by a wave emerging at a take-off angle of 45° from a focus on the surface, calculate the interval between the arrival times of the direct P- and reflected PP-waves (the PP-wave is one that is reflected at the surface at the midpoint between the focus and the point of observation).

The velocity distribution is of the type $v = v_0 \left(\frac{R}{r}\right)^\alpha$ where

$$r = R \Rightarrow v = v_0 = \frac{a}{\sqrt{R}} \Rightarrow a = v_0 \sqrt{R}$$

$$v = v_0 \left(\frac{R}{r}\right)^{\frac{1}{2}}$$

and the ray parameter p for a ray with take-off angle of 45° is

$$p = \frac{R \sin i_0}{v_0} = \frac{6000 \times \frac{1}{\sqrt{2}}}{6} = 707 \, \text{s}$$

For this type of velocity distribution the relation between the ray parameter and the epicentral distance is

$$p = \eta_0 \cos\left[\left(\frac{1+\alpha}{2}\right)\Delta\right] \qquad (161.1)$$

In this problem the value of η_0 is

$$\eta = \frac{r}{v} \Rightarrow \eta_0 = \frac{R}{v_0} = 1000 \text{ s}$$

Substituting the values in (161.1) we obtain the distance for the ray with take-off angle of 45°:

$$707 = 1000\cos\left(\frac{3}{4}\Delta\right) \Rightarrow \Delta = 60°$$

The PP-wave which arrives at $\Delta = 60°$ travels twice the distance which a P-wave does for a distance of 30°. For this type of velocity distribution the travel time for a distance Δ is given by

$$t = \frac{2\eta_0}{1+\alpha}\sin\left[\frac{(1+\alpha)\Delta}{2}\right]$$

In our case for the P- and PP-waves at distance 60° we substitute the values of the problem and find

$$t_P(60°) = 943 \text{ s}$$
$$t_{PP}(60°) = 2t_P(30°) = 1021 \text{ s}$$

The time interval between the PP- and P-waves 60° is

$$t_{PP} - t_P = 1021 - 943 = 78s$$

162. In an elastic spherical medium of radius r_0, the velocity increases with depth according to $v = ar^{-b}$. If $v_0 = 6$ km s^{-1}, $r_0 = 6000$ km, and, at a point at distance $\Delta = 90°$, the slope of the travel-time curve is 500 s, determine:

(a) The value of b.
(b) The value of r_p and of v_p of the wave reaching an epicentral distance of 90°.

(a) For this type of velocity distribution the travel time in terms of the epicentral distance is given by

$$t = \frac{2\eta_0}{1+b}\sin\left[(1+b)\frac{\Delta}{2}\right] \qquad (162.1)$$

As we know the velocity at the surface,

$$\eta = \frac{r}{v} \Rightarrow \eta_0 = \frac{r_0}{v_0} = 1000 \text{ s}$$

The ray parameter p is known, because it is equal to the slope of the travel-time curve which for $\Delta = 90°$ is given as 500 s. Using the relation between p and Δ for this type of velocity distribution,

$$p = \frac{dt}{d\Delta} = \eta_0 \cos\left[(1 + b)\frac{\Delta}{2}\right]$$

$$500 = 1000 \cos\left[(1 + b)\frac{\pi}{4}\right] \Rightarrow \cos\left[(1 + b)\frac{\pi}{4}\right] = \frac{1}{2} \Rightarrow (1 + b)\frac{\pi}{4} = \frac{\pi}{3} \Rightarrow b = \frac{1}{3}$$

(b) At the point of greatest penetration r_p for $\Delta = 90°$, we have the relation

$$p = \frac{r_p \sin 90°}{v_p} = \frac{r_p}{v_p} \Rightarrow r_p = p v_p$$

and also

$$v_p = v_0 \left(\frac{r_0}{r_p}\right)^{\frac{1}{3}}$$

From these two equations we obtain r_p and v_p:

$$r_p = v_0 p r_0^{\frac{1}{3}} r_p^{-\frac{1}{3}} \Rightarrow r_p = v_0^{\frac{3}{4}} r_0^{\frac{1}{4}} p^{\frac{3}{4}} = 3564 \, \text{km}$$

$$v_p = \frac{r_p}{p} = \frac{3564}{500} = 7.1 \, \text{km s}^{-1}$$

163. Consider a spherical Earth of radius R, the northern hemisphere with a constant speed of propagation v_0, and the southern hemisphere with a speed of propagation of
$$v = v_0 \left(\frac{R}{r}\right)^{\frac{1}{2}}.$$
(a) Calculate the travel time of seismic waves for a focus on the equator and stations on the same meridian.
(b) In which hemisphere does the wave at a distance of 60° arrive first?

(a) In the northern hemisphere the velocity is constant and the rays have straight paths and their travel time is (Fig. 163)

$$t^N = \frac{2R}{v_0} \sin\frac{\Delta}{2} \qquad 0° < \Delta < 90° \tag{163.1}$$

In the southern hemisphere the velocity increases with depth and the rays have curved paths. Their travel time is given by

$$t^S = \frac{2\eta_0}{1 + b} \sin(1 + b)\frac{\Delta}{2} \tag{163.2}$$

where

$$\eta_0 = \frac{R}{v_0} \quad \text{and} \quad v = v_0 \left(\frac{R}{r}\right)^b \Rightarrow b = \frac{1}{2}$$

Substituting in (163.2) we obtain for the travel time in the southern hemisphere

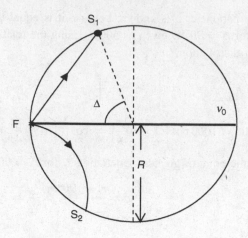

Fig. 163

$$t^S = \frac{4}{3}\frac{R}{v_0}\sin\frac{3\Delta}{4} \qquad 0 < \Delta < 90° \qquad (163.3)$$

(b) The travel times for waves in the northern and southern hemisphere are given by Equations (163.1) and (163.3). By substitution of $\Delta = 60°$ we obtain

$$t^N = \frac{R}{v_0}$$

$$t^S = \frac{R}{v_0}\frac{2\sqrt{2}}{6} = 0.47\frac{R}{v_0}$$

The waves arrive first in the southern hemisphere.

164. Consider a spherical medium of radius R consisting of two concentric regions (mantle and core), the core of radius $R/2$. The speeds of propagation are $v = ar^{-1/2}$ for the mantle and $v = aR^{-1/6}r^{-1/3}$ for the core. The surface velocity is v_0. For a wave leaving a focus with angle of incidence 14.5°, calculate the angular distance Δ at which it reaches the surface.

We calculate a by applying the boundary conditions

$$r = R \Rightarrow v = v_0 \Rightarrow v_0 = aR^{-\frac{1}{2}} \Rightarrow a = v_0 R^{\frac{1}{2}}$$

$$\text{Mantle}: v = v_0\left(\frac{R}{r}\right)^{\frac{1}{2}}$$

$$\text{Core}: v = v_0\left(\frac{R}{r}\right)^{\frac{1}{3}}$$

We determine the ray parameter corresponding to the ray with take-off angle $i_0 = 14.5°$, using Snell's law (Fig. 164a):

$$p = \frac{r\sin i}{v} = \frac{R\sin 14.5}{v_0} = \frac{R}{4v_0}$$

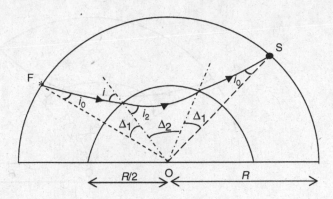

Fig. 164a

At the bottom of the mantle at depth $R/2$ the velocity is

$$v_1 = v_0 \left(\frac{R}{\frac{R}{2}} \right)^{\frac{1}{2}} = v_0 2^{\frac{1}{2}}$$

The incident angle i of this ray on the mantle–core boundary, applying Snell's law, is given by

$$\frac{\frac{R}{2} \sin i}{v_0 2^{\frac{1}{2}}} = \frac{R \sin i_0}{v_0} \Rightarrow i = 45°$$

On the top of the core the velocity is

$$v_2 = v_0 \left(\frac{R}{\frac{R}{2}} \right)^{\frac{1}{3}} = v_0 2^{\frac{1}{3}}$$

which is less than at the bottom of the mantle and there is no critical angle. Applying Snell's law again we obtain the angle i_2 of the refracted ray in the core:

$$\frac{\sin i}{v_1} = \frac{\sin i_2}{v_2} \Rightarrow i_2 = 39°$$

The take-off angle of the last ray which travels only in the mantle is given by

$$\frac{R \sin i_0}{v_0} = \frac{\frac{R}{2}}{v_0 2^{\frac{1}{2}}} \Rightarrow i_0 = 20.7°$$

In our case the angle 14.5° is less and the ray penetrates into the core.

The epicentral distance is the sum of the distances corresponding to the paths in the mantle and in the core:

$$\Delta = 2\Delta_1 + \Delta_2$$

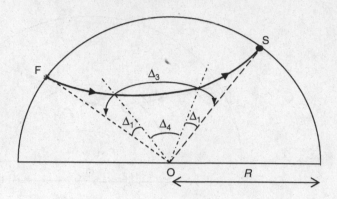

The distance corresponding to the path in the core is given by

$$\Delta_2 = \frac{2.}{1+b} \cos^{-1}\left(\frac{p}{\eta_0}\right)$$

where

$$p = \frac{R}{4v_0}$$

$$\eta_0 = \frac{\frac{R}{2}}{v_0 2^{\frac{1}{3}}}$$

and where p is the ray parameter and $\eta = r/v$. Substituting the values we obtain

$$\Delta_2 = 76.4°$$

To calculate Δ_1 we suppose that there is no core and a ray with take-off angle $i_0 = 14.5$ would arrive at distance Δ_3 which is related with Δ_1 by (Fig. 164b)

$$2\Delta_1 = \Delta_3 - \Delta_4$$

The distances Δ_3 and Δ_4 can be determined using the equation

$$\Delta = \frac{2}{1+b} \cos^{-1}\left(\frac{p}{\eta_0}\right)$$

where for Δ_3

$$\eta_0(R) = \frac{R}{v_0}$$

$$\eta\left(\frac{R}{2}\right) = \frac{\frac{R}{2}}{v_0 \left(\frac{R}{\frac{R}{2}}\right)^{\frac{1}{2}}} = \frac{R}{v_0} 2^{-\frac{3}{2}}$$

and we obtain

$$\Delta_3 = \frac{2}{1 + \frac{1}{2}} \cos^{-1} \left(\frac{\frac{R}{4v_0}}{\frac{R}{v}} \right) = 100.6°$$

and by similar substitutions for Δ_4

$$\Delta_4 = \frac{2}{1 + \frac{1}{2}} \cos^{-1} \left(\frac{\frac{R}{4v_0}}{\frac{R}{v_0} 2^{-\frac{3}{2}}} \right) = 60.0°$$

Then, $2\Delta_1 = 100.6° - 60° = 40.6°$ and the epicentral distance is

$$\Delta = 40.6 + 76.4 = 117.0°$$

Surface waves

165. A Rayleigh wave in a semi-infinite medium has a 20 s period. If the P-wave velocity is 6 km s^{-1} and Poisson's ratio is 0.25, calculate the depth at which $u_1 = 0$, and at which depth the particle movement becomes prograde.

Since Poisson's ratio is 0.25, we find the relation between P- and S-waves:

$$\sigma = \frac{1}{4} = \frac{\lambda}{2(\lambda + \mu)} \Rightarrow \lambda = \mu \Rightarrow \alpha = \sqrt{\frac{\lambda + 2\mu}{\rho}} = \sqrt{\frac{3\mu}{\rho}} = \sqrt{3}\beta$$

$$\alpha = 6 \Rightarrow \beta = \frac{6}{\sqrt{3}} = 3.4 \, \text{km s}^{-1}$$

For a half-space the velocity of Rayleigh waves is

$$c_R = 0.919\beta = 3.2 \, \text{km s}^{-1}$$

The displacement u_1 is given by

$$u_1 = \frac{\partial \varphi}{\partial x_1} - \frac{\partial \psi}{\partial x_3}$$

where the potentials are given by

$$\varphi = A \exp(-ikrx_3 + ik(x_1 - c_R t))$$
$$\psi = B \exp(-iksx_3 + ik(x_1 - c_R t))$$
$$r = i\left(1 - \frac{c_R^2}{\alpha^2}\right)^{1/2} = 0.85i$$
$$s = i\left(1 - \frac{c_R^2}{\beta^2}\right)^{1/2} = 0.39i$$

Then,

$$u_1 = 0 \Rightarrow ikA \exp(0.85kx_3) - 0.39ikB \exp(0.39kx_3) = 0 \qquad (165.1)$$

so

$$k = \frac{2\pi}{\lambda} = \frac{2\pi}{Tc_R} \cong 0.1 \, \text{km}^{-1}$$

We can write B in terms of A using the boundary condition of zero stress at the free surface:

$$\tau_{31} = 0|_{x_3=0} \Rightarrow 2rA - (1 - s^2)B = 0$$

so

$$B = 1.47iA$$

Substituting in (165.1) we obtain the value of x_3:

$$\exp(0.85kx_3) = 0.39 \times 1.47 \exp(0.39kx_3)$$
$$x_3 = -12 \, \text{km}$$

At 12 km depth u_1 is null and for greater values of depth the particle motion is prograde while for lesser values of depth it is retrograde.

166. Given a layer of thickness H and shear modulus $\mu = 0$ on top of a half-space or semi-infinite medium in which $\lambda = 0$, study (without expanding the determinant) whether there exist surface waves that propagate in the x_1-direction. Are they dispersive waves?

In the liquid layer ($\mu = 0$) the P- and S-velocities are

$$\beta' = 0 = \sqrt{\frac{\mu'}{\rho}} \Rightarrow \alpha' = \sqrt{\frac{\lambda' + 2\mu'}{\rho}} = \sqrt{\frac{\lambda'}{\rho}}$$

and in the solid half-space

$$\lambda = 0 \Rightarrow \alpha = \sqrt{\frac{2\mu}{\rho}} = \sqrt{2}\beta$$

The relation between the stress and strain is

$$\tau_{ij} = \lambda\theta\delta_{ij} + 2\mu e_{ij}$$

where $e_{ij} = \frac{1}{2}(u_{i,j} + u_{j,i})$.

In the layer: $\mu' = 0 \Rightarrow \begin{cases} \tau'_{ii} = \lambda'(e'_{11} + e'_{22} + e'_{33}) \\ \tau'_{ij} = 0 \end{cases}$

In the half-space: $\lambda = 0 \Rightarrow \tau_{ij} = 2\mu e_{ij}$

If there are surface waves propagating in the x_1-direction, their displacements in terms of the potentials are given by (Fig. 166)

$$u_1 = \varphi_{,1} - \psi_{,3}$$
$$u_2 = u_2$$
$$u_3 = \varphi_{,3} + \psi_{,1}$$

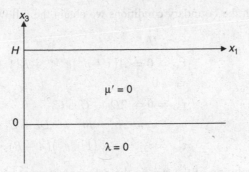

Fig. 166

The boundary conditions at the free surface are null normal stresses:

$$x_3 = H \Rightarrow \begin{cases} \tau'_{33} = 0 \\ \tau'_{31} = 0 \end{cases}$$

and at the boundary between the liquid layer and the solid half-space continuity of the normal component of the displacement and stress and zero tangential stresses,

$$x_3 = 0 \Rightarrow \begin{cases} u_3 = u'_3 \\ \tau_{33} = \tau'_{33} \\ \tau_{32} = \tau'_{32} = 0 \\ \tau'_{31} = 0 \end{cases}$$

In the liquid layer there is only the P-wave potential φ. Taking (x_1, x_3) as the incidence plane

$$\varphi' = A \exp(ikr'x_3 + ik(x_1 - ct)) + B \exp(-ikr'x_3 + ik(x_1 - ct))$$

$$r' = \sqrt{\frac{c^2}{\alpha'^2} - 1}$$

where c is the velocity of wave propagation in the x_1-direction

In the half-space

$$\psi = C \exp(-iksx_3 + ik(x_1 - ct))$$
$$u_2 = E \exp(-iksx_3 + ik(x_1 - ct))$$
$$\varphi = D \exp(-ikrx_3 + ik(x_1 - ct))$$

$$r = \sqrt{\frac{c^2}{\alpha^2} - 1}$$

$$s = \sqrt{\frac{c^2}{\beta^2} - 1}$$

In the layer we have only guided P-waves and r' is real, while in the half-space for surface waves, r and s must be imaginary. Then $\alpha > \beta > c > \alpha'$ must be satisfied.

From the boundary conditions we obtain the following equations:

$$x_3 = H$$

$$\tau'_{33} = 0 \Rightarrow A(1 + r'^2)e^{ikr'H} + B(1 + r'^2)e^{-ikr'H} = 0$$

$$x_3 = 0$$

$$\tau_{31} = 0 \Rightarrow 2Dr - C + Cs^2 = 0$$

$$u'_3 = u_3 \Rightarrow Ar' - Br' = -Dr + C$$

$$\tau'_{33} = \tau_{33} \Rightarrow -\lambda'(1 + r'^2)(A + B) = 2\mu(-Dr^2 + Cs)$$

For a solution of the system the determinant must be zero:

$$\begin{vmatrix} e^{ikr'H} & e^{-ikr'H} & 0 & 0 \\ 0 & 0 & s^2 - 1 & 2r \\ r' & -r' & -1 & r \\ -\lambda'(1 + r'^2) & -\lambda'(1 + r'^2) & -2\mu s & 2\mu r^2 \end{vmatrix} = 0$$

Expanding the determinant and working r', r, and s in terms of the variable c, we obtain c (k), the velocity of waves in the x_1-direction. They have the form of guided waves in the liquid layer and surface waves in the half-space. Since the velocity $c(k)$ is a function of the wavenumber the waves are dispersive.

167. There is a liquid layer of density ρ and speed of propagation α on top of a rigid medium (half-space). Derive the dispersion equation of waves in the layer by boundary conditions and by constructive interference in terms of ω. Plot the dispersion curve for the different modes.

Given that the layer is liquid the only potential is φ:

$$\varphi = (A \exp ikrx_3 + B \exp(-ikrx_3)) \exp ik(x_1 - ct) \tag{167.1}$$

The boundary conditions at the free surface are zero normal stress and at the boundary between the liquid layer and the rigid half-space zero normal component of displacement (Fig. 167a):

$$x_3 = H \Rightarrow \tau_{33} = 0$$

$$x_3 = 0 \Rightarrow u_3 = 0$$

where

$$\tau_{33} = \lambda\theta = \nabla^2\varphi = -\rho\omega^2\varphi = 0$$

$$u_3 = \varphi_{,3}$$

$$r = \sqrt{\frac{c^2}{\alpha^2} - 1}$$

By substitution of (167.1) we obtain

$$Ae^{ikrH} + Be^{-ikrH} = 0$$

$$A - B = 0 \Rightarrow A = B$$

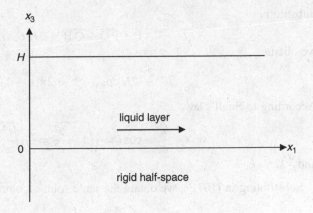

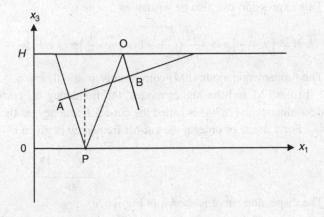

Then, $2A \cos krH = 0$. For waves propagating in the layer, r must be real and $c > a$. The solution is given by

$$krH = \left(n + \frac{1}{2}\right)\pi; \quad n = 0, 1, 2,\ldots \tag{167.2}$$

The solution can also be found by the method of constructive interference. The condition of constructive interference implies that waves coinciding at a given wavefront (AB) are in phase, that is, the distance along the ray must be an integer multiple of the wavelength, taking into account possible phase shifts (Fig. 167b). In our case on the free surface, $x_3 = H$, there is a phase shift of π ($\lambda/2$) and we write the condition as (Fig. 167b)

$$\overline{A}\,\overline{P} + \overline{P}\,\overline{Q} + \overline{Q}\,\overline{B} - \frac{\lambda_\alpha}{2} = n\lambda_\alpha$$

or

$$\frac{2\pi}{\lambda_\alpha}(\overline{A}\,\overline{P} + \overline{P}\,\overline{Q} + \overline{Q}\,\overline{B}) - \pi = 2\pi n$$

Substituting

$$\overline{AP} + \overline{PQ} + \overline{QB} = 2H\cos i$$

we obtain

$$\frac{2\pi}{\lambda_\alpha} 2H\cos i - \pi = 2\pi(n+1) \qquad (167.3)$$

According to Snell's law,

$$\sin i = \frac{\alpha}{c} \Rightarrow \cos i = \sqrt{1 - \frac{\alpha^2}{c^2}} = \frac{\alpha}{c}\sqrt{\frac{c^2}{\alpha^2} - 1} = \frac{\alpha}{c}r$$

and $\dfrac{\alpha}{c} k_\alpha = k$.

Substituting in (167.3), we obtain the same solution obtained in (167.2):

$$k_\alpha \frac{\alpha}{c} Hr = \left(n + \frac{1}{2}\right)\pi$$

This expression can also be written as

$$krH = \left(n + \frac{1}{2}\right)\pi \Rightarrow \frac{\omega H}{c}\sqrt{\frac{c^2}{\alpha^2} - 1} = \left(n + \frac{1}{2}\right)\pi \Rightarrow c = \left[\frac{1}{\alpha^2} - \left(n + \frac{1}{2}\right)^2 \frac{\pi^2}{H^2\omega^2}\right]^{-\frac{1}{2}} \qquad (167.4)$$

The fundamental mode (FM) corresponds to $n = 0$, and $n \geq 1$ to the higher modes (HM).

In the FM and the higher modes, the frequency ω_c corresponding to the zero in the denominator in (167.4) is called the cut-off frequency, as there are no values of c for $\omega < \omega_c$. For a mode of order n the cut-off frequency is given by

$$\omega_c = \frac{\pi\left(n + \frac{1}{2}\right)\alpha}{H}$$

The dispersion curve is shown in Fig. 167c.

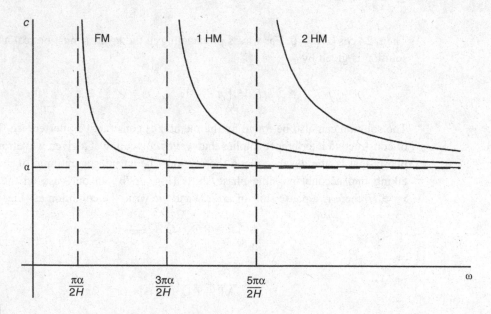

168. Consider an elastic layer of coefficients λ and μ, thickness H, and density ρ on a rigid semi-infinite medium. Derive the dispersion equation $c(\omega)$ for P-SV and SH-type channelled waves for the fundamental mode (FM) and the first higher mode (1HM). Plot the dispersion curve for the SH motion.

For a SH-wave which propagates in the x_1-direction its displacement is given by

$$u_2 = (E \exp(iksx_3) + F \exp(-iksx_3)) \exp ik(x_1 - ct)$$

The P- and SV-waves are given by their scalar potentials φ and ψ:

$$\varphi = (A \exp(ikrx_3) + B \exp(-ikrx_3)) \exp ik(x_1 - ct)$$
$$\psi = (C \exp(iksx_3) + D \exp(-iksx_3)) \exp ik(x_1 - ct)$$

where r and s were defined in Problem 166.

The boundary conditions for SH-waves are null stress at the free surface and null displacement at the boundary with the rigid medium (Fig 168a):

$$x_3 = H \Rightarrow \tau_{32} = 0 = \mu \frac{\partial u_2}{\partial x_3}$$

$$x_3 = 0 \Rightarrow u_2 = 0$$

By substitution we have

$$Ee^{iksH} - Fe^{-iksH} = 0$$
$$E + F = 0 \quad \Rightarrow F = -E$$
$$E(e^{iksH} + e^{-iksH}) = 0 \Rightarrow \cos(ksH) = 0 \Rightarrow ksh = \left(n + \frac{1}{2}\right)\pi$$

Substituting s and putting $k = \omega/c$:

$$\frac{H\omega}{c}\sqrt{\frac{c^2}{\beta^2} - 1} = \left(n + \frac{1}{2}\right)\pi \Rightarrow c = \left(\frac{1}{\beta^2} - \left(n + \frac{1}{2}\right)^2 \frac{\pi^2}{H^2\omega^2}\right)^{-\frac{1}{2}} \qquad (168.1)$$

This equation give us, for the SH component, the frequency dependence of the velocity $c(\omega)$.

The boundary conditions for P and SV are similarly

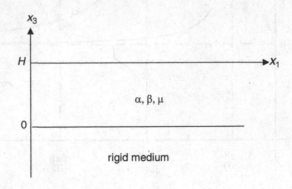

$$x_3 = H \Rightarrow \tau_{31} = 0; \tau_{33} = 0$$
$$x_3 = 0 \Rightarrow u_1 = 0; u_3 = 0$$

where

$$\tau_{33} = \lambda(e_{11} + e_{33}) + 2\mu e_{33}$$
$$\tau_{31} = 2\mu e_{31}$$
$$u_1 = \varphi_{,1} - \psi_{,3}$$
$$u_3 = \varphi_{,3} + \psi_{,1}$$

Substituting the expression for the potentials we obtain

$$(\lambda + 2\mu)(r^2 A e^{ikrH} + r^2 B e^{-ikrH}) + \lambda(A e^{ikrH} + B e^{-iksH}) + 2\mu s(C e^{iksH} - D e^{-iksH}) = 0$$
$$2r(A e^{ikrH} - B e^{-ikrH}) + (1 - s^2)(C e^{iksH} + D e^{-iksH}) = 0$$
$$(A + B) - s(C - D) = 0$$
$$r(A - B) + C + D = 0$$

For a solution we put the determinant of the system of equations equal to zero:

$$\begin{vmatrix} 1 & 1 & -s & s \\ r & -r & 1 & 1 \\ 2re^{ikrH} & -2re^{-ikrH} & (1-s^2)e^{iksH} & (1-s^2)e^{-iksH} \\ [\lambda+r^2(\lambda+2\mu)]e^{ikrH} & [\lambda+r^2(\lambda+2\mu)]e^{-ikrH} & 2\mu s e^{iksH} & -2\mu s e^{iksH} \end{vmatrix} = 0 \quad (168.2)$$

Expanding the determinant and putting it equal to zero, we obtain the dependence with frequency of the velocity $c(\omega)$ which gives us the dispersion curve.

For the wave with SH component the dispersion curve is given in Fig. 168b:

$$c = \left(\frac{1}{\beta^2} - \left(n + \frac{1}{2} \right)^2 \frac{\pi^2}{H^2\omega^2} \right)^{-\frac{1}{2}}$$

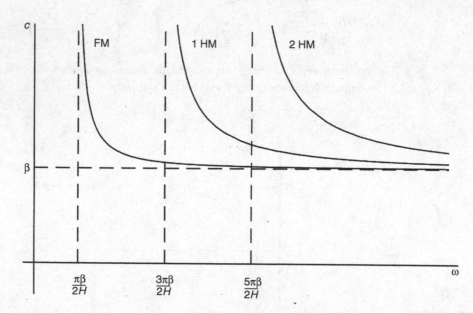

Fig. 168b

For $n = 0$ the curves correspond to the fundamental mode and for $1 \leq n$ to the higher modes. For all modes, including the fundamental mode, there is a cut-off frequency $\omega_c = (n+1)\pi\beta/2H$, with $n = 0$ for the fundamental mode and $n \geq 1$, for higher-order modes.

169. For a liquid layer of thickness H with a rigid medium above and below, derive the dispersion equation $c(\omega)$ of the fundamental and higher modes. For the FM, at what height above the layer is the motion circular?

Given that the medium is a liquid, motion is represented only by the scalar potential ϕ:

$$\varphi = (A \exp ikrx_3 + B \exp(-ikrx_3)) \exp ik(x_1 - ct) \tag{169.1}$$

where $r = \sqrt{\dfrac{c^2}{\alpha^2} - 1}$.

The boundary condition at the two boundaries between the liquid and rigid solid is that the normal component of the displacement is null (Fig. 169):

$$x_3 = 0 \Rightarrow u_3 = 0$$
$$x_3 = H \Rightarrow u_3 = 0$$

Substituting $u_3 = \varphi_{,3}$ we have

$$A - B = 0$$
$$Ae^{ikrH} - Be^{-ikrH} = 0 \tag{169.2}$$

which leads to the equation

$$A\left[e^{ikrH} - e^{-ikrH}\right] = 0 \tag{169.3}$$

Consider first that r is real, that is, $c > \alpha$. Then, from (169.1)

$$2iA \sin krH = 0 \Rightarrow krH = n\pi, \ n = 0, 1, 2, \ldots$$

with $n = 0$, fundamental mode (FM), and $n \geq 1$ for higher modes.

For the FM, $n = 0$ and $r = 0$, and then

$$Hk\sqrt{\frac{c^2}{\alpha^2} - 1} = 0 \Rightarrow c = \alpha$$

The displacements from (169.1) and (169.2) are

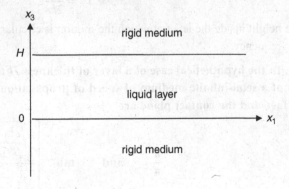

Fig. 169

$$u_3 = \frac{\partial \varphi}{\partial x_3} = Aikr(\exp ikrx_3 - \exp -ikrx_3) \exp ik(x_1 - ct)$$

$$u_1 = \frac{\partial \varphi}{\partial x_1} = Aik(\exp ikrx_3 + \exp -ikrx_3) \exp ik(x_1 - ct)$$

For the FM $r = 0$, then $u_3 = 0$ and this is a P-wave, with only a u_1 component, which propagates in the x_1-direction. For all HM the displacements have both components

For the first higher mode (1HM), $n = 1$:

$$H\frac{\omega}{c}\sqrt{\frac{c^2}{\alpha^2} - 1} = \pi$$

$$c^2 = \frac{1}{\dfrac{1}{\alpha^2} - \dfrac{\pi^2}{\omega^2 H^2}}$$

If

$$\frac{1}{\alpha^2} - \frac{\pi^2}{\omega^2 H^2} = 0 \qquad \text{then} \qquad \omega = \frac{\alpha\pi}{H} \Rightarrow c \to \infty$$

The cut-off frequency is $\omega_c > \alpha\pi/H$. For each higher mode there is a cut-off frequency $\omega''_c > n\alpha\pi/H$.

If $r = i\hat{r}$ is imaginary, then $c < \alpha$ and this implies that $-2\sinh(k\hat{r}H) = 0$ which is impossible ($1 < \sinh x < \infty$).

The particle motion inside the layer is circular when

$$u_1 = u_3 \tag{169.4}$$

so

$$u_1 = u_3 \Rightarrow (1 - r)\exp ikrx_3 + (1 + r)\exp -ikrx_3 = 0$$

Taking only the amplitudes of the displacements,

$$(1 - r)\cos krx_3 + (1 + r)\cos krx_3 = 0 \Rightarrow \cos krx_3 = 0$$

$$krx_3 = \left(n + \frac{1}{2}\right)\pi \Rightarrow x_3 = \left(n + \frac{1}{2}\right)\frac{\pi}{kr}$$

For the FM, we have seen that $u_3 = 0$, so there is no circular motion. For the 1HM, the height in the layer at which the motion is circular is

$$x_3 = \frac{3\pi}{2kr}$$

The height inside the layer at which the motion is circular depends in each higher mode of the frequency.

170. In the hypothetical case of a layer of thickness H and speed of propagation β' on top of a semi-infinite medium of speed of propagation β, the phase shifts at the free surface and the contact plane are

$$\frac{\pi}{4} \qquad \text{and} \qquad \tan^{-1}\left[\frac{\sqrt{\dfrac{c^2}{\beta'} - 1}}{\sqrt{1 - \dfrac{c^2}{\beta^2}}}\right].$$

Determine:

(a) The dispersion equation using constructive interference.
(b) The cut-off frequency of the fundamental mode and first higher mode.
(c) Plot the dispersion curve of the FM and 1HM using units of c/β and H/λ for $\beta = 2\beta'$.

(a) The distance from A to B along the ray path is (Fig. 170a)

$$AB = 2H \cos i$$

According to Snell's law

$$\sin i = \frac{\beta'}{c}$$

$$\cos i = \frac{\beta'}{c} \sqrt{\frac{c^2}{\beta'^2} - 1}$$

and the wavenumber k associated with velocity c is

$$k_{\beta'} \sin i = k_{\beta'} \frac{\beta'}{c} = k$$

As explained in Problem 167, the condition for constructive interference is that the distance AB along the ray path be an integer multiple of the wavelength, taking into account the phase shift at the free surface and the boundary surface between the two media:

$$2k_{B'} H \cos i + \frac{\pi}{4} - \tan^{-1}\left(\frac{s'}{s}\right) = 2\pi n \qquad (170.1)$$

where

$$s' = \sqrt{\frac{c^2}{\beta'^2} - 1}$$

$$s = \sqrt{1 - \frac{c^2}{\beta^2}}$$

By substitution in (170.1) we obtain the dispersion equation

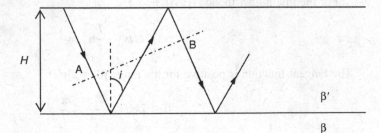

$$\tan\left[2kHs' - \left(2n - \frac{1}{4}\right)\pi\right] = \sqrt{\frac{\dfrac{c^2}{\beta'^2} - 1}{1 - \dfrac{c^2}{\beta^2}}} \qquad (170.2)$$

(b) For the fundamental mode (FM), $n = 0$:

$$\tan\left[2kH\sqrt{\frac{c^2}{\beta'^2} - 1} + \frac{\pi}{4}\right] = \frac{\sqrt{\dfrac{c^2}{\beta'^2} - 1}}{\sqrt{1 - \dfrac{c^2}{\beta^2}}} > 0 \qquad \text{where} \qquad \beta' < c < \beta$$

Given that $\tan\varsigma > 0 \Rightarrow \dfrac{\pi}{4} \le \varsigma \le \dfrac{\pi}{2}$ so that $\beta' < c < \beta$, then
If $k = 0$ then

$$\tan\left(\frac{\pi}{4}\right) = 1 = \frac{\sqrt{\dfrac{c^2}{\beta'^2} - 1}}{\sqrt{1 - \dfrac{c^2}{\beta^2}}} \Rightarrow c = \frac{\sqrt{2}\beta\beta'}{\sqrt{\beta^2 + \beta'^2}}$$

so

$$\tan\left(2kHs' + \frac{\pi}{4}\right) = \tan(\varsigma) = \infty \Rightarrow \varsigma = \frac{\pi}{2}$$

If $c = \beta$ then

$$2kH\sqrt{\frac{c^2}{\beta'^2} - 1} + \frac{\pi}{4} = \frac{\pi}{2} \Rightarrow k = \frac{\pi}{8Hs'} \Rightarrow s' = \sqrt{\frac{\beta^2}{\beta'^2} - 1}$$

so

$$\tan\left(2kHs' + \frac{\pi}{4}\right) = 0$$

If $c = \beta'$ then

$$s' = \sqrt{\frac{c^2}{\beta'^2} - 1} = 0 \Rightarrow \tan\left(kHs' + \frac{\pi}{4}\right) = 0 \Rightarrow k = \infty \Rightarrow kHs' = -\frac{\pi}{4}$$

But kHs' must be positive, so this last solution is not possible.
For the first higher mode (1HM), $n = 1$:

$$\tan\left[2kHs' - \frac{7}{4}\pi\right] = \frac{s'}{s}$$

The tangent function is positive for the range $[0, \pi/2]$:

$$0 \le 2kHs' - \frac{7\pi}{4} \le \frac{\pi}{2}$$

For $c = \beta'$

$$\tan\left[2kHs' - \frac{7\pi}{4}\right] = 0 \Rightarrow kHs' = \frac{7\pi}{8} \Rightarrow k = \infty$$

For $c = \beta$

$$\tan\left[2kHs' - \frac{7\pi}{4}\right] = \infty \Rightarrow 2kHs' - \frac{7\pi}{4} = \frac{\pi}{2} \Rightarrow k = \frac{9\pi}{8Hs'}$$

(c) Taking $\beta' = \beta/2$, the dispersion equation for the FM is

$$\tan\left[2kH\sqrt{\frac{4c^2}{\beta^2} - 1} + \frac{\pi}{4}\right] = \frac{\sqrt{\dfrac{4c^2}{\beta^2} - 1}}{\sqrt{1 - \dfrac{c^2}{\beta^2}}} \Rightarrow \left(\frac{c}{\beta}\right)^2 = \frac{2}{5 - 3\sin\left[8\pi\left(\dfrac{H}{\lambda}\right)\sqrt{4\left(\dfrac{c}{\beta}\right)^2 - 1}\right]}$$

But we have seen that

$$2kHs' + \frac{\pi}{4} \le \frac{\pi}{2} \Rightarrow \frac{H}{\lambda} \le \frac{1}{16s'}$$

$$c = \beta \Rightarrow s' = \sqrt{3}$$

$$\frac{H}{\lambda} \le \frac{1}{16\sqrt{3}}$$

Giving values to H/λ we can calculate the corresponding values of c/β by means of a numerical method. For example, we obtain,

H/λ	c/β
0.0	0.63
0.01	0.68
0.02	0.78
0.036	1

For the first higher mode (1HM), we arrive at the same equation, given that $\tan(a + \pi/4)$ = $\tan(a + 7\pi/4)$, but vary the intervals of H/λ and kHs':

$$\frac{7\pi}{8} \le kHs' \le \frac{9\pi}{8}$$

$$\frac{7}{16s'} \le \frac{H}{\lambda} \le \frac{9}{16s'}$$

For $\dfrac{H}{\lambda} = \dfrac{7}{16s'}$ at one limit we have the value $s' = 0$ which corresponds to $c = \beta'$ and $c/\beta = 0.5$.

For $\dfrac{H}{\lambda} = \dfrac{9}{16s'}$ we have $s = 0$ and $c = \beta$, and consequently $\dfrac{H}{\lambda} = \dfrac{9}{16\sqrt{3}}$

The dispersion curves for the fundamental mode and the first higher mode are shown in Fig.170b.

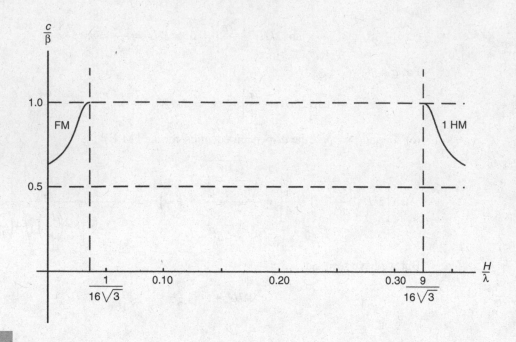

Fig. 170b

171. In a structure with a layer of thickness H and speed of propagation β' on top of a medium of speed of propagation β, the phase shifts at the free surface and the contact plane are:

$$\delta_1 = -\frac{\pi}{2} \text{ and } \delta_2 = -\sin^{-1}\left\{\frac{\sqrt{1-\dfrac{c^2}{\beta^2}}}{\sqrt{\dfrac{c^2}{\beta'^2}-1}}\right\}$$

Calculate:

(a) The dispersion equation of the Love wave.

(b) For the fundamental and first higher mode, and the minimum and maximum frequencies as functions of H, β, and β'.

(c) For this mode, given $\beta' = \beta/2$, the maximum and minimum frequencies, and the corresponding values of c.

(a) As in Problem 170, the condition of constructive interference with the phase shifts given in this problem results in (Fig. 171)

$$\frac{4\pi}{\lambda} H \cos i - \frac{\pi}{2} - \sin^{-1}\left(\frac{s}{s'}\right) = 2\pi n \qquad (171.1)$$

where

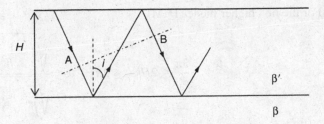

Fig. 171

$$s = \sqrt{1 - \frac{c^2}{\beta^2}}$$

$$s' = \sqrt{\frac{c^2}{\beta'^2} - 1}$$

According to Snell's law

$$\sin i = \frac{\beta'}{c}$$

$$\cos i = \frac{\beta'}{c}\sqrt{\frac{c^2}{\beta'^2} - 1} \quad .$$

Substituting in (171.1) we find the dispersion equation for Love waves:

$$2Hks' - \frac{\pi}{2} - 2\pi n = \sin^{-1}\left(\frac{s}{s'}\right) \Rightarrow -\cos\left(2Hk\sqrt{\frac{c^2}{\beta'^2} - 1}\right) = \frac{\sqrt{1 - \frac{c^2}{\beta^2}}}{\sqrt{\frac{c^2}{\beta'^2} - 1}}$$

(b) The fundamental mode (FM) corresponds to the values

$$0 \leq 2Hks' \leq \frac{\pi}{2} \Rightarrow 1 \geq \frac{\sqrt{1 - \frac{c^2}{\beta^2}}}{\sqrt{\frac{c^2}{\beta'^2} - 1}} \geq 0$$

The velocity at the limit of lowest frequencies, $k = 0$, is given by

$$k = 0 \Rightarrow \frac{s}{s'} = \frac{\sqrt{1 - \frac{c^2}{\beta^2}}}{\sqrt{\frac{c^2}{\beta'^2} - 1}} = 1 \Rightarrow c^2 = \frac{2\beta'^2\beta^2}{\beta'^2 + \beta^2}$$

$$k = \frac{\pi}{4Hs'} \quad \Rightarrow \frac{s}{s'} = 0 \Rightarrow s = 0 \Rightarrow c = \beta \Rightarrow s' = \frac{1}{\beta'}\sqrt{\beta^2 - \beta'^2}$$

For the first higher mode (1HM)

$$\frac{3\pi}{2} \le 2Hks' \le 2\pi \Rightarrow 0 \le \frac{\sqrt{1 - \dfrac{c^2}{\beta^2}}}{\sqrt{\dfrac{c^2}{\beta'^2} - 1}} \le 1$$

In the lowest limit

$$k_1 = \frac{3\pi}{4Hs'} \Rightarrow \frac{s}{s'} = 0 \Rightarrow s = 0 \Rightarrow c = \beta \Rightarrow s' = \frac{1}{\beta'} \sqrt{\beta^2 - \beta'^2}$$

The minimum value of the frequency is

$$k_1 = \frac{3\pi\beta'}{4H\sqrt{\beta^2 - \beta'^2}}$$

In the highest frequency limit

$$k_2 = \frac{\pi}{Hs'} \Rightarrow \frac{s}{s'} = 1 \Rightarrow s = s' \Rightarrow c^2 = \frac{2\beta'^2\beta^2}{\beta^2 + \beta'^2} \Rightarrow s' = \sqrt{\frac{\beta^2 - \beta'^2}{\beta^2 + \beta'^2}}$$

The maximum value of the frequency is given by

$$k_2 = \frac{\pi\sqrt{\beta^2 + \beta'^2}}{H\sqrt{\beta^2 - \beta'^2}}$$

(c) If we put $\beta' = \beta/2$ in the 1MS,

$$k_1 = \frac{3\pi\dfrac{\beta}{2}}{4H\sqrt{\beta^2 - \dfrac{\beta^2}{4}}} = \frac{\pi\sqrt{3}}{H4} = \frac{2\pi}{\lambda} \Rightarrow \frac{H}{\lambda} = \frac{\sqrt{3}}{8}, \ c = \beta$$

$$k_2 = \frac{\pi\sqrt{\beta^2 + \dfrac{\beta^2}{4}}}{4H\sqrt{\beta^2 - \dfrac{\beta^2}{4}}} = \frac{\pi\sqrt{5}}{H\sqrt{3}} = \frac{2\pi}{\lambda} \Rightarrow \frac{H}{\lambda} = \frac{\sqrt{5}}{2\sqrt{3}}, \ c = \beta\sqrt{\frac{2}{5}}$$

172. **Consider a layer of thickness H and parameters μ' and ρ' on top of a semi-infinite medium of parameters $\mu = 4\mu'$ and $\rho = \rho'$. If $\dfrac{c}{\beta} = a$ and $\dfrac{H}{\lambda} = b$:**

(a) **Write the dispersion equation of the Love wave in terms of a and b.**

(b) **Calculate the values of b corresponding to $a = \dfrac{1}{2}, \dfrac{3}{4}$ and 1 for the fundamental mode and first higher mode.**

(c) **For which values of b is the node of the amplitude of the first higher mode at a depth of $H/2$?**

(a) The dispersion equation for Love waves is

$$\tan\left\{ kH\sqrt{\frac{c^2}{\beta'^2} - 1} \right\} = \frac{\mu\sqrt{1 - \frac{c^2}{\beta^2}}}{\mu'\sqrt{\frac{c^2}{\beta'^2} - 1}} \tag{172.1}$$

In this problem,

$$\mu = 4\mu'$$
$$\rho = \rho'$$

then,

$$\beta = \sqrt{\frac{\mu}{\rho}} \rightarrow \beta' = \sqrt{\frac{\mu}{4\rho}} = \frac{\beta}{2}$$

We now introduce a and b:

$$a = \frac{c}{\beta} \Rightarrow \frac{c}{\beta'} = 2a$$

$$b = \frac{H}{\lambda} \Rightarrow k = \frac{2\pi}{\lambda} = \frac{2\pi b}{H}$$

Substituting in (172.1):

$$\tan\left(2\pi b\sqrt{4a^2 - 1}\right) = 4\frac{\sqrt{1 - a^2}}{\sqrt{4a^2 - 1}}$$

(b) For the FM

$$a = 1, \ b = 0 \text{ and } c = \beta$$

$$a = \frac{1}{2}, \ b \rightarrow \infty \text{ and } c = \beta'$$

$$a = \frac{3}{4} \Rightarrow \tan\left[2\pi b\sqrt{4\frac{9}{16} - 1}\right] = 4\frac{\sqrt{1 - \frac{9}{16}}}{\sqrt{4\frac{9}{16} - 1}} \Rightarrow b = 0.17$$

For the 1HM,

$$a = \frac{1}{2}, \ b \rightarrow \infty$$

$$a = 1 \Rightarrow \tan\left(2\pi b\sqrt{4a^2 - 1}\right) = 0 \Rightarrow 2\pi b\sqrt{3} = \pi \Rightarrow b = \frac{1}{2\sqrt{3}} = 0.29$$

$$a = \frac{3}{4} \Rightarrow \tan\left(2\pi b\sqrt{4\left(\frac{3}{4}\right)^2 - 1}\right) = 4\sqrt{\frac{7}{20}} \Rightarrow \pi b\sqrt{5} = \tan^{-1}\left(4\sqrt{\frac{7}{20}}\right) + \pi \Rightarrow b = 0.61$$

(c) Inside the layer the amplitude of the displacements of the Love wave are given by

$$u_2' = 2A'\cos\left[ks'H\left(1 - \frac{x_3}{H}\right)\right]\cos k(s'H + x_1 - ct)$$

The nodes are the points where the amplitude is zero. For the 1HM the node is located at the value of x_3 which satisfies the relation

$$ks'H\left(1 - \frac{x_3}{H}\right) = \frac{3\pi}{2}$$

If we want a node located at $x_3 = H/2$ then

$$k = \frac{3\pi}{s'H} \Rightarrow b = \frac{3}{2\sqrt{4a^2 - 1}}$$

If we substitute the values of a, ½, 1, and ¾ we obtain for b infinity, 0.86, and 1.34. The infinite value of b corresponds to $\lambda = 0$.

Focal parameters

173. Consider three stations with coordinates:

St1 = 36.2° N, 4.8° E;   St2 = 37.0° N, 2.4° E; St3 = 38.6° N, 4.0° E

An earthquake is recorded at the three stations with the following respective S-P time intervals:

$$t_1^{S-P} = 26.7 \, \text{s}$$

$$t_2^{S-P} = 27.0 \, \text{s}$$

$$t_3^{S-P} = 22.5 \, \text{s}$$

Given that the focus is at the surface, the P-wave velocity is constant and equal to 6 km s^{-1}, and Poisson's ratio is 1/3, calculate the coordinates of the epicentre.

Given that Poisson's ratio is 1/3,

$$\sigma = \frac{1}{3} = \frac{\lambda}{2(\lambda + \mu)} \Rightarrow \lambda = 2\mu \Rightarrow \alpha = \sqrt{\frac{\lambda + 2\mu}{\rho}} = 2\beta$$

where α and β are the velocities of the P- and S-waves, respectively.

From the S-P time intervals, calling x the epicentral distance (Fig. 173):

$$t^{S-P} = \frac{x}{\beta} - \frac{x}{\alpha} = \frac{x}{2\beta} \Rightarrow x = 2t^{S-P}\beta$$

Given that $\alpha = 6$ km s^{-1} then $\beta = 3$ km s^{-1}.

The epicentral distances corresponding to each station in kilometres and degrees are

$$x_1 = 26.7 \times 6 = 160 \, \text{km} = \frac{160}{111.11} = 1.44°$$

$$x_2 = 27 \times 6 = 162 \, \text{km} = \frac{162}{111.11} = 1.46°$$

$$x_3 = 22.5 \times 6 = 135 \, \text{km} = \frac{135}{111.11} = 1.22°$$

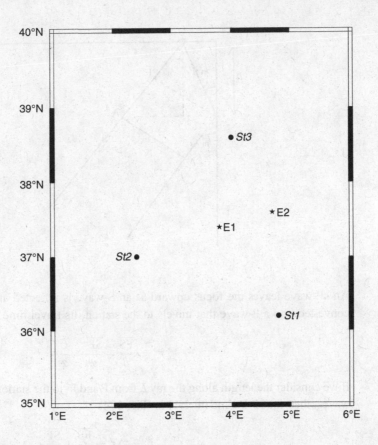

Fig. 173

If x_e and y_e are the geographical coordinates of the epicentre,

$$(36.2 - x_e)^2 + (4.8 - y_e)^2 = 1.44^2$$
$$(37.0 - x_e)^2 + (2.4 - y_e)^2 = 1.46^2$$
$$(38.6 - x_e)^2 + (4.0 - y_e)^2 = 1.22^2$$

Solving the system we find two possible solutions:

$$(x_e, y_e) = \begin{cases} (37.40° \, \text{N}, 3.83° \, \text{E}) \\ (37.63° \, \text{N}, 4.71° \, \text{E}) \end{cases}$$

174. A focus is at a depth h, in a medium with constant P- and S-wave speeds of propagation α and β. Calculate an expression for the depth h in terms of the time interval between the P and sP phases.

If L is the distance along the ray, the travel-time of the P-wave from the focus to the station is given by (Fig. 174)

$$t^P = \frac{L}{\alpha}$$

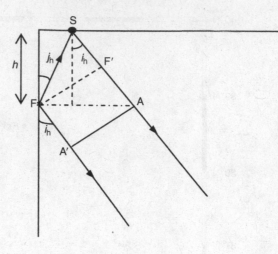

Fig. 174

An sP-wave leaves the focus upward as an S-wave is reflected at the Earth's surface and converted into a P-wave that travels to the station. Its travel time is (Fig. 174)

$$t^{sP} = \frac{\overline{FS}}{\beta} + \frac{\overline{SF'}}{\alpha} + \frac{L}{\alpha}$$

If we consider the length along the ray L from F and F′ to the station to be the same for both waves, then the sP-P time interval is (Fig. 174)

$$t^{sP} - t^{P} = \Delta t' = \frac{\overline{FS}}{\beta} + \frac{\overline{SF'}}{\alpha} \tag{174.1}$$

At the focus at depth h, the take–off angle of the direct P-wave is i_h and the take-off angle of the sP-wave is j_h. We can write

$$\overline{FS} = \frac{h}{\cos j_h}$$

$$\cos(j_h + i_h) = \frac{\overline{SF'}}{\overline{FS}} \Rightarrow \overline{SF'} = \frac{h}{\cos j_h}(\sin j_h \sin i_h - \cos j_h \cos i_h)$$

$$\sin i_h = \frac{\alpha \sin j_h}{\beta}$$

By substitution in (174.1) we obtain

$$\Delta t' = h\left(\frac{\cos j_h}{\beta} + \frac{\cos i_h}{\alpha}\right)$$

175. The displacement vector l of an earthquake is (0, 1, 0) and the vector normal to the plane of displacement n is (0, 1, 0). Determine:

(a) The components of the P-wave displacement at the point of azimuth 45° and angle of incidence 30°.

(b) The kind of mechanism it represents.

(a) The elastic displacements due to a dislocation Δu in the direction of l_i on a plane of normal n_i (Fig. 175) are given by

$$u_k^P = \Delta u \left[\lambda n_k l_k \delta_{ij} + \mu \left(l_i n_j + n_i l_j\right)\right] G_{ki,j}$$

where Green's function corresponding to the P-wave in the far field for an infinite medium is given by

$$G_{ki} = \frac{1}{4\pi\rho\alpha^2 r} \gamma_i \gamma_k \delta\left(t - \frac{r}{\alpha}\right)$$

and its derivative is

$$G_{ki,j} = \frac{1}{4\pi\rho\alpha^3 r} \gamma_i \gamma_k \gamma_j \dot{\delta}\left(t - \frac{r}{\alpha}\right)$$

where γ_i are the direction cosines of the line from the focus to the observation point.
The amplitude of the displacement is then

$$u_k^P = \frac{\Delta u}{4\pi\rho\alpha^3 r} \left[\lambda n_s l_s \delta_{ij} + \mu \left(l_i n_j + l_j n_i\right)\right]\gamma_i \gamma_k \gamma_j \qquad (175.1)$$

In our problem the direction cosines of the ray of the waves arriving at the point are

$$\gamma_1 = \sin i \cos a_z = \frac{\sqrt{2}}{4}$$

$$\gamma_2 = \sin i \sin a_z = \frac{\sqrt{2}}{4}$$

$$\gamma_3 = \cos i = \frac{\sqrt{3}}{2}$$

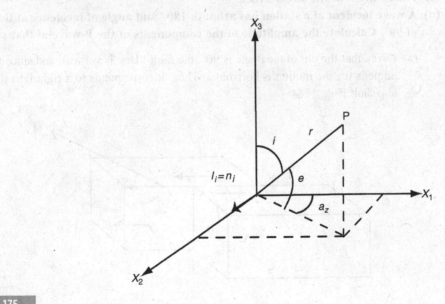

Fig. 175

The orientation of the source is given by $l_i = (0, 1, 0)$ and $n_i = (0, 1, 0)$, and substituting in (175.1) gives

$$u_1^P = A(\lambda + 2\mu\gamma_2^2)\gamma_1$$
$$u_2^P = A(\lambda + 2\mu\gamma_2^2)\gamma_2$$
$$u_3^P = A(\lambda + 2\mu\gamma_2^2)\gamma_3$$
$$A = \frac{\Delta u}{4\pi\alpha^3 \rho r}$$

Substituting the direction cosines of the ray we obtain,

$$u_1^P = A\left(\lambda + \frac{1}{4}\mu\right)\frac{\sqrt{2}}{4}$$

$$u_2^P = A\left(\lambda + \frac{1}{4}\mu\right)\frac{\sqrt{2}}{4}$$

$$u_3^P = A\left(\lambda + \frac{1}{4}\mu\right)\frac{\sqrt{3}}{2}$$

(b) The mechanism corresponds to a fault on the (x_1, x_3) plane which opens in the direction of its normal, x_2, under tensional forces in that direction.

176. The focal mechanism of an earthquake can be represented by a double-couple (DC) model. The orientation of the fault plane is azimuth 30°, dip 90°, and slip angle 0°. Calculate:

(a) **What kind of fault it is. Sketch it, indicating the direction of motion.**
(b) **The auxiliary plane.**
(c) **The azimuth of the stress axis.**
(d) **A wave incident at a station has azimuth 180° and angle of incidence at the focus of 90°. Calculate the amplitude of the components of the P-wave at that station.**

(a) Given that the dip of the plane is 90°, the fault plane is vertical, and since the slip angle is 0°, the motion is horizontal. Thus, it corresponds to a right lateral strike-slip fault (Fig. 176).

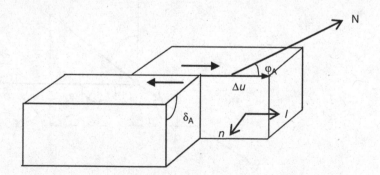

Fig. 176

(b) From the azimuth ($\varphi = 30°$), dip ($\delta = 90°$), and slip ($\lambda = 0°$) we can calculate the unit vectors l_i and n_i which give the direction of the fracture and of the normal to the fault plane:

$$n_1 = -\sin \delta \sin \varphi = -\frac{1}{2} = \sin \Theta_n \cos \Phi_n$$

$$n_2 = \sin \delta \cos \varphi = \frac{\sqrt{3}}{2} = \sin \Theta_n \cos \Phi_n$$

$$n_3 = -\cos \delta = 0 = \cos \Theta_n$$

$$l_1 = \cos \lambda \cos \varphi + \cos \delta \sin \lambda \sin \varphi = \frac{\sqrt{3}}{2} = \sin \Theta_l \cos \Phi_l$$

$$l_2 = \cos \lambda \sin \varphi - \cos \delta \sin \lambda \cos \varphi = \frac{1}{2} = \sin \Theta_l \cos \Phi_l$$

$$l_3 = -\sin \lambda \sin \delta = 0 = \cos \Theta_l$$

$$\varphi_A = \Phi_n + 90 \Rightarrow \Phi_n = 120°$$
$$\delta_A = \Theta_n = 90°$$
$$\lambda_A = \sin^{-1}\left(\frac{\cos \Theta_l}{\sin \Theta_n}\right) = 0 \Rightarrow \Theta_i = 90°$$

where Θ and Φ are the spherical coordinates for the vectors $\boldsymbol{n}$ and $\boldsymbol{l}$ ($r = 1$, unitary vectors) and for the auxiliary plane,

$$\varphi_B = 30 + 90 = 120°$$
$$\delta_B = 90°$$
$$\lambda_B = \sin^{-1}\frac{\cos \Theta_n}{\sin \Theta_l} = 0°$$

(c) The T-axis is on the same plane as n_i and l_i at $45°$ between them, and the direction cosines are

$$\begin{pmatrix} T_1 \\ T_2 \\ T_3 \end{pmatrix} = \begin{pmatrix} n_1 & n_2 & n_3 \\ l_1 & l_2 & l_3 \\ Z_1 & Z_2 & Z_3 \end{pmatrix} \begin{pmatrix} \frac{1}{\sqrt{2}} \\ \frac{1}{\sqrt{2}} \\ 0 \end{pmatrix} \tag{176.1}$$

where Z_i is the axis normal to n_i and l_i, that is, $Z_i = n_i \times l_i$ which results in $Z_i = (0, 0, -1)$. Substituting in (176.1) we obtain $T_i = \left(\frac{\sqrt{3}-1}{2\sqrt{2}}, \frac{\sqrt{3}+1}{2\sqrt{2}}, 0\right)$. The azimuth of the T-axis is

$$\Phi_T = \tan^{-1}\left(\frac{T_2}{T_1}\right) = \tan^{-1}\left(\frac{\sqrt{3}+1}{\sqrt{3}-1}\right) = 75°$$

(d) The direction cosines of the direction from the focus to the station are

$$\gamma_1 = \sin i_h \cos a_z = -1$$
$$\gamma_2 = \sin i_h \sin a_z = 0$$
$$\gamma_3 = \cos i_h = 0$$

The amplitude of the displacements for a shear fracture or double-couple (DC) source in an infinite medium is given by

$$R(i_h, a_z) \Rightarrow u_j^P = A(n_i l_k + n_k l_i)\gamma_i\gamma_k\gamma_j$$
$$u_1^P = -0.84A$$
$$u_2^P = 0$$
$$u_3^P = 0$$
$$A = \frac{M_0}{4\pi\alpha^3\rho r}$$

177. An earthquake is caused by a shear fracture. The vectors n and l (normal and direction of travel) are, in terms of the angles Φ and Θ,

$$n = (57.13°, 66.44°)$$
$$l = (305.96°, 50.09°)$$

Calculate the orientation of the fault plane, the auxiliary plane, and the tension (T) and pressure (P) stress axes.

The orientation of the fault plane and auxiliary plane in terms of the angles φ, δ, and λ (azimuth, dip, and slip) are found directly from the given values (Fig. 177), using the following relations:

$$\varphi_A = \Phi_n + 90° = 147.13°$$
$$\delta_A = \Theta_n = 66.64°$$
$$\lambda_A = \sin^{-1}\left(\frac{\cos\Theta_l}{\sin\Theta_n}\right) = 135.68°$$
$$\varphi_B = \Phi_l + 90° = 35.96°$$
$$\delta_B = \Theta_l = 50.09°$$
$$\lambda_B = \sin^{-1}\left(\frac{\cos\Theta_n}{\sin\Theta_l}\right) = 31.12°$$

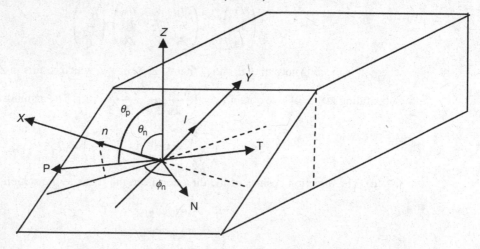

Fig. 177

To calculate the T and P axes, we calculate first the direction cosines of the l and n axes from the given angles:

$$x_1 = \sin \Theta \cos \Phi \qquad n_1 = 0.50, l_1 = 0.45$$
$$x_2 = \sin \Theta \sin \Phi \Rightarrow n_2 = 0.77, l_2 = -0.62$$
$$x_3 = \cos \Theta \qquad n_3 = 0.40, l_3 = 0.64$$

Given that the T and P axes are on the same plane as n and l at $45°$ between them, we can write, as in Problem 176,

$$\begin{pmatrix} T_1 \\ T_2 \\ T_3 \end{pmatrix} = \begin{pmatrix} l_1 & n_1 & Z_1 \\ l_2 & n_2 & Z_2 \\ l_3 & n_3 & Z_3 \end{pmatrix} \begin{pmatrix} \frac{1}{\sqrt{2}} \\ \frac{1}{\sqrt{2}} \\ 0 \end{pmatrix} \qquad (177.1)$$

where the Z-axis is normal to n and l and is found by $Z = n \times l$. Its direction cosines are $(Z_1, Z_2, Z_3) = (0.72, -0.14, -0.66)$. By substitution of n_i, l_i, and Z_i in (177.1) we obtain

$$\left. \begin{aligned} T_1 &= 0.67 = \sin \Theta_T \cos \Phi_T \\ T_2 &= 0.11 = \sin \Theta_T \sin \Phi_T \\ T_3 &= 0.74 = \cos \Theta_T \end{aligned} \right\} \Rightarrow T(\Theta_T = 42.27°, \Phi_T = 9.32°)$$

In the same way for the axis P

$$\begin{pmatrix} P_1 \\ P_2 \\ P_3 \end{pmatrix} = \begin{pmatrix} l_1 & n_1 & Z_1 \\ l_2 & n_2 & Z_2 \\ l_3 & n_3 & Z_3 \end{pmatrix} \begin{pmatrix} \frac{1}{\sqrt{2}} \\ -\frac{1}{\sqrt{2}} \\ 0 \end{pmatrix} \rightarrow P(\Theta_P = 80.02°, \Phi_P = 268.03°)$$

178. The seismic moment tensor relative to the geographical axes (X_1, X_2, X_3) (north, east, nadir) is

$$M_{ij} = \begin{pmatrix} 2 & -1 & 1 \\ -1 & 0 & 1 \\ 1 & 1 & 2 \end{pmatrix}$$

Find the values of the principal stresses, and the orientation of the tension and pressure stress axes.

First we calculate the eigenvalues of M_{ij}. Since M_{ij} is a symmetric tensor its eigenvalues are real and the corresponding eigenvectors mutually orthogonal (Problem 111):

$$\begin{vmatrix} 2 - \sigma & -1 & 1 \\ -1 & 0 - \sigma & 1 \\ 1 & 1 & 2 - \sigma \end{vmatrix} = 0 \Rightarrow \sigma = \begin{pmatrix} 3 \\ 2 \\ -1 \end{pmatrix}$$

Ordered by magnitude, the three eigenvalues are

$$\sigma_1 = 3, \sigma_2 = 2, \sigma_3 = -1$$

The diagonalized matrix is

$$M_{ij} = \begin{pmatrix} 3 & 0 & 0 \\ 0 & 2 & 0 \\ 0 & 0 & -1 \end{pmatrix}$$

In this form M_{ij} is referred to the coordinate system formed by the eigenvectors or principal axes. Given that the sum of the elements of the principal diagonal is not zero, the source has net volume changes. Then, we can separate M_{ij} into two parts: an isotropic part with volume changes (ISO) and a deviatoric part without volume changes. The second part can be separated into two parts: a part corresponding to a double-couple or shear fracture (DC) and a part corresponding to a non-double-couple source usually expressed as a compensated linear vector dipole (CLVD). Thus the moment tensor is separated into three parts, namely

$$M = M^{\mathrm{ISO}} + M^{\mathrm{DC}} + M^{\mathrm{CLVD}}$$

The isotropic part is given by

$$M^{\mathrm{ISO}} = \sigma_0 = \frac{1}{3}(\sigma_1 + \sigma_2 + \sigma_3) = \frac{4}{3}$$

The deviatoric part (DC+CLVD) is given by

$$M'_{ij} = M_{ij} - \delta_{ij}\sigma_o$$

and in our case

$$M'_{ij} = \begin{pmatrix} \sigma_1 & 0 & 0 \\ 0 & \sigma_2 & 0 \\ 0 & 0 & \sigma_3 \end{pmatrix} = \begin{pmatrix} \dfrac{5}{3} & 0 & 0 \\ 0 & \dfrac{2}{3} & 0 \\ 0 & 0 & -\dfrac{7}{3} \end{pmatrix}$$

Now we separate this part into two parts, DC and CLVD:

$$M'_{ij} = M^{\mathrm{DC}} + M^{\mathrm{CLVD}}$$

$$M'_{ij} = \begin{pmatrix} \dfrac{1}{2}(\sigma_1 - \sigma_3) & 0 & 0 \\ 0 & 0 & 0 \\ 0 & 0 & -\dfrac{1}{2}(\sigma_1 - \sigma_3) \end{pmatrix} + \begin{pmatrix} -\dfrac{\sigma_2}{2} & 0 & 0 \\ 0 & \sigma_2 & 0 \\ 0 & 0 & -\dfrac{\sigma_2}{2} \end{pmatrix}$$

$$M'_{ij} = \begin{pmatrix} 2 & 0 & 0 \\ 0 & 0 & 0 \\ 0 & 0 & -2 \end{pmatrix} + \begin{pmatrix} -\dfrac{1}{3} & 0 & 0 \\ 0 & \dfrac{2}{3} & 0 \\ 0 & 0 & -\dfrac{1}{3} \end{pmatrix}$$

The orientation of the P and T axes is calculated from the double-couple part M^{DC}:

$$M_{ij}^{DC} = 2 \begin{pmatrix} 1 & 0 & 0 \\ 0 & 0 & 0 \\ 0 & 0 & -1 \end{pmatrix}$$

We can find the n and l axes:

$$M_{ij}^{DC} = M_0(l_i n_j - l_j n_i)$$

$$\vec{n} : \left(\frac{1}{\sqrt{2}}, 0, -\frac{1}{\sqrt{2}} \right) \quad \Rightarrow \quad n(\Theta_n = -45°, \Phi_n = 0°)$$

$$\vec{l} : \left(\frac{1}{\sqrt{2}}, 0, \frac{1}{\sqrt{2}} \right) \quad \Rightarrow \quad l(\Theta_l = 45°, \Phi_l = 0°)$$

In the same way as in Problems 176 and 177, we determine T and P from n and l, finding first $Z = n \times l$:

$$\begin{pmatrix} T_1 \\ T_2 \\ T_3 \end{pmatrix} = \begin{pmatrix} l_1 & n_1 & Z_1 \\ l_2 & n_2 & Z_2 \\ l_3 & n_3 & Z_3 \end{pmatrix} \begin{pmatrix} \frac{1}{\sqrt{2}} \\ \frac{1}{\sqrt{2}} \\ 0 \end{pmatrix}$$

$$T_1 = 1 = \sin \Theta_T \cos \Phi_T$$
$$T_2 = 0 = \sin \Theta_T \sin \Phi_T \quad \to \quad T(\Theta_T = 90°, \Phi_T = 0°)$$
$$T_3 = 0 = \cos \Theta_T$$

For the P-axis,

$$\begin{pmatrix} P_1 \\ P_2 \\ P_3 \end{pmatrix} = \begin{pmatrix} l_1 & n_1 & Z_1 \\ l_2 & n_2 & Z_2 \\ l_3 & n_3 & Z_3 \end{pmatrix} \begin{pmatrix} \frac{1}{\sqrt{2}} \\ -\frac{1}{\sqrt{2}} \\ 0 \end{pmatrix} \quad P(\Theta_P = 0°, \Phi_P = 0°)$$

179. The magnitude M_s of an earthquake as calculated for surface waves of period 20 s is 6.13.

(a) Calculate the amplitude of these waves at a station 3000 km away. If the instrument's amplification is 1500, what will be the amplitude of the seismogram's waves and the seismic energy?

(b) If $M_s = M_w$, and the area of the fault is 12 km $\times$ 8 km with $\mu = 4.4 \times 10^4$ MPa, find the fault slip Δu.

(a) The surface wave magnitude M_s is given by

$$M_s = \log \frac{A}{T} + 1.66 \log \Delta + 3.3$$

where A is the ground motion amplitude, T is the period of the wave, and Δ is epicentral distance in degrees. Knowing the magnitude and period of the waves we can calculate the wave amplitude:

$$6.13 = \log\frac{A}{T} + 1.66\log\frac{3000}{111.11} + 3.3$$

$$\Rightarrow \log\frac{A}{T} = 0.454 \Rightarrow A = 2.84 \times 20 \times 1500 = 8.5\,\text{cm}$$

We have reduced the ground motion to the amplitude of the seismogram using the amplification of the instrument (1500).

Knowing the magnitude we can calculate the seismic energy:

$$\log E_s = 11.8 + 1.5M_s \Rightarrow E_s = 10^{21}\,\text{ergs} = 10^{14}\,\text{J}$$

(b) $M_w = 6.13 = 2/3 \log M_0 - 6.1$; $M_0 = 2.19 \times 10^{18}$ Nm

If $M_0 = \mu \Delta u\, S$, with $S = 12 \times 8 = 9.6 \times 10^7$ m^2, then

$$\Delta u = \frac{M_0}{\mu S} = \frac{2.19 \times 10^{18}\,\text{Nm}}{4.4 \times 10^{10}\,\text{N\,m}^{-2} \times 9.6 \times 10^7\,\text{m}^2} = 0.52\,\text{m}$$

Heat flow and geochronology

Heat flow

180. Assume that the temperature variation within the Earth is caused by gravitational forces under adiabatic conditions. Knowing that the coefficient of thermal expansion at constant pressure is $\alpha_P = 2 \times 10^{-5}$ K^{-1} and the specific heat at constant pressure is $c_P = 1.3$ kJ kg^{-1} K^{-1}, determine an expression for the gradient of the temperature with depth. Compare it with the value observed at the surface which is 30 K km^{-1}, knowing that, at 200 km depth, $T = 1600$ K.

Under adiabatic conditions, there is no heat flow and the variation of pressure with depth z is a function of gravity g and density ρ:

$$dP = \rho g dz \qquad (180.1)$$

Using the first and second laws of thermodynamics

$$dU = \delta Q - PdV$$
$$\delta Q = TdS$$

where Q is the heat, U is the internal energy, S is the entropy, T is the absolute temperature, P is the pressure, and V is the volume.

If we use the specific variables (variables divided by mass) we can write

$$du = \delta q - pdv$$
$$\delta q = Tds$$

Considering that

$$dS = \left(\frac{\partial S}{\partial T}\right)_P dT + \left(\frac{\partial S}{\partial P}\right)_T dP$$

we can write

$$\delta q = Tds = T\left(\frac{\partial s}{\partial T}\right)_P dT + T\left(\frac{\partial s}{\partial P}\right)_T dP$$

According to the definition of specific heat at constant pressure, c_P and the increase in heat δq are given by

$$c_P = T\left(\frac{\partial S}{\partial T}\right)_P$$

$$\delta q = c_P dT + T\left(\frac{\partial s}{\partial P}\right)_T dP$$

(180.2)

The Gibbs function G is defined as

$$G = u - Ts + pv$$

Taking the differential in this expression, and taking into account the second law of thermodynamics

$$du = Tds - pdv$$

we obtain

$$dG = vdp - sdT$$

If we compare this expression to the differential of the Gibbs function

$$dG = \left(\frac{\partial G}{\partial p}\right)_T dp + \left(\frac{\partial G}{\partial T}\right)_p dT$$

we differentiate again and using the Schwartz theorem we obtain

$$\left(\frac{\partial v}{\partial T}\right)_P = -\left(\frac{\partial s}{\partial P}\right)_T$$

But the coefficient of thermal expansion is defined as

$$\alpha_p = \frac{1}{v}\left(\frac{\partial v}{\partial T}\right)_p$$

In consequence we can write Equation (180.2) as

$$\delta q = c_p dT - Tv\alpha_p dp$$

In our case the process is adiabatic and in consequence using this equation and Equation (180.2) we obtain

$$\frac{dT}{dz} = \frac{T\alpha_p g}{c_p}$$

(180.3)

where we have taken into account that the variables are by unit mass so $\rho v = 1$, and substituting the values we obtain:

$$\frac{dT}{dz} = \frac{1600 \times 2 \times 10^{-5} \times 10\,\mathrm{K\,K^{-1}\,ms^{-2}}}{1.3 \times 10^3\,\mathrm{J\,kg^{-1}\,K^{-1}}} = 0.25\,\mathrm{K\,km^{-1}}$$

We observe that this result is two orders of magnitude lower than the observed values. This shows that observations correspond to heat flow at the lithosphere and are not satisfied by purely adiabatic conditions.

181. If the Earth's temperature gradient is 1 °C/30 m, calculate the heat loss per second due to conduction from its core. Compare this with the average power received from the Sun.

Data:

Thermal conductivity $K = 4 \text{ W m}^{-1} \text{ °C}^{-1}$.

Earth's radius $R = 6370$ km.

Solar constant: 1.35 kW m^{-2}.

The heat flow is given by

$$\dot{q} = -KA \frac{dT}{dr}$$

where A is the area of the Earth's surface

$$A = 4\pi R^2 = 5.10 \times 10^{14} \text{ m}^2$$

and

$$\frac{dT}{dr} = -\frac{1 \text{ °C}}{30 \text{ m}}$$

Substituting the values

$$\dot{q} = 6.80 \times 10^{13} \text{ J s}^{-1} = 1.63 \times 10^{13} \text{ cal s}^{-1}$$

the average power received on the Earth's surface by the radiation from the Sun is

$$1.35 \times 10^3 \frac{\text{W}}{\text{m}^2} 5.10 \times 10^{14} \text{ m}^2 = 6.89 \times 10^{17} \text{ J s}^{-1} = 1.65 \times 10^{17} \text{ cal s}^{-1}$$

In consequence, from these values we can see that on the Earth's surface the average solar power is much larger than that due to the heat flow from inside the Earth.

182. At the Earth's surface, the heat flow is 60 mW m^{-2} and $T_0 = 0$ °C. If all the heat is generated by the crust at whose base the thermal conductivity is $K = 4 \text{ W m}^{-1} \text{ °C}^{-1}$, and T is 1000 °C, determine the thickness of the crust and the heat production per unit volume

If we assume that all the heat is generated at the crust and there is no heat flow from the mantle at the crust base, then we can write

$$\dot{q}|_{z=H} = 0$$

Using the temperature equation for a flat Earth for one-dimensional heat-flow and the stationary case we can write

$$T = -\frac{\varepsilon}{2K} z^2 + \frac{\dot{q}_0}{K} z + T_0$$

$$\varepsilon = \frac{\dot{q}_0}{H}$$

For $z = H$:

$$T_H = \frac{\dot{q}_0 H}{2K} + T_0 \Rightarrow H = \frac{(T_H - T_0)2K}{\dot{q}_0}$$

Substituting the values given in the problem we obtain

$$H = 133.3 \, \text{km}$$

and the heat production by unit of volume is

$$\varepsilon = \frac{\dot{q}_0}{H} = 4.5 \times 10^{-4} \, \text{mW} \, \text{m}^{-3}$$

183. Consider the crust to be $H = 30$ km thick and the heat flow at the surface to be 60 mW m^{-2}.

(a) **If all the heat is generated in the crust, what is the value of the heat generated per unit volume? (Take $K = 3$ W m^{-1} K^{-1})**

(b) **If all the heat is generated in the mantle with a distribution $Ae^{-z/H}$ mW m^{-3}, what is the value of A? What is the temperature at 100 km depth?**

(a) We solve the heat equation for a stationary one-dimensional case, assuming a flat Earth with one-dimensional flow in the z-direction (vertical) positive downward. In this case the solution of the heat equation is given by

$$T = -\frac{\varepsilon}{2K}z^2 + \frac{\dot{q}_0}{K}z + T_0$$

where ε is the heat generated by unit volume and time, K is the thermal conductivity, q_0 and T_0 are the heat flow and temperature at the surface of the Earth, respectively

If all the heat is generated at the crust we can write (Fig. 183)

$$z = H \rightarrow \dot{q}_0 = 0$$

The heat generated by unit volume is

$$\dot{q}|_{z=H} = 0 = -K\frac{dT}{dz}\bigg|_{z=H} = -K\left(-\frac{\varepsilon 2z}{2K} + \frac{\dot{q}_0}{K}\right)\bigg|_{z=H} = 0$$

$$\varepsilon = \frac{\dot{q}_0}{H} = \frac{60 \times 10^{-3}}{30 \times 10^3} = 2 \times 10^{-6} \, \text{W} \, \text{m}^{-3}$$

(b) If all heat is generated in the mantle with distribution

$$\varepsilon = Ae^{-z/H}$$

the heat equation is

$$\frac{d^2T}{dz^2} = -\frac{\varepsilon}{K} = -\frac{A}{K}e^{-z/H}$$

and the solution is given by

$$T = -\frac{A}{K}H^2 e^{-z/H} + Cz + D \tag{183.1}$$

where C and D are constants of integration. They may be estimated from the boundary conditions at the surface

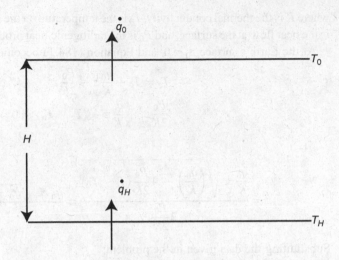

Fig. 183

$$z = 0 \rightarrow T = T_0 = -\frac{A}{K}H^2 + D \rightarrow D = T_0 + H^2\frac{A}{K}$$

$$z = 0 \rightarrow \dot{q} = -\dot{q}_0 = 0 \rightarrow C = -\frac{A}{K}H$$

Substituting in (183.1) we obtain

$$T = -\frac{A}{K}H^2 e^{-z/H} - \frac{AH}{K}z + T_0 + H^2\frac{A}{K} = \frac{AH}{K}\left(-He^{-z/H} - z + H\right) + T_0 \qquad (183.2)$$

If the heat has its origin in the mantle, the flow at the base of the crust is

$$\dot{q}|_{z=H} = -\dot{q}_o = -K\frac{dT}{dz}\Big|_{z=H} \Rightarrow A = \frac{\dot{q}_0}{H(e^{-1} - 1)}$$

The temperature at $z = 100$ km may be estimated from (183.2) assuming that $T_0 = 0$:

$$T(z) = \frac{\dot{q}_0}{K(e^{-1} - 1)}\left(-He^{-z/H} - z + H\right)$$

$$T(z = 100) = \frac{60 \times 10^{-3}}{3(e^{-1} - 1)}\left(-30 \times 10^3 e^{-100/30} - 100 \times 10^3 + 30 \times 10^3\right) = 2249\,\text{K}$$

184. Calculate the thickness of the continental lithosphere if its boundary coincides with the 1350 °C geothermal, knowing that the surface temperature is 15 °C, the heat flow at the surface is $\dot{q}_0 = 46\,\text{mW m}^{-2}$, the lithospheric mantle's thermal conductivity is $K = 3.35$ W m^{-1} K^{-1}, and the radiogenic heat production is $P = 0.01 \times 10^{-3}$ mW m^{-3}

The geothermal equation at depth z is given by:

$$T_z = T_0 + \frac{\dot{q}_0}{K}(z - z_0) - \frac{P_0}{2K}(z - z_0)^2 \qquad (184.1)$$

where K is the thermal conductivity, T_0 is the temperature at the surface of the Earth (in K), $\dot{q}_0$ is the heat flow at the surface, and P_0 is the radiogenic heat production at the Earth's surface.

At the Earth's surface $z_0 = 0$, and Equation (184.1) becomes:

$$T_z = T_0 + \frac{\dot{q}_0}{K}z - \frac{P_0}{2K}z^2$$

$$\frac{P_0}{2K}z^2 - \frac{\dot{q}_0}{K}z + (T_z - T_0) = 0$$

So

$$z = \frac{\frac{\dot{q}_0}{K} \pm \sqrt{\left(\frac{\dot{q}_0}{K}\right)^2 - 4\frac{P_0}{2K}(T_z - T_0)}}{2\frac{P_0}{2K}} = \frac{\dot{q}_0 \pm \sqrt{\dot{q}_0^2 - 2P_0K(T_z - T_0)}}{P_0}$$

Substituting the data given in the problem:

$K = 3.35 \text{ W m}^{-1}\text{ K}^{-1}$

$\dot{q}_0 = 46 \times 10^{-3} \text{ W m}^{-2}$

$P_0 = 0.01 \times 10^{-6} \text{ mW m}^{-3}$

$T_z = 1623 \text{ K}$

$T_0 = 288 \text{ K}$

we obtain two solutions, but only $z = 98.27$ km is realistic (the second one gives a depth larger than the Earth's radius).

185. On the surface of an Earth of radius 6000 km, the temperature is 300 K, the heat flow is 6.7 mW m^{-2}, and the thermal conductivity is 3 W m^{-1} K^{-1}. If the heat production per unit volume inside the Earth is homogeneously distributed, what is the temperature at the centre of the planet?

We begin solving the problem of heat conduction inside a sphere with constant internal heat generation per unit volume ε and conductivity K. The differential equation for heat conduction with spherical symmetry is

$$\frac{1}{r^2}\frac{d}{dr}\left(r^2\frac{dT}{dr}\right) = -\frac{\varepsilon}{K} \tag{185.1}$$

Integrating twice and using the boundary conditions:

Surface: $r = R \rightarrow T = T_0$

Center: $r = 0 \rightarrow T$ finite

we obtain the solution

$$T = T_0 + \frac{\varepsilon}{6K}\left(R^2 - r^2\right) \tag{185.2}$$

The heat flow is given by

$$\dot{q} = -K\frac{dT}{dr} \Rightarrow \dot{q}_0(r = R) = 6.7 \text{ mW m}^{-2}$$

$$= -K\frac{d}{dr}\left(T_0 + \frac{\varepsilon}{6K}\left(R^2 - r^2\right)\right) = \frac{\varepsilon r}{3}$$

Solving for the heat production ε

$$\varepsilon = 3.35 \times 10^{-9}\,\mathrm{Wm^{-3}}$$

From Equation (185.1) the temperature at the Earth's centre is

$$T_{r=0} = 300 + \frac{3.35 \times 10^{-9}}{6 \times 3} \times 6^2 \times 10^{12} = 7000\,\mathrm{K}$$

186. Consider a spherical Earth of radius $R = 6000$ km and a core at $R/2$, in which there is a uniform and stationary distribution of ε heat sources per unit volume. The heat flow at the surface is 5 mW m^{-2}, the thermal conductivity is 3 W m^{-1} K^{-1}, and the temperature at the core–mantle boundary is 4000 K. Calculate the temperature at the Earth's surface.

We consider the problem as one of heat conduction inside a sphere with conductivity K and constant heat generation per unit volume ε inside the core (radius $R/2$). We begin with Equation (185.1)

$$\frac{1}{r^2}\frac{d}{dr}\left(r^2\frac{dT}{dr}\right) = -\frac{\varepsilon}{K}$$

The boundary conditions at the core–mantle boundary and its centre are

$$r = R/2 \rightarrow T = T_N$$
$$r = 0 \rightarrow T \quad \text{finite}$$

where T_N is the temperature at the core–mantle boundary
  Integrating twice we obtain

$$T(r) = T_N + \frac{\varepsilon}{6K}\left(\left(\frac{R}{2}\right)^2 - r^2\right)$$

The heat flow is given by

$$\dot{q}_0(r = R) = -K\frac{dT}{dr}\bigg|_{r=R} = \frac{\varepsilon R}{3} \Rightarrow \varepsilon = \frac{\dot{q}_0 \times 3}{R} = 2.5 \times 10^{-9}\,\mathrm{W\,m^{-3}}$$

Then, the temperature at the Earth's surface is

$$T(r = R) = 4000 + \frac{2.5 \times 10^{-9}}{6 \times 3}\left(\left(\frac{6000 \times 10^3}{2}\right)^2 - 6000^2 \times 10^6\right) = 250\,\mathrm{K}$$

187. Consider the Earth of radius $R_0 = 6000$ km formed by a spherical crust with its base at 500 km and constant thermal conductivity K. If the temperature at the base of the crust is T_1 and at the surface of the Earth is $T_0 = 0\ °C$, determine:

(a) An expression for the heat flow through the crust.

(b) **An expression for the temperature distribution within the Earth.**

(c) **The temperature at the base of the crust if, at that depth, $\dot{q} = 5.5 \times 10^{13}$ W and $K = 4$ W m^{-1} °C^{-1}.**

(a) We assume a spherical Earth where the temperature varies only in the radial direction. Then we can solve the problem as one of spherical unidirectional flow. For the stationary case, when the conductivity and heat generation are constant, the Fourier law may be written as

$$\dot{q} = -KA\frac{dT}{dr} \qquad (187.1)$$

where $A = 4\pi r^2$ is the area in the normal direction to the heat flow. Integrating this equation:

$$\frac{\dot{q}}{4\pi}\int_{r_1}^{R_0} \frac{dr}{r^2} = -K\int_{T_1}^{T_0} dT$$

where the conditions at the Earth's surface are, $r = R_0 \rightarrow T = T_0$ and at the base of the crust, $r = r_1 \rightarrow T = T_1$.

Solving Equation (187.1), assuming that K is constant, we obtain

$$\dot{q} = -\frac{4\pi K}{\frac{1}{r_1} - \frac{1}{R_0}}(T_0 - T_1) = \frac{-4\pi K r_1 R_0}{R_0 - r_1}(T_0 - T_1) \qquad (187.2)$$

(b) The temperature distribution inside of the Earth may be obtained by integration of Equation (187.1):

$$\frac{\dot{q}}{4\pi}\int_{r_1}^{r} \frac{dr}{r^2} = -K\int_{T_1}^{T} dT$$

$$T(r) = T_1 + \frac{R_0(r - r_1)}{(R_0 - r_1)r}(T_0 - T_1)$$

(c) The radial distance to the base of the crust is $r_2 = 5500$ km, so, using expression (187.2), we obtain

$$T_1 = T_0 + \frac{\dot{q}(R_0 - r_1)}{4\pi K r_1 R_0} = 0 + \frac{5.5 \times 10^{13} \times 500 \times 10^3}{4\pi \times 4 \times 5500 \times 10^3 \times 6000 \times 10^3} = 16579\,°C$$

This result implies a constant increase of temperature from the Earth's surface of 1 °C each 33.2 m similar to the observed gradient in the real Earth of 1 °C per 30 m

188. Assume that the heat flow inside the Earth is due to solar heating of the Earth's surface. Calculate the maximum penetration of this flow in the diurnal and annual cycles. Take as typical values for the Earth $K = 3$ Wm^{-1} K^{-1}, $\rho = 5.5$ g cm^{-3}, $C_v = 1$ kJ kg^{-1} K^{-1}.

We assume the heat propagation inside the Earth coming from the solar radiation on its surface as unidirectional flow thermal diffusion (inside the Earth) with periodic variation of surface temperature. The diffusivity equation is

$$\kappa \frac{\partial^2 T}{\partial z^2} = \frac{\partial T}{\partial t} \tag{188.1}$$

where the thermal diffusivity is $\kappa = \dfrac{K}{\rho C_v}$, K is the thermal conductivity, ρ is the density, and C_v is the specific heat at constant volume. We solve Equation (188.1) using the separation of variables

$$T(z,t) = Z(z)\theta(t)$$

Substituting in (188.1) we obtain the solution

$$Z(z) = Ae^{-\alpha z} + Be^{\alpha z}$$
$$\theta(t) = Ce^{\kappa \alpha^2 t}$$

where α is the constant of separation of variables. Using the boundary condition of periodic flow and the temperature T_0 at the Earth's surface,

$$z = 0 \Rightarrow T = T_0 e^{i\omega t}$$

and as $Z(z)$ exists only inside the Earth, $B = 0$. At the surface, $z = 0$, so

$$ACe^{\kappa \alpha^2 t} = T_0 e^{i\omega t}$$

Then

$$AC = T_0$$
$$\kappa \alpha^2 = i\omega$$

But putting, $i = \frac{1}{2}(1 + i)^2$, we have

$$\alpha = (1 + i)\sqrt{\frac{\omega}{2\kappa}}$$

Then, we can write the temperature variation inside of the Earth as:

$$T(z,t) = T_0 \exp\left[-\sqrt{\frac{\omega}{2\kappa}}z + i\left(-\sqrt{\frac{\omega}{2\kappa}}z + \omega t \right) \right]$$

This equation corresponds to a periodic wave, with angular frequency ω propagating for positive z values (to the Earth's interior) and with the amplitude decreasing with depth. The propagation velocity and wavelength are given by

$$v = \sqrt{\frac{2\kappa}{\omega}}$$

$$\lambda = 2\pi v = \pi\sqrt{\frac{8\kappa}{\omega}}$$

The values of λ corresponding to the daily and annual cycles give their maximum penetration:
Daily cycle:

$$\omega = \frac{2\pi}{24 \times 60 \times 60} = 7.2 \times 10^{-5}\,\text{s}^{-1}$$

$$\kappa = \frac{K}{\rho C_v} = \frac{3\,\text{W}\,\text{m}^{-1}\,\text{K}^{-1}}{5.5 \times 10^3\,\text{K}\,\text{gm}^{-3} \times 10^3\,\text{J}\,\text{Kg}^{-1}\,\text{K}^{-1}} = 0.5 \times 10^{-6}\,\text{m}^2\,\text{s}^{-1}$$

Then

$$\lambda = \pi\sqrt{\frac{8 \times 0.5 \times 10^{-6}}{7.2 \times 10^{-5}}} = 0.74\,\text{m}$$

Annual cycle:

$$\omega = \frac{2\pi}{365 \times 24 \times 60 \times 60} = 2 \times 10^{-7}\,\text{s}^{-1}$$
$$\lambda = 14\,\text{m}$$

The penetration of the solar radiation as periodic heat conduction inside the Earth is very shallow due to the poor heat conduction.

189. Consider a lithospheric plate of 100 km thickness created from asthenospheric material originating from a ridge in the asthenosphere with constant temperature T_a and in which no heat is generated. Given that $\kappa = 10^{-6}\,\text{m}^2\,\text{s}^{-1}$, that the temperature at the base of the lithosphere is 1100 °C, and in the asthenosphere is 1300 °C, calculate the age of the plate, and, if the velocity of drift is 2 cm yr^{-1}, how far it has moved away from the ridge.

The heat propagation inside the plate is given by:

$$\frac{K}{\rho c_v}\left(\frac{\partial^2 T}{\partial x^2} + \frac{\partial^2 T}{\partial z^2}\right) = u\frac{\partial T}{\partial x} \tag{189.1}$$

where T is the temperature, ρ is the density, c_v is the specific heat at constant volume, and u is the horizontal velocity of the plate in the x-direction (normal to the plate front). If we assume that the horizontal conduction of heat is insignificant in comparison with the horizontal advection and vertical conduction, we can write, using the following change of variable $t = x/u$,

$$\frac{K}{\rho c_v}\frac{\partial^2 T}{\partial z^2} = \frac{\partial T}{\partial t}$$

Integrating this equation and using the boundary conditions at the ridge and surface:

$$x = 0 \rightarrow T = T_a$$
$$z = 0 \rightarrow T = 0$$

we obtain for the temperature distribution

$$T(z,t) = T_a\,\text{erf}\left(\frac{z}{2\sqrt{\kappa t}}\right)$$

where

$$\kappa = \frac{K}{\rho c_v}$$

$$\text{erf}(x) = \frac{2}{\sqrt{\pi}} \int_0^x e^{-y^2} dy$$

Substituting the data of the problem,

$$1100 = 1300 \text{erf}\left(\frac{L}{2\sqrt{\kappa t}}\right) \Rightarrow \text{erf}\left(\frac{L}{2\sqrt{\kappa t}}\right) = 0.846$$

Values of the error function, erf(x), may be obtained from tables. If erf(x) = 0.846, x = 1.008, then

$$\left(\frac{L}{2\sqrt{\kappa t}}\right) = 1.008 \Rightarrow t = \frac{L^2}{4\kappa \times 1.008^2} = \frac{10^{10}}{4 \times 10^{-6} \times 1.008^2}$$

$$= 2.5 \times 10^{15}\,\text{s} = 79\,\text{Myr}$$

If the displacement velocity is 2 cm yr^{-1}, the plate has moved 1580 km.

190. If the concentrations of ^{235}U and ^{235}Th in granite are 4 ppm and 17 ppm, respectively, and the respective values of heat production are 5.7×10^{-4} W kg^{-1} and 2.7×10^{-5} W kg^{-1}, respectively, calculate the heat flow at the base of a granite column of 1 m² cross-section and 30 km height (the density of granite is 2.65 g cm^{-3}).

We estimate first the mass of the granite column:

$$M = \rho V = 2.65 \times 10^3 \times 1 \times 30 \times 10^3 = 7.95 \times 10^7\,\text{kg}$$

If the concentration of ^{235}U in the granite is 4 ppm, its quantity in the column is

$$^{235}\text{U}: \frac{4 \times 7.95 \times 10^7}{10^6} = 318\,\text{kg}$$

Then the heat flow due to the ^{235}U is

$$\dot{q} = 5.7 \times 10^{-4} \times 318 = 181.26\,\text{mW m}^{-2}$$

For ^{235}Th, the heat flow is

$$^{235}\text{Th}: 7.95 \times 10^7 \times 17 \times 10^{-6} = 1351.5\,\text{kg}$$

$$\dot{q} = 1351.5 \times 2.7 \times 10^{-5} = 36.49\,\text{mW m}^{-2}$$

Geochronology

191. The mass of 1 millicurie of ^{214}Pb is 3×10^{14} kg. Calculate the value of the decay constant of ^{214}Pb.

The mass of the sample is

$$3 \times 10^{14} \, \text{kg} = 3 \times 10^{17} \, \text{g} = \frac{N}{N_0} M$$

where N is the number of atoms in the sample, N_0 is Avogadro's number $= 6.02 \times 10^{23}$, and M is atomic number $= 214$. Solving for N we obtain

$$N = \frac{3 \times 10^{17} \times 6.02 \times 10^{23}}{214} = 8.44 \times 10^{38} \text{ atoms}$$

The correspondence of a curie is

$$1 \, \text{curie} = \frac{dN}{dt} = \lambda N = 3.7 \times 10^{10} \text{ disintegrations s}^{-1}$$

where λ is the decay constant and t is the time. Then

$$\lambda = \frac{3.7 \times 10^7}{8.44 \times 10^{38}} = 0.44 \times 10^{-31} \, \text{s}^{-1}$$

192. The isotope ^{40}K decays by emission of β particle with a half-life of 1.83×10^9 years. How many β decays occur per second in one gram of pure ^{40}K?

The average life $\bar{t}$ of a radioactive material is a function of the decay constant λ:

$$\bar{t} = \frac{1}{\lambda} \rightarrow \lambda = \frac{1}{1.83 \times 10^9} = 0.55 \times 10^{-9} \, \text{yr}^{-1} = 1.73 \times 10^{-2} \, \text{s}^{-1}$$

The number of atoms N contained in 1 g of ^{40}K may be estimated from Avogadro's number N_0 and the atomic number M:

$$1 = \frac{N}{N_0} M \rightarrow N = \frac{1 \times 6.02 \times 10^{23}}{40} = 0.15 \times 10^{23} \text{ atoms}$$

The rate of disintegration is given by

$$\frac{dN}{dt} = \lambda N = 0.26 \times 10^{21} \, \text{Bq}$$

193. The half-life of ^{238}U is 4468×10^6 yr and of ^{235}U is 704×10^6 yr. The ratio ^{235}U/^{238}U in a sample is 0.007257. Given that the ratio was 0.4 at the time of formation, calculate the sample's age.

From the half-life $T_{1/2}$ we can obtain the decay constant λ:

$$T_{1/2} = \frac{0.693}{\lambda}$$

$$\lambda_{238} = \frac{0.693}{4468 \times 10^6} = 1.5510 \times 10^{-10} \, \text{yr}^{-1}$$

$$\lambda_{235} = \frac{0.693}{704 \times 10^6} = 9.8434 \times 10^{-10} \, \text{yr}^{-1}$$

The number N of disintegrating atoms at time t is given by

$$N = N_0 \, e^{-\lambda t} \tag{193.1}$$

where N_0 is the number of atoms at time $t = 0$. Then

$$N_{235} = N_0^{235} e^{-\lambda_{235}t}$$
$$N_{238} = N_0^{238} e^{-\lambda_{238}t}$$

If we divide these equations:

$$\frac{N_{235}}{N_{238}} = \frac{N_0^{235}}{N_0^{238}} e^{(\lambda_{238}-\lambda_{235})t}$$

Substituting the values given in the problem we obtain

$$-4.0095 = -8.2924 \times 10^{-10}t \rightarrow t = 4.8 \times 10^9 \, \text{yr}$$

194. Date a meteorite which contains potassium knowing that its content of ^{40}K is 1.19×10^{14} atoms g^{-1}, of ^{40}Ar is 4.14×10^{17} atoms g^{-1}, and that the half-life of $^{40}K \rightarrow ^{40}Ar$ is 1.19×10^9 years.

We obtain the decay constant of the ^{40}K from its half-life $T_{1/2}$:

$$\lambda = \frac{0.693}{T_{1/2}}$$
$$\lambda = \frac{0.693}{1.19 \times 10^9} = 0.58 \times 10^{-9} \, \text{yr}^{-1}$$

We can solve the problem considering it as a case of radioactive 'parent' atom disintegrating to a 'daughter' stable atom. At time $t = 0$ we have n_0 'parent' atoms at the sample, and at time t there remain NR radioactive atoms in the sample and NE daughter atoms, from the disintegration of the n_0 parent atoms:

$$n_0 = \text{NR} + \text{NE}$$

But

$$\frac{n_0}{n_t} = \frac{\text{NR} + \text{NE}}{\text{NR}} = 1 + \frac{\text{NE}}{\text{NR}}$$

From Equation (193.1) we obtain the age of the sample:

$$n_t = n_0 \, e^{-\lambda t} \rightarrow \frac{n_0}{n_t} = e^{-\lambda t} = 1 + \frac{\text{NR}}{\text{NR}} \qquad (194.1)$$

so

$$t = \frac{1}{\lambda} \ln\left(1 + \frac{\text{NE}}{\text{NR}}\right)$$

NE, the number of atoms of ^{40}K in the sample, can be estimated from the number of atoms contained in 1 g of potassium:

$$1 = \frac{N}{6.023 \times 10^{23}} 40 \rightarrow N = 1.506 \times 10^{22} \, \text{atoms} \, g^{-1}$$

so

$$NE = 1.506 \times 10^{22} \times 1.19 \times 10^{14} = 1.792 \times 10^{36} \text{ atoms g}^{-1}$$

Then the age of the meteorite from (194.1) is

$$t = \frac{1}{\lambda} \ln\left(1 + \frac{1.792 \times 10^{36}}{4.41 \times 10^{17}}\right) = 7.4 \times 10^{10} \text{ yr}$$

195. At an archaeological site, human remains were found and assigned an age of 2000 years. One wants to confirm this with ^{14}C dating whose half-life is 5730 yr. If the proportion of ^{14}C/^{12}C in the remains is 6×10^{-13}, calculate their age. (Assume that at the initial time the ^{14}C/^{12}C ratio was 1.2×10^{-12}.)

The decay constant λ may be obtained from the half-life:

$$T_{1/2} = \frac{0.693}{\lambda} \rightarrow \lambda = \frac{0.693}{5730} = 1.2094 \times 10^{-4} \text{ yr}^{-1}$$

The activity in a sample is given by

$$R = R_0 \, e^{-\lambda t}$$

where $R_0 = (^{14}\text{C}/^{12}\text{C})_{t=0}$ and $R = (^{14}\text{C}/^{12}\text{C})$.

Then the age of the remains is

$$t = \frac{1}{\lambda} \ln\left(\frac{R_0}{R}\right) = \frac{1}{1.2094 \times 10^{-4}} \ln\left(\frac{1.2 \times 10^{-12}}{9 \times 10^{-13}}\right) = 2379 \text{ yr}$$

196. Mass spectrometry of the different minerals in an igneous rock yielded the following table of values for the concentrations of ^{87}Sr originating from the radioactive decay of ^{87}Rb and of ^{87}Rb, with the concentration expressed relative to the concentrations of ^{86}Sr of non-radioactive origin.

Mineral	^{87}Sr /^{86}Sr	^{87}Rb /^{86}Sr
A	0.709	0.125
B	0.715	0.418
C	0.732	1.216
D	0.755	2.000
E	0.756	2.115
F	0.762	2.247

Express on a ^{87}Sr/^{86}Sr–^{87}Rb/^{86}Sr diagram the isochron corresponding to the formation of the rock, and calculate the age of the rock. Take $\lambda = 1.42 \times 10^{-11} \text{ yr}^{-1}$.

For the decay of ^{87}Rb

$$[^{87}\text{Sr}]_{\text{now}} = [^{87}\text{Sr}]_0 + [^{87}\text{Rb}]_{\text{now}}\left(e^{\lambda t} - 1\right) \tag{196.1}$$

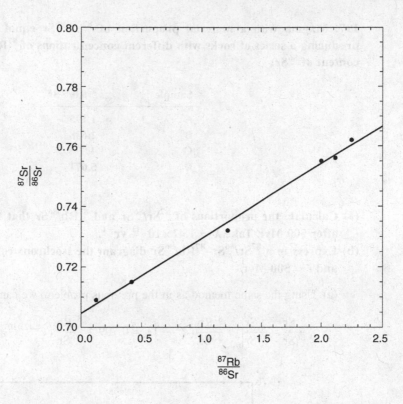

Fig. 196

where $[^{87}\text{Sr}]_{\text{now}}$ and $[^{87}\text{Rb}]_{\text{now}}$ are the number of atoms of each isotope at time t, $[^{87}\text{Sr}]_0$ is the amount of original number of atoms of the isotope ^{87}Sr $[^{87}\text{Sr}]_{\text{now}}$, and λ is the decay constant.

Equation (196.1) may be written as

$$\frac{^{87}\text{Sr}}{^{86}\text{Sr}}\bigg|_{\text{now}} = \frac{^{87}\text{Sr}}{^{86}\text{Sr}}\bigg|_0 + \frac{^{87}\text{Rb}}{^{86}\text{Sr}}\bigg|_{\text{now}}\left(e^{\lambda t} - 1\right) \qquad (196.2)$$

This equation corresponds to a line ($y = a + bx$) with intercept $\dfrac{^{87}\text{Sr}}{^{86}\text{Sr}}\bigg|_0$ and slope $(e^{\lambda t} - 1)$, which is called an isochron.

If we plot the values given in the problem (Fig. 196) we can obtain the equation of the line by least-squares fitting:

$$y = 0.025x + 0.705$$

The age of the sample can be obtained from the slope $b = 0.025$:

$$e^{-\lambda t} - 1 = b$$

$$t = \frac{\ln(1 + b)}{\lambda}$$

Substituting the values of b and λ:

$$t = 1.72 \times 10^9 \text{ yr.}$$

197. Magma with a material proportion of $^{87}Sr/^{86}Sr$ equal to 0.709 crystallizes producing a series of rocks with different concentrations of ^{87}Rb with respect to the content of ^{86}Sr:

Sample	$^{87}Rb/^{86}Sr$
A	1.195
B	2.638
C	4.892
D	5.671

(a) Calculate the proportions of $^{87}Sr/^{86}Sr$ and $^{87}Rb/^{86}Sr$ that these rocks will have after 500 Myr. Take $\lambda = 1.42 \times 10^{-11}$ yr^{-1}.

(b) Express in a $^{87}Sr/^{86}Sr-^{87}Rb/^{86}Sr$ diagram the isochrons corresponding to $t = 0$ and $t = 500$ Myr.

(a) Using the same method as in the previous problem, we can write

$$\frac{^{87}Sr}{^{86}Sr} = \frac{^{87}Sr}{^{86}Sr}\bigg|_0 + \frac{^{87}Rb}{^{86}Sr}\left(e^{\lambda t} - 1\right) = 0.709 + \frac{^{87}Rb}{^{86}Sr}\left(e^{1.42\times10^{-11}\times5\times10^8} - 1\right)$$

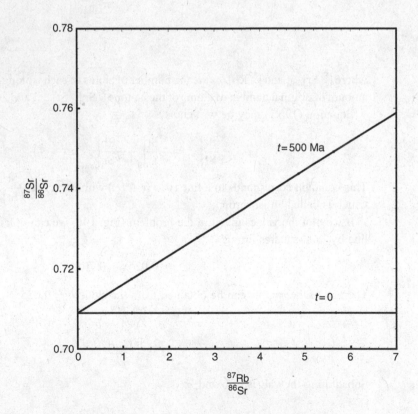

Fig. 197

The results for each rock are given in the following table

Sample	$^{87}Sr/^{86}Sr$	$^{87}Rb/^{86}Sr$
A	0.717	1.187
B	0.728	2.619
C	0.744	4.857
D	0.749	5.631

(b) For t = 500 Myr, we carry out a least-squares fitting to obtain the isochron, which results in

$$y = 0.007x + 0.709$$

In Fig. 197 the isochrones corresponding to $t = 0$ and $t = 500$ Myr are shown.

Bibliography

General geophysics

Berckhemer, H. (1990). *Grundlagen der Geophysik*. Wissenschaftlische Buchgeselschaft, Darmstadt.

Buforn, E., C. Pro, and A. Udías (2010). *Problemas Resueltos de Geofísica*. Pearson, Madrid.

Cara, M. (1989). *Geophysique*. Dunod, Paris.

Coulomb, J. and G. Jobert (1973 and 1976). *Traité de Géophysique Interne (I and II)*. Masson, Paris.

Fowler, C.M.R. (2005). *The Solid Earth: An Introduction to Global Geophysics*. (2nd edn). Cambridge University Press, Cambridge.

Garland, G.D. (1979). *Introduction to Geophysics*. W.B. Saunders, Philadelphia.

Kaufman, A.A. (1992). *Geophysical Field Theory and Method*. Academic Press, San Diego.

Larroque, C. and J. Virieux (2001). *Physique de la Terre solide. Observations et Théories*. Gordon and Breach, Paris.

Lillie, R.J. (1999). *Whole Earth Geophysics. An Introductory Textbook for Geologists and Geophysicists*. Prentice Hall, New Jersey.

Lowrie, W. (2007). *Fundamentals of Geophysics* (2nd edn). Cambridge University Press, Cambridge.

Lowrie, W. (2011). *A Student's Guide to Geophysical Equations*. Cambridge University Press, Cambridge.

Officer, C.B. (1974). *Introduction to Theoretical Geophysics*. Springer, New York.

Poirier, J.P. (2000). *Introduction to the Physics of the Earth's Interior*. (2nd edn). Cambridge University Press, Cambridge.

Schick, R. and G. Schneider (1973). *Physik des Erdkörpers. Eine Einfürung für Naturwissenschafler und Ingeníeure*. Ferdinand Enke, Stuttgart.

Sleep, N.H. and K. Fujita (1997). *Principles of Geophysics*. Blackwell Science, London.

Stacey, F.D. (1992). *Physics of the Earth* (3rd edn). Brookfield Press, Brisbane.

Udías, A. and J. Mezcua (1997). *Fundamentos de Geofísica*. Alianza Universidad, Madrid.

Gravimetry

Blakely, R.J. (1995). *Potential Theory in Gravity and Magnetic Applications*. Cambridge University Press, Cambridge.

Bomford, G. (1980). *Geodesy* (4th edn). Clarendon Press, Oxford.

Heiskanen, W.A. and H. Moritz (1985). *Physical Geodesy*. Freeman, San Francisco.

Hofmann-Wellenhof, B. and H. Moritz. (2006). *Physical Geodesy*. Springer, New York.

Lambeck, K. (1988). *Geophysical Geodesy*. Clarendon Press. Oxford.

Levallois, J.J. (1970). *Géodesie Générale (Vol. 3, Le Champ de la Pesanteur)*. Edition Eyrolles, Paris.

Pick, M., J. Picha, and V. Vyskocil (1973). *Theory of the Earth's Gravity Field*. Elsevier, Amsterdam.

Torge, W. (1989). *Gravimetry*. Walter Gruyter, Berlín.

Geomagnetism

Backus, G., R. Ladislav, and C. Constable (1996). *Foundations of Geomagnetism*. Cambridge University Press, Cambridge.

Basavaiah, N. (2010). *Geomagnetism: Solid Earth and Atmospheric Perspectives*. Springer, Berlin.

Blakely, R. J. (1995). *Potential Theory in Gravity and Magnetic Applications*. Cambridge University Press, Cambridge.

Campbell, W.H. (2003). *Introduction to Geomagnetic Fields*. Cambridge University Press, Cambridge.

Campbell, W.H. (2001). *Earth Magnetism: A Guided Tour through Magnetic Fields*. Cambridge University Press, Cambridge.

Delcourt, J.J. (1990). *Magnétisme Terrestre. Introduction*. Masson, Paris.

Jacobs, J.A. (1987–1991). *Geomagnetism* (4 volumes). Academic Press, London.

Miguel, L. de (1980). *Geomagnetismo*. Instituto Geográfico Nacional, Madrid.

Parkinson, W.D. (1983). *Introduction to Geomagnetism*. Scottish Academic Press, Edinburgh.

Rikitake, T. and Y. Honkura (1985). *Solid Earth Geomagnetism*. Terra Scientific Publishing, Tokyo.

Wait, J.R. (1982). *Geo-electromagnetism*. Academic Press, New York.

Seismology

Aki, K. and P.G. Richards (2002). *Quantitative Seismology*. (2nd edn). University Science Books, Sausalito, California.

Bullen, K.E. and B.A. Bolt (1985). *An Introduction to the Theory of Seismology* (4th edn). Cambridge University Press, Cambridge.

Chapman, C. (2004). *Fundamentals of Seismic Wave Propagation*. Cambridge University Press, Cambridge.

Dahlen, F.A. and J. Tromp (1998). *Theoretical Global Seismology*. Princeton University Press, Princeton

Gubbins, D. (1990). *Seismology and Plate Tectonics*. Cambridge University Press, Cambridge.

Lay, T. and T.C. Wallace (1995). *Modern Global Seismology*. Academic Press, San Diego.

Kennnett, B.L.N. (2001). *The Seismic Wavefield*. Cambridge University Press, Cambridge.

Pujol, J. (2003). *Elastic Wave Propagation and Generation in Seismology*. Cambridge University Press, Cambridge.

Shearer, P.M. (1999). *Introduction to Seismology*. Cambridge University Press, Cambridge.

Stein, S. and M. Wyssesion (2003). *An Introduction to Seismology, Earthquakes and Earth Structure*. Blackwell Publishers, Oxford.

Udías, A. (1999). *Principles of Seismology*. Cambridge University Press, Cambridge.

Heat flow and geochronology

Bunterbarth, G. (1984). *Geothermics*. Springer, Berlin.

Faure, G. (1986). *Principles of Isotope Geology*. John Wiley, New York.

Jaupart, C. and J.-C. Mareschal (2011). *Heat Generation and Transport in the Earth*. Cambridge University Press, Cambridge.

Printed in the United States
By Bookmasters